INTRODUCTION TO STRATEGIES FOR ORGANIC SYNTHESIS

INTRODUCTION TO STRATEGIES FOR ORGANIC SYNTHESIS

SECOND EDITION

Laurie S. Starkey
California State Polytechnic University, Pomona
CA, USA

Registered Office
John Wiley & Sons, Inc., 111 River Street, Hoboken, NJ 07030, USA

Editorial Office
111 River Street, Hoboken, NJ 07030, USA

For details of our global editorial offices, customer services, and more information about Wiley products visit us at www.wiley.com.

Wiley also publishes its books in a variety of electronic formats and by print-on-demand. Some content that appears in standard print versions of this book may not be available in other formats.

Library of Congress Cataloging-in-Publication Data

Names: Starkey, Laurie Shaffer, 1969– author.
Title: Introduction to strategies for organic synthesis / Laurie S. Starkey.
Description: Second edition. | Hoboken, NJ : John Wiley & Sons, 2018. |
    Identifiers: LCCN 2017046041 (print) | LCCN 2017052512 (ebook) |
    ISBN 9781119347231 (pdf) | ISBN 9781119347217 (epub) |
    ISBN 9781119347248 (pbk.)
Subjects: LCSH: Organic compounds–Synthesis.
Classification: LCC QD262 (ebook) | LCC QD262 .S812 2018 (print) | DDC 547/.2–dc23
LC record available at https://lccn.loc.gov/2017046041

Cover design by Wiley
Cover image: © pixhook/Getty Images

Set in 10/12pt Times Ten by SPi Global, Pondicherry, India

Printed in the United States of America

10 9 8 7 6 5 4 3 2 1

■ CONTENTS

I really could have used this book when I started graduate school! I became fascinated with organic synthesis ever since running my first Grignard reaction as an undergraduate student at the University of Connecticut. As I watched the magnesium metal disappear into the solvent in my round-bottom flask, I was intrigued by the thought of making new molecules. Although my interest and enthusiasm continued as I entered graduate school at UCLA, I quickly found myself being thrown into the proverbial deep end when I took my first graduate course in organic synthesis. I had never taken a synthesis course at UConn, and my undergraduate organic chemistry course seemed like a foggy memory. I scoured every textbook I could find in an effort to stay afloat, but it was a struggle to work through the advanced material I found there. I benefited immensely from the mentorship and patience of my research advisor, and I eventually earned my Ph.D. in organic chemistry. Although I was able to make progress on my graduate research projects, I didn't truly appreciate the strategies of organic synthesis until I taught the course myself as a faculty member. As I embarked on my teaching career at Cal Poly Pomona, I was eager to share my passion for organic synthesis, but I found that most of my students experienced the same difficulties that I had encountered as a student. The quantum leap from sophomore-level organic chemistry to senior-level organic synthesis is nearly insurmountable for some students. I did my best to bridge this gap with my teaching, but it was a challenge since all of the available textbooks were written at the graduate level (or beyond!). Throughout many years of teaching the organic synthesis course, I gradually developed a teaching strategy that seemed to foster student success. My approach involves a significant amount of review of the sophomore-level material (functional group transformations, reagents, and reaction mechanisms) before changing the perspective and attempting to *plan a synthesis* (functional group analysis and making strategic disconnections: the retrosynthesis of a target molecule). Simply put, taking a year of organic chemistry does not make you an organic chemist, so this review is an essential element for most students. Through practice and experience, envisioning a reaction both in the forward direction and in the reverse direction eventually becomes a routine exercise, but it cannot be assumed to be a trivial matter from the beginning. Such an assumption is made when little to no distinction is made between undergraduate-level and graduate-level organic synthesis courses, and it can result in a frustrating experience for the student. This book is designed as an intermediate-level introduction to the tools and skills needed to study organic synthesis. It contains worked-through examples and detailed solutions to the end-of-chapter problems, so it is ideal for any student who is interested in pursuing research in the field of organic chemistry, including beginning graduate students. For the second edition, over 130 mid-level, in-chapter problems have been added to provide opportunities for practice and self-assessment. With its thorough review of the reactions of organic chemistry and its Study Guide approach, this book aims to build confidence as it deepens students' knowledge. More challenging topics are also explored, with the goal of raising students to a new skill level and preparing them for advanced coursework. To the students studying organic chemistry, I offer this book along with the same advice that I give to my kids (a quote from Mahatma Gandhi), "Live as if you were to die tomorrow. Learn as if you were to live forever."

■■■■ ACKNOWLEDGMENTS

I want to thank my students for making this the greatest job in the world, my graduate advisor, Mike Jung, for providing a supportive environment and for believing in a struggling student, and my mentor, Phil Beauchamp, who is both a passionate teacher and an organic synthesis junkie. The preparation of the manuscripts would be impossible without my tireless reviewers, including Phil Beauchamp, Chantal Stieber, Joe Casalnuovo, Chris Nichols, Phil Lukeman, Richard Johnson, and especially Michael O'Donnell, an avid lifelong student of chemistry who somehow found the first edition of my book before it was even finished! Thanks to my extended family for their love, support, and friendship...and for making me the person that I am today. Most of all, I am grateful to my wolfpack, Mike, Ellie, and Andy, because the most important things in life aren't things.

CHAPTER 1

SYNTHETIC TOOLBOX 1: RETROSYNTHESIS AND PROTECTIVE GROUPS

This book will demonstrate how to synthesize target molecules (TMs) that contain various functional groups (FGs) such as C≡C (alkyne), OH (alcohol or carboxylic acid), and C=O (aldehyde, ketone, and many others). The process of planning a synthesis, called a retrosynthesis, is one of the most critical tools within the "toolbox" needed to solve synthesis problems. The method of retrosynthetic analysis is introduced in this chapter and is used throughout the book. This first chapter will also review the use of protective groups in organic synthesis. The second chapter provides additional useful tools needed by the beginning synthesis student, by reviewing common nucleophiles and electrophiles, as well as some general reagents for oxidation and reduction reactions.

Introduction to Strategies for Organic Synthesis, Second Edition. Laurie S. Starkey.
© 2018 John Wiley & Sons, Inc. Published 2018 by John Wiley & Sons, Inc.

RETROSYNTHETIC ANALYSIS

Every organic synthesis problem actually begins at the end of the story, a target molecule (TM). The goal is to design a reasonable synthesis that affords the TM as the major product. In the interest of saving both time and money, an ideal synthesis will employ readily available starting materials and will be as efficient as possible. The planning of a synthesis involves imagining the possible reactions that could give the desired product; this process is called doing a retrosynthesis or performing a retrosynthetic analysis of a TM. A special arrow is used to denote a retrosynthetic step. The ⇒ arrow leading away from the TM represents the question "What starting materials could I use to make this product?" and points to an answer to that question. The analysis begins by identifying a functional group present on the target molecule and recalling the various reactions that are known to give products containing that functional group (or pattern of FGs). The process is continued by analyzing the functional groups in the proposed starting material and doing another retrosynthetic step, continuing to work backward toward simple, commercially available starting materials. Once the retrosynthetic analysis is complete, then the forward multistep synthesis can be developed, beginning with the proposed starting materials and treating them with the necessary reagents to eventually transform them into the desired TM.

Introduction to Strategies for Organic Synthesis, Second Edition. Laurie S. Starkey.
© 2018 John Wiley & Sons, Inc. Published 2018 by John Wiley & Sons, Inc.

Retrosynthesis (planning a synthesis)

target molecule ⟹ possible starting
TM   *"What starting*   material(s)
material is
needed?"

Synthesis (making the TM)

starting   reagents   target molecule
material(s) ⟶   TM

Retrosynthesis and Synthesis of a Target Molecule (TM)

A *retrosynthesis* involves working backward from the given target molecule (work done in our minds and on paper), while the *synthesis* is the forward path leading to the target molecule (experimental work done in the lab). Performing a retrosynthetic analysis is challenging since it requires not only knowledge of an enormous set of known organic reactions but also the ability to imagine the experimental conditions necessary to produce a desired product. This challenge becomes more manageable by developing a systematic approach to synthesis problems.[*]

When evaluating a given target molecule, it is important to consider how the functional groups present in the TM can be formed. There are two possibilities for creating a given functional group: by conversion from a different functional group (called a functional group interconversion or FGI), or as a result of a carbon—carbon bond-forming reaction (requiring a retrosynthetic "disconnection"). In order to synthesize a target molecule (or transform a given starting material into a desired product), a combination of FGIs and carbon—carbon bond-forming reactions will typically be required. While the key to the "synthesis" of complex organic molecules is the formation of new carbon—carbon bonds, the synthetic chemist must also be fully capable of swapping one functional group for another.

1.1.1 RETROSYNTHESIS BY FUNCTIONAL GROUP INTERCONVERSION (FGI)

Each functional group has a characteristic reactivity; for example, it might be electron-rich, electron-deficient, acidic, or basic. In order to synthesize organic compounds, we must construct the desired carbon framework while locating the required functional groups in the appropriate

---

[*] For the classic textbook on such an approach, see Stuart Warren and Paul Wyatt, *Organic Synthesis: The Disconnection Approach*, 2nd ed. (Wiley, 2009).

positions. This necessitates that the chemist is familiar not only with the reactivities of each functional group but also with the possible interconversions between functional groups. Such functional group interconversions (FGIs) enable the chemist to move along a synthetic pathway toward a desired target.

Selected "FGI" Reactions	Examples
oxidation, reduction	
addition, elimination	
substitution, hydrolysis	

Examples of Functional Group Interconversions (FGI)

Let us consider a carboxylic acid target molecule (RCO_2H). There are many ways to generate a carboxylic acid functional group, so there are many possible syntheses to consider (often, there may be more than one good solution to a given synthesis problem!). One reaction that gives a carboxylic acid product is the hydrolysis of a carboxylic acid derivative, such as a nitrile. Therefore, a possible retrosynthesis of a carboxylic acid TM (What starting materials are needed?) is to consider an FGI and imagine a nitrile starting material. In other words, if we had a nitrile in our hands, we could convert it to a carboxylic acid, leading to a synthesis of the target molecule.

	FGI	
TM	"What starting material is needed?"	$R-C\equiv N$
target molecule (a carboxylic acid)		a nitrile is a possible starting material

Retrosynthesis of a TM via FGI

$$R-C\equiv N \xrightarrow{H_3O^+} R-C(=O)OH$$

TM

Synthesis of the TM

Choice of Reagents

There is almost always more than one reagent that can be used to achieve any given transformation. In fact, a quick look at a book such as *Comprehensive Organic Transformations* by Richard Larock[*] reveals that there may be dozens of possibilities. Why have so many methods been developed over the years for organic reactions? Because not every molecule—or every chemist—has the same needs. The most obvious reason any "one size fits all" approach fails is that complex synthetic targets contain a wide variety of functional groups. The molecule as a whole must tolerate the reaction conditions used, and side reactions with other functional groups must be kept to a minimum. For example, chromic acid oxidation ($Na_2Cr_2O_7$, H_2SO_4) of a 2° alcohol to give a ketone would not be useful if the starting material contains any functional groups that are sensitive to acidic conditions. In such a case, the Swern oxidation might be preferred (DMSO, ClCOCOCl, Et_3N). New reagents, catalysts, and methods are continuously being developed, with goals of having better selectivity, better tolerance for certain functional groups, being "greener" with less waste or lower toxicity, requiring fewer steps, being more efficient and/or less expensive, and so on.

The focus of this book is on the *strategies* of organic synthesis; it is not intended to be comprehensive in the treatment of modern reagents.[†] Instead, reagents used are those that are typically found in undergraduate organic chemistry textbooks. Hopefully, these reagents will be familiar to the reader, although they would not necessarily be the ones selected when the synthesis moves from paper to the laboratory. Furthermore, experimental details[‡] have largely been omitted from this book. For example, osmium tetroxide oxidation of an alkene is given simply as "OsO_4." In reality, this expensive and toxic reagent is used in catalytic amounts in conjunction with some other oxidizing agent (e.g., NMO), so the precise reagents and experimental reaction conditions are much more complex than what is presented herein.

1.1.2   RETROSYNTHESIS BY MAKING A DISCONNECTION

Rather than being created via an FGI, a functional group (or pattern of FGs) may be created as a result of a reaction that also forms a carbon—carbon sigma bond. In that case, the retrosynthesis involves the disconnection of that

---

[*] Richard C. Larock, *Comprehensive Organic Transformations: A Guide to Functional Group Preparations*, 2nd ed. (Wiley-VCH, 1999).
[†] Tse-Lok Ho, *Fieser and Fieser's Reagents for Organic Synthesis Volumes 1–26, and Collective Index for Volumes 1–22, Set*, 1st ed. (Wiley, 2011); Leo A. Paquette et al., *Encyclopedia of Reagents for Organic Synthesis, 14 Volume Set*, 2nd ed. (Wiley, 2009); George Zweifel and Michael Nantz, *Modern Organic Synthesis*, 1st ed. (W. H. Freeman, 2006).
[‡] A. I. Vogel et al., *Vogel's Textbook of Practical Organic Chemistry*, 5th ed. (Prentice Hall, 1996).

bond. In a typical carbon–carbon bond-forming reaction, one of the starting material carbons must have been a nucleophile (Nu:, electron-rich), and the other must have been an electrophile (E+, electron-deficient). While this is certainly not the *only* way to make a carbon–carbon bond (e.g., the organo-metallic coupling reactions explored in Chapter 8 offer an alternate approach), the pairing of appropriate nucleophiles and electrophiles serves as an important foundation to the logic of organic synthesis, and such strategies will solve a wide variety of synthetic problems. Therefore, the disconnection of the carbon–carbon bond is made heterolytically to give an anion (nucleophile) and a cation (electrophile). These imaginary fragments, called "synthons," are then converted into reasonable starting materials. By being familiar with common nucleophiles and electrophiles, we can make logical disconnections. The example below shows the logical disconnection of an ether TM, affording recognizable alkyl halide E+ and alkoxide Nu: starting materials.

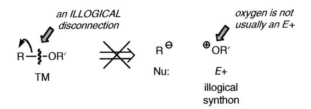

A Logical Disconnection of a TM

Disconnecting that same carbon–oxygen bond in the other direction (with both electrons going to the carbon) would be an illogical disconnection, since it leads to an electrophilic oxygen synthon for which there is no reasonable equivalent reagent.

An Illogical Disconnection of a TM

Let us consider once again a carboxylic acid TM. We have seen that a carboxylic acid can be prepared by an FGI if the carbon chain is already in place, but it is also possible to create new carbon–carbon bonds in a carboxylic acid synthesis. For example, the reaction of a Grignard reagent with carbon dioxide generates a carboxylic acid functional group, so this presents a possible disconnection for the target molecule's retrosynthesis. The logical disconnection is the one that moves the electrons away from the carbonyl, giving reasonable synthons and recognizable starting materials (RMgBr Nu: and CO_2 E+).

Retrosynthesis via Disconnection of a TM

Synthesis of the TM

What Makes a Good Synthesis?

The fact that multiple retrosynthetic strategies usually exist means that there will often be more than one possible synthesis of a desired target molecule. How can we determine which synthesis is best? This depends on many factors, but there are some general rules that can help us devise a good plan to synthesize the simple target molecules found in this book.

1. Start with reasonable starting materials and reagents. A good synthesis begins with commercially available starting materials. Most of these starting materials will have a small number of functional groups (just one or two), although some complex natural products are readily available and inexpensive (e.g., sugars and amino acids). A quick check in any chemical supplier catalog can confirm whether a starting material is ordinary (i.e., available and inexpensive) or exotic (i.e., expensive or not listed).

2. Propose a reaction with a reasonable reaction mechanism. Look for familiar nucleophiles and electrophiles to undergo predictable reactions. A poor choice for a bond disconnection can lead to impossible synthons (and impossible reagents). However, we will learn that certain seemingly impossible synthons are, in fact, possible with the use of synthetic equivalents.

3. Strive for disconnections that lead to the greatest simplification. It is bad practice to put together a 10-carbon target molecule one carbon at a time

(an example of a *linear* synthesis). Remember, the synthetic schemes drawn on paper represent reactions that will be performed in the lab. While this book will not be focusing on experimental details, we should recognize that the more steps in a reaction sequence, the lower the overall yield of product will be. Starting with a 9-carbon starting material, which is nearly as big and possibly as complicated as a 10-carbon target molecule, also would not be a good synthesis. The most efficient synthesis would be one that links together two 5-carbon structures, or perhaps one that combines a 4-carbon with a 6-carbon compound (described as a *convergent* synthesis). The more nearly equal the resulting pieces, the better the bond disconnection. One useful strategy is to look for branch points in a target molecule for good places to make a disconnection. In the example below, the starting materials resulting from disconnection "a" are not only more simple molecules, but also the butanal starting material (butyraldehyde) is one-tenth the price of the aldehyde in disconnection "b" (2-methylbutyraldehyde).

Good Disconnections Lead to Simple, Inexpensive Starting Materials

PROTECTIVE GROUPS

If a target molecule contains more than one functional group, then its synthesis becomes increasingly challenging. The synthesis of a complex natural product is difficult not only because there are many transformations that must be accomplished but also because care must be taken to ensure that the functional groups do not interfere with each other. Those functional groups not involved in a given reaction sequence must be stable to the various reagents and reaction conditions being employed. One way to achieve this stability is by using a protective group to temporarily mask (or hide) the functional group's reactivity. The strategy involves installing a protective group, conducting a reaction elsewhere in the molecule, and then removing the protective group (called "deprotection"). Protective groups are usually denoted using abbreviations, which can make a natural product synthetic scheme seem like alphabet soup to the beginning student! However, as you spend more time with the literature you will quickly become familiar with the more widely used protective groups and you will likely be able to recognize certain transformations as a protection or deprotection step, even if you do not know a particular abbreviation.

Introduction to Strategies for Organic Synthesis, Second Edition. Laurie S. Starkey.
© 2018 John Wiley & Sons, Inc. Published 2018 by John Wiley & Sons, Inc.

Example of General Protective Group (PG) Strategy

Similar functional groups may be differentiated by selective protection, such as protecting a more reactive aldehyde in the presence of a ketone or a less hindered primary alcohol in the presence of a tertiary alcohol. Protective groups can be used to hide the acidic proton of an alcohol or the electrophilic carbonyl of a ketone. Protection of the functional groups found in amino acids (carboxylic acids, amines, and thiols) finds significant applications in the synthesis of peptides and proteins. While hundreds of protective groups have been developed for use in organic synthesis,* only a brief sampling is provided here. A wide variety of protective groups is available since each has its own advantages and disadvantages; factors such as the reactions conditions needed to install and remove the protective group, as well as the stability of the protective group to various reaction conditions are taken into consideration when planning a given synthesis.

1.2.1 PROTECTION OF KETONES AND ALDEHYDES

Ketones and aldehydes will be attacked by strong nucleophiles, such as Grignard reagents, and can be deprotonated at the alpha carbon with strong bases. To protect a ketone or aldehyde from reacting with strong nucleophiles and bases, it can be converted to an acetal. Reaction with ethylene glycol in the presence of acid under dehydrating conditions converts an aldehyde or a ketone to its corresponding cyclic acetal (called a 1,3-dioxolane). It is possible to protect an aldehyde in the presence of a ketone, or even the less hindered carbonyl of a diketone. Removal of the protective group is achieved by aqueous acidic hydrolysis.

---

* Peter G. M. Wuts and Theodora W. Greene, *Greene's Protective Groups in Organic Synthesis*, 4th ed. (Wiley-Interscience, 2006).

formation of acetal
requires removal of H_2O

acetal PGs are removed
by acidic hydrolysis

Formation/Removal of 1,3-Dioxolane Protective Group

1.2.2 PROTECTION OF ALCOHOLS

Alcohols have acidic protons that can interfere with strongly basic species such as a Grignard reagent. Protection of the alcohol involves replacing the hydrogen with some groups that can later be removed ($RO-H \rightarrow RO-PG$). A variety of protective groups are available for alcohols, including ethers, esters, and acetals.

Ether Protective Groups for Alcohols

An ether is a very stable functional group that generally resists reactions with nucleophiles, bases, and oxidizing agents. In fact, ethers are so unreactive that many are unsuitable as protective groups since it would be nearly impossible to deprotect and get rid of the ether once it is created! A variety of special ethers with simple deprotection strategies have been developed and are regularly used in synthesis. For example, the benzyl ether protective group ($-CH_2Ph$, or Bn) is useful since it is very stable to most reaction conditions, but it can be removed by catalytic hydrogenation (H_2, Pd). The highly reactive benzylic carbon can be reduced, thus "deprotecting" the alcohol.

add benzyl PG

H_2, Pd

reduce to
remove Bn PG

Protecting an Alcohol as a Benzyl Ether

Similarly, the p-methoxybenzyl ether ($-CH_2C_6H_4OCH_3$, abbreviated as PMB or MPM for methoxyphenylmethyl) is simple to put on and can be removed by *oxidation* of the benzylic position (usually with 2,3-dichloro-5,6-dicyanobenzoquinone, or DDQ).

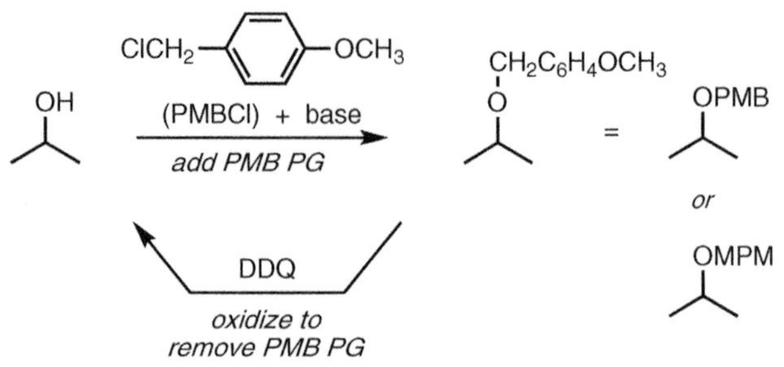

Protecting an Alcohol as a *p*-Methoxybenzyl Ether

The triphenylmethyl, or trityl, ether ($-CPh_3$, or Tr) can be cleaved by reduction or with acid, via the stabilized trityl carbocation. Derivatives of the trityl group are widely used as protective groups in the laboratory synthesis of oligonucleotides (DNA).

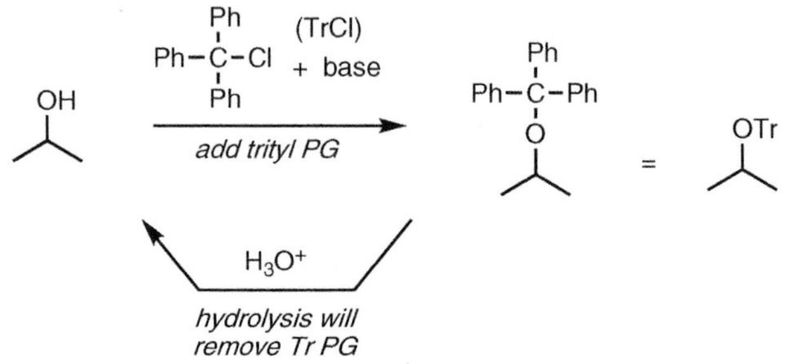

Protecting an Alcohol as a Trityl Ether

Silyl ethers are widely used protective groups because they also have a convenient method of deprotection: treatment with tetrabutylammonium fluoride (*n*-Bu$_4$NF, or TBAF, pronounced "t-baff"). Reaction of an alcohol (ROH) with trimethylsilyl chloride (($CH_3)_3$SiCl, or TMSCl) and base makes the protected trimethylsilyl ether (ROTMS). Variation of the alkyl groups on the silicon (e.g., triisopropylsilyl TIPS, or *t*-butyldimethylsilyl TBDMS/TBS) increases the silyl ether's stability, especially toward acidic conditions, and also allows for selective protection of less hindered alcohols.

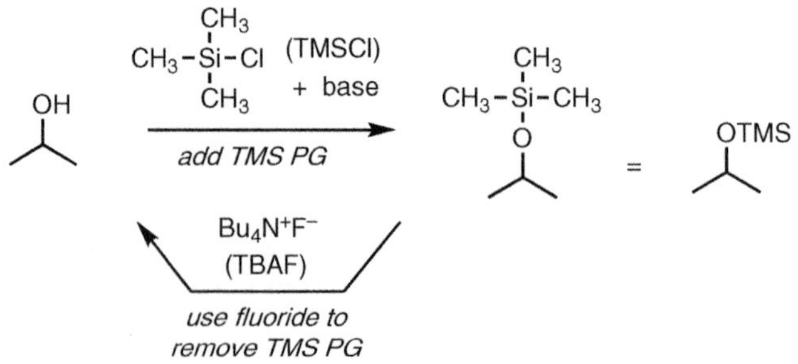

Protecting an Alcohol as a Silyl Ether

Ester Protective Groups for Alcohols

Esters (ROCOR') can be easily prepared by reacting an alcohol (ROH) with an acid chloride (R'COCl) and base. Deprotection is generally accomplished by basic hydrolysis or alcoholysis of the ester. Commonly used ester groups include acetate (—COCH$_3$, or Ac) and benzoate (—COPh, or Bz). The pivaloate ester (—COt-Bu, or Pv) is useful for selective acylation of a primary alcohol in the presence of more hindered secondary or tertiary alcohols.

Protecting an Alcohol as an Ester

Acetal Protective Groups for Alcohols

Since an acetal is produced by the reaction of a carbonyl with a diol, an acetal can serve as a protective group for either functional group. Deprotection is accomplished as usual, by treatment with acid and water; hydrolysis of the acetal regenerates both the carbonyl and the diol. Both 1,2- and 1,3-diols can be protected by reaction with acetone and acid. The resulting cyclic acetal (called an acetonide) is widely used in carbohydrate chemistry to selectively mask pairs of hydroxyl groups in sugars.

Protecting a 1,2- or 1,3-Diol as an Acetonide

The methoxymethyl (CH$_3$OCH$_2$—, or MOM) and tetrahydropyranyl (THP) groups are called "ether" groups (e.g., the hydroxyl group was protected to give the MOM ether), but they are actually examples of acetals that are cleaved by acidic hydrolysis. The MOM group is introduced by treatment of the alcohol with the very reactive chloromethyl methyl ether (CH$_3$OCH$_2$Cl, or MOMCl) and base (less toxic reagents are also available). The THP group is formed by reaction of an alcohol with dihydropyran (DHP).

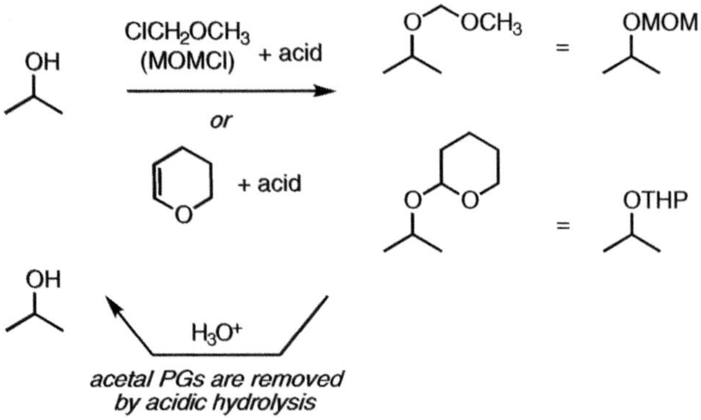

Protecting an Alcohol as an Acetal

1.2.3   PROTECTION OF CARBOXYLIC ACIDS

Carboxylic acids have acidic protons that can interfere with any basic species, including amines. Carboxylic acids are usually protected as esters. A methyl ester (RCO_2CH_3) can be prepared in a variety of ways, including reaction of an acid chloride with methanol (nucleophilic acyl substitution), reaction of a carboxylate nucleophile with methyl iodide (S_N2), or treatment of the carboxylic acid with diazomethane (CH_2N_2). Deprotection involves hydrolysis, usually under basic conditions (saponification). Bulky esters that inhibit nucleophilic attack of the carbonyl and esters with unique deprotection strategies are also commonly used. Examples include t-butyl esters ($RCO_2C(CH_3)_3$), which are stable to base but can be removed with trifluoroacetic acid (TFA), and benzyl esters (RCO_2CH_2Ph) that can be removed by catalytic hydrogenation.

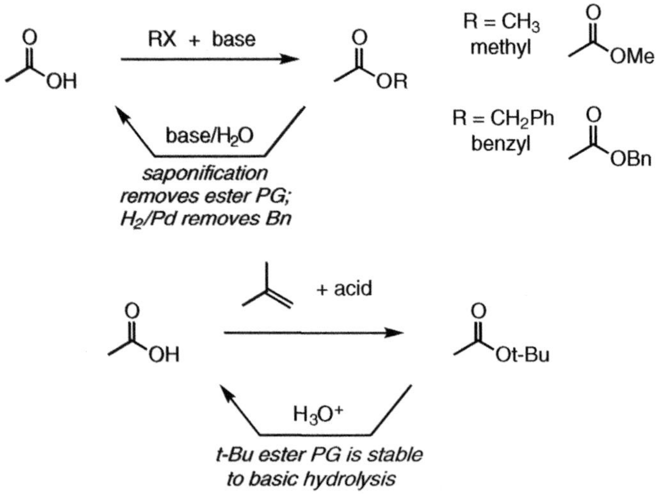

Protecting a Carboxylic Acid as an Ester

1.2.4 PROTECTION OF AMINES

Amines are good bases and strong nucleophiles. Protection of amines, typically as an amide or a carbamate, works by introducing a carbonyl that can delocalize the nitrogen's lone pair of electrons by resonance, thus rendering the nitrogen much less reactive.

Amide Protective Groups for Amines

Like esters, amides (RNHCOR') can be easily prepared by reacting an amine (RNH$_2$) with an acid chloride (R'COCl) and base. Commonly used acyl groups include acetyl (—COCH$_3$, or Ac) and benzoyl (—COPh, or Bz). Amide protective groups are generally removed by hydrolysis, but removal is often difficult since amides are fairly unreactive and stable.

Protecting an Amine as an Amide

Carbamate Protective Groups for Amines

Carbamate protective groups (RNHCO$_2$R') are more widely used in the protection of amines, especially for amino acids and in the synthesis of peptides. The *t*-butoxycarbonyl (—CO$_2$*t*-Bu, or BOC) group is added using di-*t*-butyl dicarbonate (BOC$_2$O) and base, and it is readily removed with acidic hydrolysis. The carboxybenzyl (—CO$_2$CH$_2$Ph, or Cbz or Z) group is introduced by treatment of the amine with benzyl chloroformate (BnOCOCl) and base; it can be removed by catalytic hydrogenation.

Protecting an Amine as a *t*-Butyl Carbamate

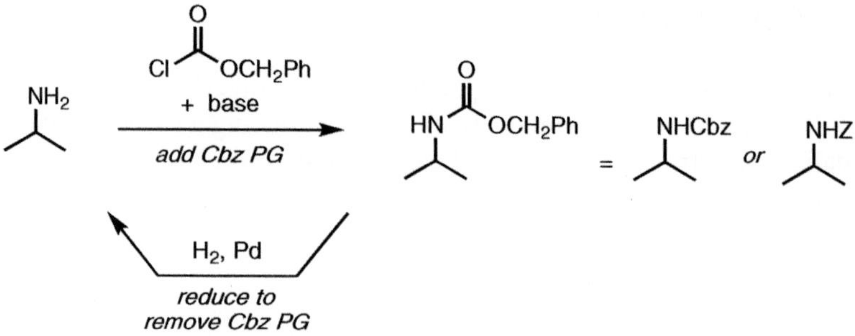

Protecting an Amine as a Benzyl Carbamate

PROTECTIVE GROUPS

1-1. Predict the major product expected when the given compound is treated with each of the following reagent(s). If no reaction is expected, write N.R.

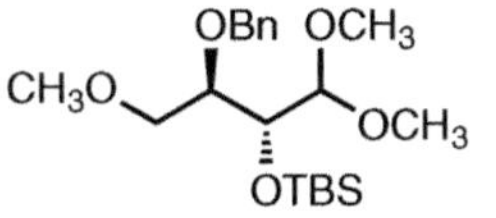

(a) H_2, Pd

(b) 1. PhMgBr 2. aq. NH_4Cl

(c) H_3O^+ (pH 1)

(d) $Bu_4N^+F^-$

(e) LDA (strong base)

1-2. A key strategy in organic synthesis is the ability to selectively protect and deprotect multiple functional groups. The compound shown below contains two protective groups, and the goal is to remove one protective group while leaving the other intact, to form either product **A** or product **B**. Of the following protective groups (PG = TMS, MOM, Tr, MPM, THP, Bz, Bn, Ac), which could be used to produce product **A** in the transformation shown below? For each of the suitable protective groups, what reaction conditions are needed to form **A**? Which of the protective groups listed are appropriate if the transformation to product **B** is desired, and what reaction conditions for each protective group are required to form **B**?

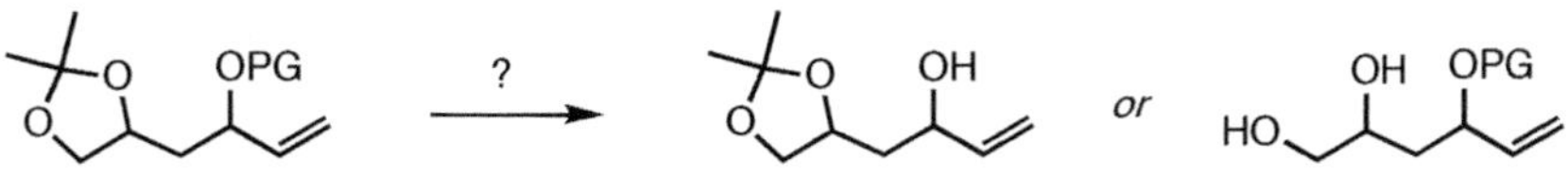

A

B

Introduction to Strategies for Organic Synthesis, Second Edition. Laurie S. Starkey.
© 2018 John Wiley & Sons, Inc. Published 2018 by John Wiley & Sons, Inc.

1-3. Provide the steps needed to accomplish each of the following transformations, using the starting materials and reagents given, along with any reagents needed for the installation and removal of required protective groups.

A

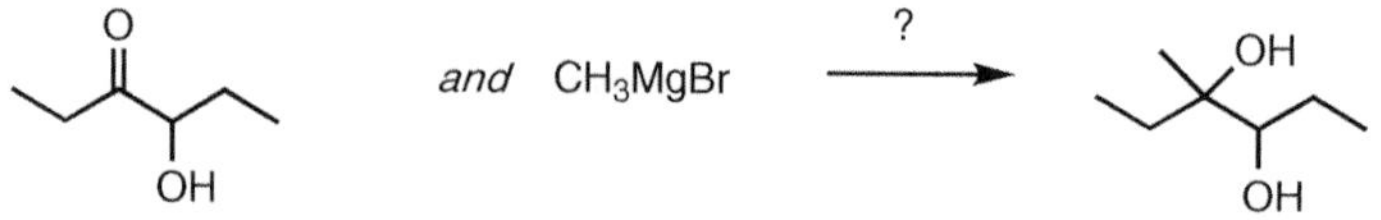
O
OH
and CH$_3$MgBr
?
OH
OH

B

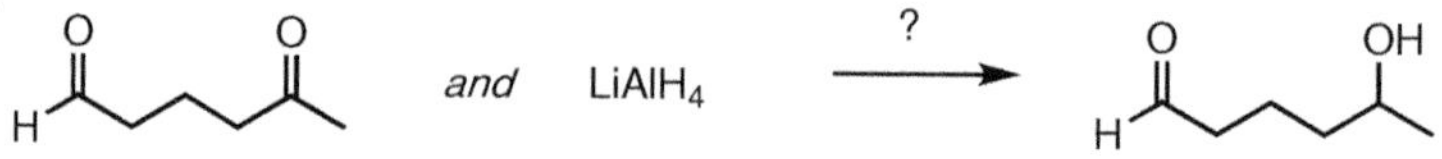
O
H
O
and LiAlH$_4$
?
O
H
OH

C

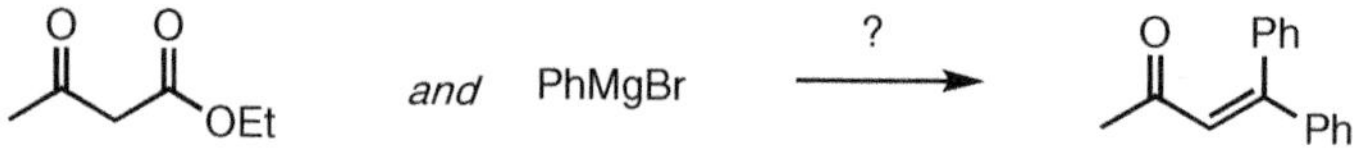
O
O
OEt
and PhMgBr
?
O
Ph
Ph

D

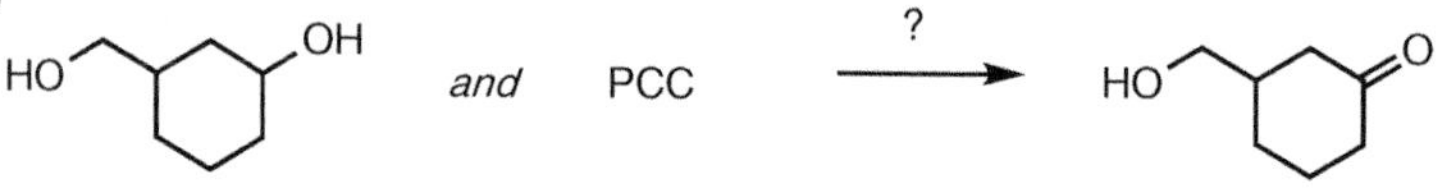
HO
OH
and PCC
?
HO
O

E

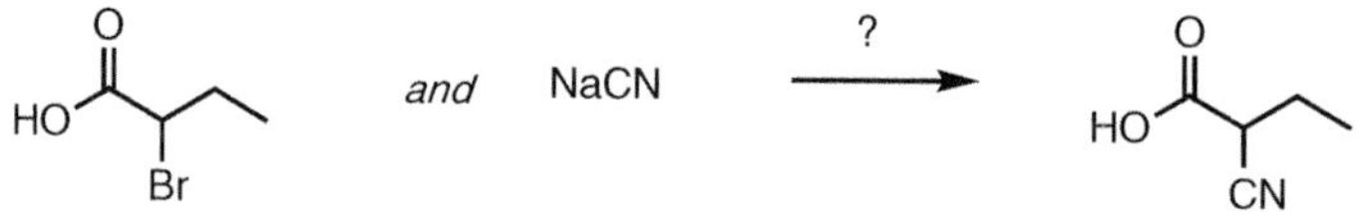
O
HO
Br
and NaCN
?
O
HO
CN

F

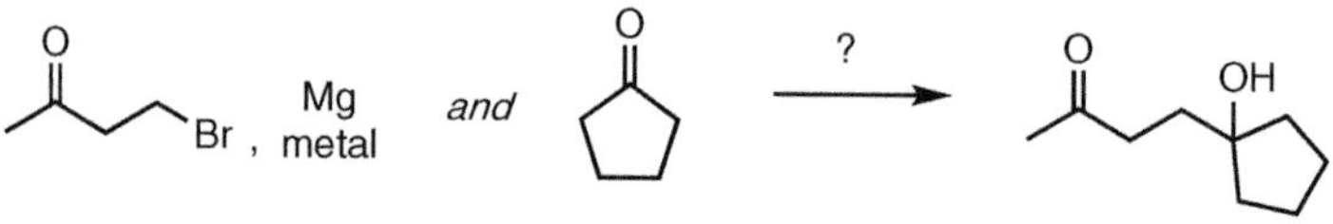
O
Br
Mg
, metal
and
O
?
O
OH

1-4. Show how protective groups can be used to prepare the following dipeptide from the given amino acids. Note: dicyclohexylcarbodiimide (DCC) is a reagent commonly used in the formation of amides from carboxylic acids and amines.

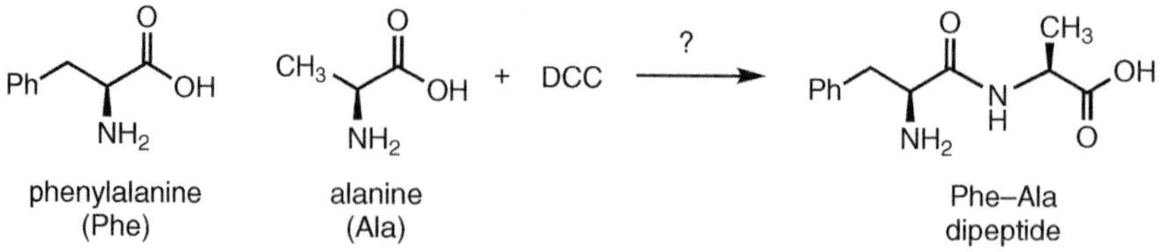
Ph
O
OH
NH$_2$
phenylalanine
(Phe)
CH$_3$
O
OH
NH$_2$
alanine
(Ala)
+ DCC
?
Ph
O
NH$_2$
N
H
CH$_3$
OH
O
Phe–Ala
dipeptide

SYNTHETIC TOOLBOX 2: OVERVIEW OF ORGANIC TRANSFORMATIONS

In order to learn the *strategies* of organic synthesis, one first needs to be knowledgeable about the various organic reactions and transformations that are possible. Toward that goal, this book will systematically review the many reactions that have been explored during a typical yearlong organic chemistry lecture sequence before using those reactions in the context of synthesis problems. In Chapter 2, our "Synthetic Toolbox" will be filled by exploring common nucleophiles and electrophiles and also by reviewing common oxidizing and reducing agents.

Introduction to Strategies for Organic Synthesis, Second Edition. Laurie S. Starkey.
© 2018 John Wiley & Sons, Inc. Published 2018 by John Wiley & Sons, Inc.

NUCLEOPHILES AND ELECTROPHILES

Most organic reactions result from the union of an electron-rich nucleophile (Nu:) with an electron-poor electrophile ($E+$). (Exceptions to this generalization include radical reactions, pericyclic reactions, and reactions mediated by organometallic species.) In order to properly plan for a typical organic synthesis, one must be familiar with commonly used electrophiles and nucleophiles. Presented in this chapter is an overview of such species, all of which are either commercially available or readily prepared. These nucleophiles and electrophiles will be employed throughout this book as their reactions and uses in synthesis are presented in detail in subsequent chapters.

2.1.1 COMMON NUCLEOPHILES

When a new bond is being formed, it is the nucleophile that is providing the electrons. A nucleophile is a species with either a lone pair of electrons or a pi bond that can be used to attack an electrophile. One can imagine a set of strong nucleophiles by considering those that would initiate an S_N2 (backside attack) substitution mechanism. The more electron-rich a nucleophile is, the stronger and more reactive it is, so most good nucleophiles have a negative charge. A negative charge is reasonable on an electronegative atom such as oxygen, or on a large atom such as iodine. Carbon is too electropositive to handle a negative charge by itself, so every carbanion nucleophile has some significant source of stabilization, such as being sp-hybridized or having resonance if it is ionic (e.g., RC≡C:⁻ and enolate, respectively), or being complexed with a metal such as in a Grignard reagent (RMgBr). Good nucleophiles that are neutral include atoms that are

Introduction to Strategies for Organic Synthesis, Second Edition. Laurie S. Starkey.
© 2018 John Wiley & Sons, Inc. Published 2018 by John Wiley & Sons, Inc.

not too electronegative, such as the nitrogen in an amine ($:NH_2R$), and those that are large and polarizable, such as the phosphorus in a phosphine ($Ph_3P:$).

Ionic and Other Commercially Available Nucleophiles

Amines and phosphines are among the few commonly used nucleophiles that are not negatively charged. Many anionic nucleophiles are listed below as stable salts that are commercially available. Note that the only carbanions shown here are cyanide and acetylide; these anions are relatively stable because the carbon bearing the negative charge is sp-hybridized. Other alkynyl anions ($RC{\equiv}C:^-$) can be readily prepared by deprotonation of a terminal alkyne ($RC{\equiv}CH$) with a strong base such as sodium amide ($NaNH_2$).

Nucleophile (Nu:)	Name	Reagents (notes)
$HO^{\ominus}$	hydroxide	NaOH
$RO^{\ominus}$	alkoxide	ROH + NaH
$RS^{\ominus}$	alkylthiolate	RSH + NaOH
RNH_2	amine	($RNH^{\ominus}$ is a strong base, not a Nu:)
R_3P	phosphine	(phosphorus is large and polarizable)
$NO_2^{\ominus},\ N_3^{\ominus}$	nitrite, azide	$NaNO_2$, NaN_3 (useful for amine synthesis)
$I^{\ominus}$	iodide	NaI (largest, best halide Nu:)
$N{\equiv}C:^{\ominus}$	cyanide	NaCN
$RC{\equiv}C:^{\ominus}$	acetylide	$RC{\equiv}CH$ + $NaNH_2$

Common Nucleophiles (Commercially Available or Readily Prepared)

Metal-Stabilized Nucleophiles

When an atom center is complexed with a metal, that atom will have some characteristics of an anion and can often behave as a nucleophile. An organometallic Grignard reagent (RMgBr) is an example of such a nucleophile that often behaves as if it were a carbanion. It is convenient to draw a Grignard reagent as a carbanion, but since it is not an ionic species, quotation marks are typically drawn around the carbanion when it is used in a mechanism ("R:⁻"). This same convention is sometimes used for the nucleophilic hydride species ("H:⁻") that is available when using lithium aluminum hydride ($LiAlH_4$).

Nucleophile (Nu:)	Name	Reagents
H:$^\ominus$	hydride	$LiAlH_4$ or $NaBH_4$
R:$^\ominus$	Grignard	$RX + Mg \longrightarrow RMgX$
	organolithium	$RX + Li \longrightarrow RLi$
	organocuprate	$2\,RLi + Cu \longrightarrow R_2CuLi$

Metal-Stabilized Nucleophiles

Resonance-Stabilized Nucleophiles

Certain carbanions can be prepared by deprotonation. This deprotonation will be favored only if the resulting conjugate base is more stable than the attacking base. The conjugate base carbanion is usually stabilized by resonance, but inductive stabilization can also enable deprotonation. Protons alpha to an electron-withdrawing group, such as a carbonyl (C=O), cyano (—C≡N), or nitro (—NO$_2$) group, are acidic because the resulting carbanion is resonance-stabilized (an enolate). If a carbon is alpha to two electron-withdrawing groups, it is especially acidic and will be readily deprotonated to give a "stabilized" enolate that has extra resonance delocalization of the negative charge. Such enolates are even more stable and, therefore, less basic, making them excellent nucleophiles.

Enolate (Nu:)		Reagents	Notes
	Nu: here	+ LDA	ketone pK$_a$ ~20
	extra resonance (2 EWGs)	+ NaOR	diketone pK$_a$ ~9
		$O_2N–CH_3$ + NaOH	α to nitro pK$_a$ ~10
$N≡C–\overset{..}{\underset{\ominus}{C}H_2}$	$^\ominus N=C=CH_2$	$NC–CH_3$ + LDA	nitrile pK$_a$ ~25

Resonance-Stabilized Nucleophiles (Enolates)

2.1.2 COMMON ELECTROPHILES

The natural partner to an electron-rich nucleophile is an electron-poor species that is seeking electrons: an electrophile. Unlike the anionic nucleophiles seen in the previous sections, cationic electrophiles are not readily available. For example, a carbocation would be a great electrophile, but because it is missing an octet it is highly unstable and very reactive. This is not something one can easily put in a bottle or get from the stockroom! Instead, commercially available electrophiles typically have electron-deficient, *partial* positive (δ^+) reactive sites that will combine with nucleophiles. If a nucleophile is going to form a bond with the electrophile, then the electrophile must somehow be able to accept those additional electrons. How can this happen if the electrophilic atom already has a filled octet? In order for a nucleophile to create a new bond, an old bond must also be broken. The broken bond will be either a sigma bond to a leaving group or a pi bond to an electron acceptor, so these are common features in the electrophiles shown below.

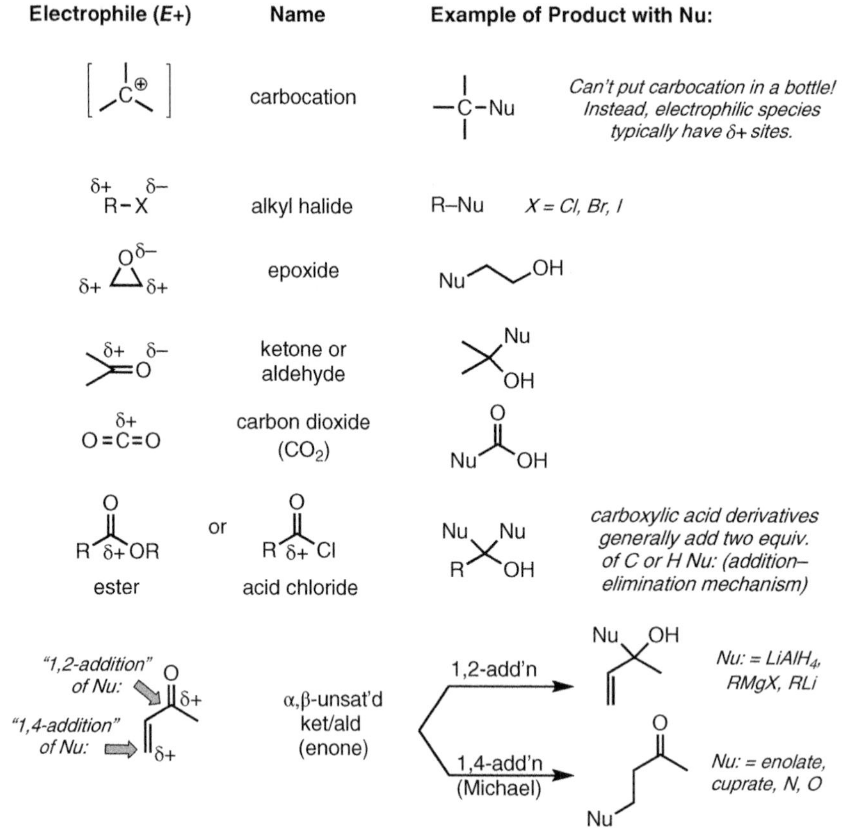

Electrophile (E+)	Name	Example of Product with Nu:	
[carbocation structure]	carbocation	—C—Nu	*Can't put carbocation in a bottle! Instead, electrophilic species typically have $\delta+$ sites.*
δ^+ δ^- R–X	alkyl halide	R–Nu	$X = Cl, Br, I$
epoxide structure (δ^+, δ^-, δ^+)	epoxide	Nu⌒OH	
δ^+ δ^- C=O	ketone or aldehyde	Nu / OH structure	
δ^+ O=C=O	carbon dioxide (CO_2)	Nu—C(=O)—OH	
ester (R–$\delta+$–OR) or acid chloride (R–$\delta+$–Cl)	ester or acid chloride	Nu, Nu / R, OH structure	*carboxylic acid derivatives generally add two equiv. of C or H Nu: (addition–elimination mechanism)*

"1,2-addition" of Nu: ↘
"1,4-addition" of Nu: ⟹

α,β-unsat'd ket/ald (enone)

1,2-add'n → [Nu, OH structure]   *Nu: = LiAlH₄, RMgX, RLi*

1,4-add'n (Michael) → [Nu structure]   *Nu: = enolate, cuprate, N, O*

Common Electrophiles

OXIDATION AND REDUCTION REACTIONS

2.2.1   OVERVIEW OF OXIDATIONS AND REDUCTIONS

General Chemistry Examples of Redox Reactions

When an atom gets oxidized, it experiences an increase in its oxidation number. A reagent causing an oxidation is known as an oxidizing agent. When an atom gets reduced (by reacting with a reducing agent), that atom's oxidation number is "reduced" as it accepts electrons. In a typical General Chemistry introduction to the topic involving the reactions of metals and metal ions, one can readily observe the change in oxidation number, along with the gain or loss of electrons associated with redox reactions.

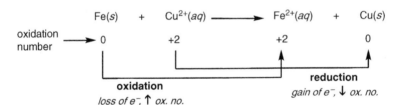

General Chemistry Description of Redox Reactions

Organic Chemistry Examples of Redox Reactions

When dealing with carbon atoms, oxidations and reductions are often easy to recognize without calculating oxidation numbers. Any reaction that *increases the number of C–O bonds* (while decreasing the number of C–H or C–C bonds) is described as an *oxidation*. Any reaction that *decreases the number*

Introduction to Strategies for Organic Synthesis, Second Edition. Laurie S. Starkey.
© 2018 John Wiley & Sons, Inc. Published 2018 by John Wiley & Sons, Inc.

of C—O bonds (while also increasing the number of C—H or C—C bonds) is described as a *reduction*. Since stable carbon atoms contain four bonds, an example of the most oxidized state of carbon can be found in CO_2 (product of the oxidative combustion process) where all four bonds are to oxygen, and an example of the most reduced form of carbon can be found in CH_4, where all four bonds are to hydrogen. While organic reductions and oxidations do not always involve oxygen bonds, these general patterns can still be helpful. For example, it is possible to identify the catalytic hydrogenation of an alkene as a reduction since it involves an increase in the number of C—H bonds (while decreasing the number of C—C bonds in this case).

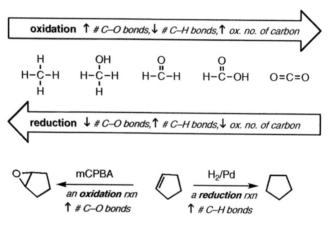

Organic Chemistry Description of Redox Reactions

As you replace a bond to hydrogen (ox. no. +1) with a bond to oxygen (ox. no. −2) the oxidation number of the carbon increases as a result. However, a calculation is not required and the determination that a given reaction is an oxidation (or a reduction) can be done by inspection. If it is recognized that a given transformation is an oxidation, the task then becomes the simple one of selecting an appropriate oxidizing agent.

Effect of Heteroatoms on the Oxidation State of Carbon

If a carbon atom is bonded to something other than hydrogen and oxygen, determination of its oxidation state becomes more challenging. Still, it can typically be done by inspection. The term heteroatom is used to describe any atom other than carbon or hydrogen; those commonly encountered in organic molecules include O, N, S, Cl, Br, and I. Since each of these atoms is more electronegative than carbon, they all have the same effect on the oxidation state of a carbon to which they are attached.

While the halogenation of CH_4 to give CH_3Br is an oxidation reaction (note that a C–H bond has been replaced by a C–Br bond), it is usually described simply as a halogenation rather than an oxidation, since a new *oxygen* bond has not been introduced.

Not all Redox Reactions Involve Oxygen

However, the conversion of HC≡N to HCO_2H is not an oxidation, since the carbon started out with three bonds to a heteroatom and still has three bonds to a heteroatom in the product. While the transformation involves an increase in the number of C–O bonds, it is only by replacement of C–N bonds, so there is no change in the oxidation number of carbon, and the reaction is not described as an oxidation (in this instance, it is an example of a hydrolysis reaction).

Not all Reactions Involving Heteroatoms are Redox Reactions

2.2.2 COMMON OXIDATION REACTIONS AND OXIDIZING AGENTS

A wide variety of oxidizing agents exist, and they can look deceivingly similar to the beginning student. For example, various chromium(VI)-based oxidizing agents can have quite different reactivities depending on the exact experimental reaction conditions (e.g., addition of pyridine–HCl, water, heat, etc.), and most textbooks point to distinctly different results when using "hot" and "cold" $KMnO_4$ conditions (or acidic/basic). The purpose of this section is to review common oxidation reactions, along with some representative reagents to achieve these transformations.

Oxidation of Alcohols

The oxidation of an alcohol (contains one C—O bond) can generate either an aldehyde or a ketone (a C=O has two C—O bonds) or a carboxylic acid (RCO_2H has three C—O bonds). In each case, the functional group formed depends on the type of alcohol being oxidized and the oxidizing agent being used. It is the carbon bearing the OH group that is being oxidized, and a primary alcohol (1° ROH, e.g., RCH_2OH) has two hydrogens on that carbon. If a strong oxidizing agent is used (such as potassium permanganate or chromic acid, called a Jones oxidation), then both of these C—H bonds will be replaced with C—O bonds, resulting in a carboxylic acid product.

Oxidation of 1° Alcohols to Carboxylic Acids

If a more selective oxidizing agent is used, then the alcohol will be only partially oxidized to give an aldehyde product (only one C—H bond is replaced). A wide variety of oxidizing agents/conditions are available for this transformation, including pyridinium chlorochromate (PCC), Dess–Martin periodinane (DMP), and Swern oxidation, just to name a few. Typically, the motivation for development of new reagents and reaction conditions may be driven by any number of goals including of increased selectivity, lower cost, or being more environmentally friendly. Although commonly found in textbooks, PCC is no longer widely used since hexavalent chromium compounds are toxic (as a result, they are hazardous to work with and waste disposal is expensive).

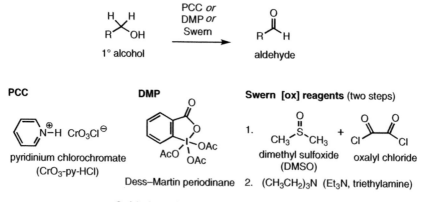

Oxidation of 1° Alcohols to Aldehydes

Because a secondary alcohol (2° ROH, e.g., R_2CHOH) has only one hydrogen atom on the carbon bearing the OH group, it can give only one possible oxidation product. Treatment of a secondary alcohol with a variety of oxidizing agents (e.g., any of the oxidizing agents listed earlier) will afford a ketone product. Tertiary alcohols (3° ROH, e.g., R_3COH) have no hydrogens on the alcohol carbon, so no useful oxidation reaction will take place.

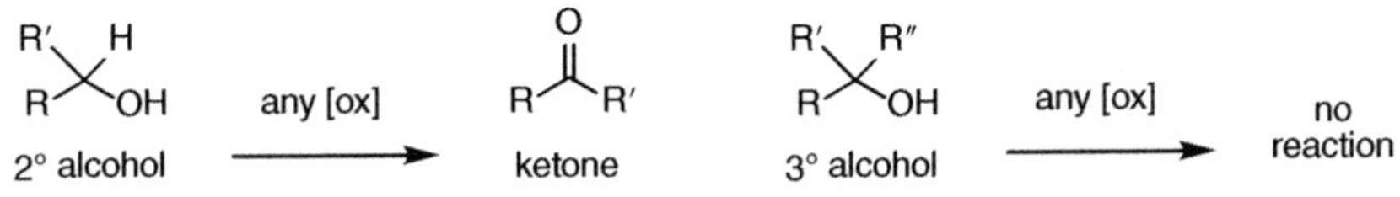

Oxidation of 2° and 3° Alcohols

Oxidation of Diols

While most oxidation reactions in organic synthesis involve loss of a C–H bond, it is sometimes possible to break a C–C sigma bond instead. For example, 1,2-diols (also called vicinal diols or glycols) undergo oxidative cleavage when treated with a strong oxidizing agent such as periodic acid, HIO_4. This reaction breaks the bond between the alcohol carbons and replaces the C–C bond with a new C–O bond on each, resulting in a carbonyl at each carbon.

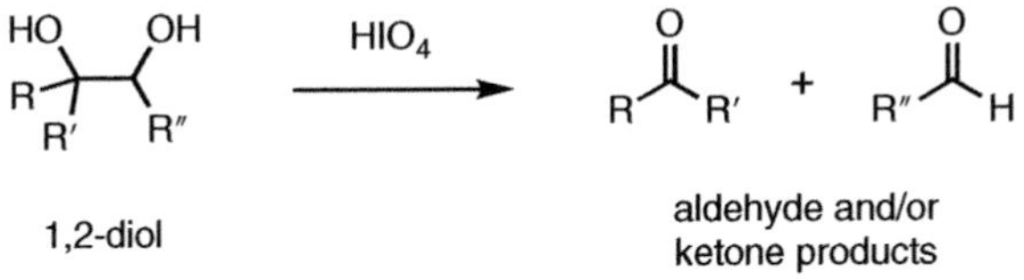

Periodate Oxidation of 1,2-Diols

Oxidation of Aldehydes

Since the carbonyl carbon of an aldehyde contains a C–H bond, it is subject to oxidation. Treatment with any strong oxidizing agent (e.g., Jones, $KMnO_4$, etc.) results in the formation of a carboxylic acid.

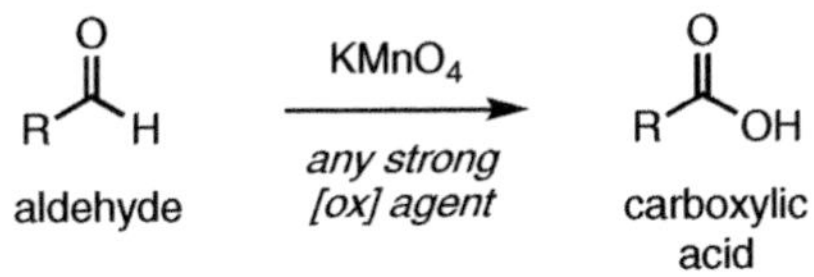

Oxidation of Aldehydes

Oxidation of Ketones

Reaction of a ketone with a peroxyacid such as *meta*-chloroperoxybenzoic acid (mCPBA) or peracetic acid (CH_3CO_3H), results in the insertion of an oxygen atom on one side or the other of the ketone to give an ester product. This reaction is called a Baeyer–Villiger oxidation and results in the migration of one of the carbon groups attached to the carbonyl carbon. The mechanism involves a nucleophilic addition to the carbonyl by the peroxyacid, followed by a "collapse" of the tetrahedral intermediate to reform a carbonyl and break a C—C sigma bond. The carbon group that is "ejected" in this step does not leave, however; it simply shifts over and bonds with the peroxy oxygen (while displacing a carboxylate leaving group attached to that oxygen). The order of preference for this migration is as follows:

3° alkyl > 2° alkyl > phenyl > 1° alkyl > methyl
(most likely to migrate)   (least likely)

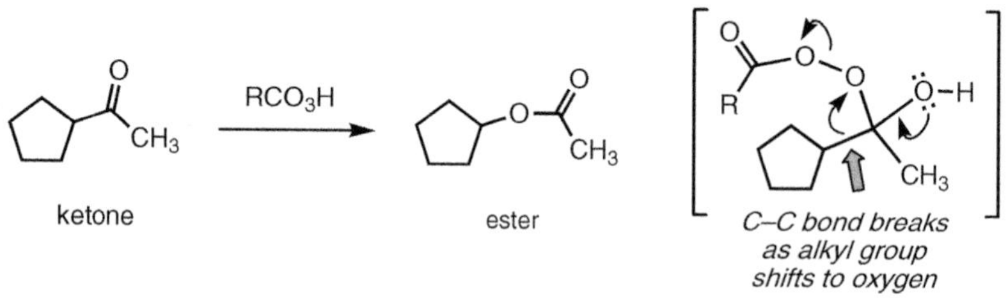

Baeyer–Villiger Oxidation of Ketones

Baeyer–Villiger oxidation of a cyclic ketone results in a cyclic ester (a lactone) with a ring that has been expanded. Note that the stereochemistry of the carbon center involved in the migration is retained during the oxidation.

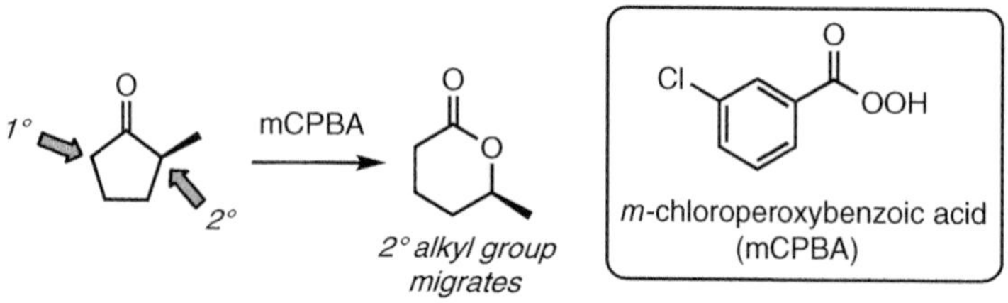

Baeyer–Villiger Oxidation to Prepare Lactones

Oxidation of Alkenes

The three most common alkene oxidation reactions are epoxidation, dihydroxylation, and ozonolysis. Epoxides are formed when an alkene is treated with a peroxyacid such as mCPBA. Since both C–O bonds are formed in the same step (described as a concerted mechanism), the stereochemistry of the starting alkene is preserved in the product.

concerted, syn addition, so cis R groups remain cis

Epoxidation of Alkenes

Alkenes can undergo a dihydroxylation with either syn or anti addition. Treatment with basic $KMnO_4$ or OsO_4 results in the syn addition of two OH groups. (The oxidizing strength of $KMnO_4$ varies widely with pH and other reaction conditions, so it would also be possible to form other oxidation products, including cleavage products like those generated in the HIO_4 oxidation of diols.) Since both C–O bonds are formed in the same step in this concerted mechanism, they are added to the same face of the alkene. An alternate dihydroxylation sequence involves a ring-opening reaction of an epoxide with water (or hydroxide) as a nucleophile, resulting in a product with two hydroxyl groups that are trans to each other. The epoxidation and subsequent ring-opening reaction can be achieved in a "one-pot" method by treatment of an alkene with a peroxide, water, and acid. Reaction with peracetic acid (CH_3CO_3H) and water causes *in situ* formation of the epoxide, followed by acid-catalyzed opening of the epoxide with backside attack by water, resulting in overall anti addition of the two hydroxyl groups.

syn dihydroxylation
*concerted, syn addition
gives cis hydroxyl groups*

anti dihydroxylation
*epoxide ring opening (S_N2)
gives trans hydroxyl groups*

Dihydroxylation of Alkenes

As the name implies, ozonolysis causes molecules to "lyse" or break up by reaction with ozone (O_3). Ozonolysis is a two-step process: treatment of an alkene with ozone creates an ozonide intermediate that breaks down upon workup. A "reductive" workup uses a reducing agent such as zinc or dimethylsulfide (DMS, CH_3SCH_3). If there were any hydrogens attached to the original C=C double bond, they are preserved in a reductive workup and aldehyde products would be formed. The use of $NaBH_4$ in the workup reduces the carbonyls formed *in situ*, affording alcohol products. If an "oxidative" workup were used instead, the added oxidizing agent (H_2O_2) would replace any aldehyde C–H bonds with C–O bonds to give carboxylic acid products. Ozonolysis can be selective for the more electron-rich double bond of a diene (e.g., a trisubstituted double bond will be cleaved preferentially over a disubstituted one). The variety of workup options makes the ozonolysis reaction a versatile synthetic tool. Rather than using ozonolysis, an alkene can also be converted to a ketone/aldehyde by using a combination of OsO_4 (makes a 1,2-diol) and $NaIO_4$ (cleaves and oxidizes the 1,2-diol *in situ*).

Ozonolysis of Alkenes

Oxidation of Alkynes

Upon oxidation of an alkyne, the C≡C triple bond can be cleaved completely. All three C–C bonds are oxidized to form three new C–O bonds, resulting in carboxylic acid products. This oxidation can be carried out either by ozonolysis or with $KMnO_4$ (dihydroxylation occurs, followed by additional oxidative cleavage).

Oxidation of Alkynes

Oxidation of Allylic and Benzylic Carbons

In reviewing the oxidation reactions presented thus far, it can be observed that in order for a carbon to be oxidized, it must have either a pi bond or at least one bond to oxygen already in place. However, introduction of a new C–O bond onto an ordinary alkyl group can be possible if that group is attached to an alkene or an aromatic ring. Such carbons, called allylic and benzylic carbons, respectively, are unusually reactive due to the neighboring π molecular orbitals. Treatment of an alkene with selenium dioxide (SeO_2) oxidizes the allylic carbon to give an allylic alcohol product. Allylic bromination (NBS) would also be described as an oxidation reaction.

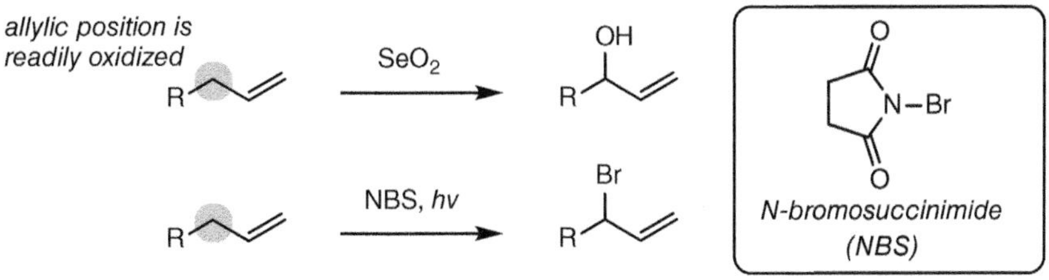

Oxidation of Allylic Carbons

When an aromatic compound is subjected to a strong oxidizing agent (e.g., Jones), almost any alkyl side chain will be completely oxidized to give a carboxylic acid. It is interesting to note that all bonds to the benzylic carbon will be broken (including C—C sigma bonds) to give a benzoic acid product. The only exception is that the benzylic carbon cannot be a quaternary carbon, as these will not be oxidized.

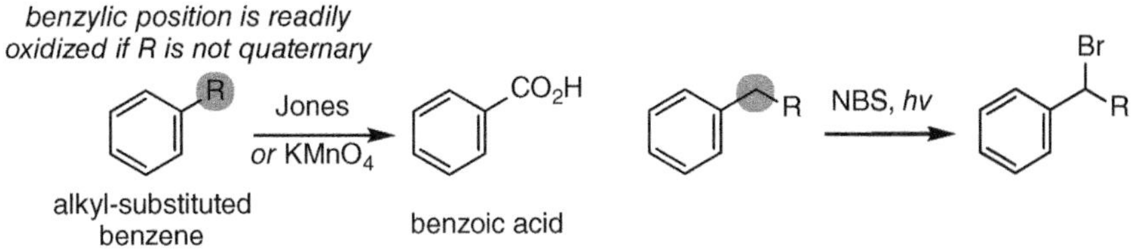

Oxidation of Benzylic Carbons

Oxidation of Ketone α-Carbons

The introduction of any heteroatom alpha to a carbonyl can be described as an oxidation of the α-carbon. In one method, the ketone is first converted to the corresponding silyl enol ether, by treatment with trimethylsilyl chloride and base (TMSCl, Et$_3$N). Reaction of this enol ether with bromine (Br$_2$) gives an α-bromo product. Oxidation of the enol ether with OsO$_4$ or mCPBA gives the α-hydroxy ketone product.

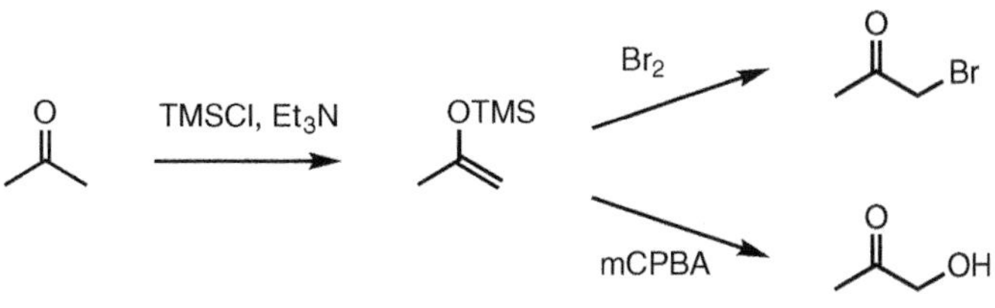

Oxidation of α-Carbons

2.2.3 COMMON REDUCTION REACTIONS AND REDUCING AGENTS

In organic synthesis, reduction reactions are typically characterized as those reactions that add hydrogen (i.e., increase the number of C—H bonds).

Catalytic Hydrogenation

Treatment of an alkene or alkyne with hydrogen gas in the presence of a metal catalyst such as palladium affords an alkane product. Syn addition of hydrogen is observed. If a "poisoned" catalyst is used (such as Lindlar's catalyst) in the hydrogenation of an alkyne, only one equivalent of H_2 is added to give a cis alkene product. Selective reduction of the less hindered double bond of a diene can be achieved with the use of an organometallic catalyst such as $Rh(PPh_3)_3Cl$ instead of Pd.

The unusual reactivity of the benzylic position is evident when benzylic ethers are reduced with catalytic hydrogenation (ordinary ethers have no reaction). This reactivity makes the benzyl group ($PhCH_2-$ or Bn) an especially useful protecting group for alcohols, since hydrogenation could be used to remove the protecting group and regenerate the alcohol. Catalytic hydrogenation is also useful in the synthesis of amines, since reduction of nitrile, imine, nitro, and azide functional groups all give amine products.

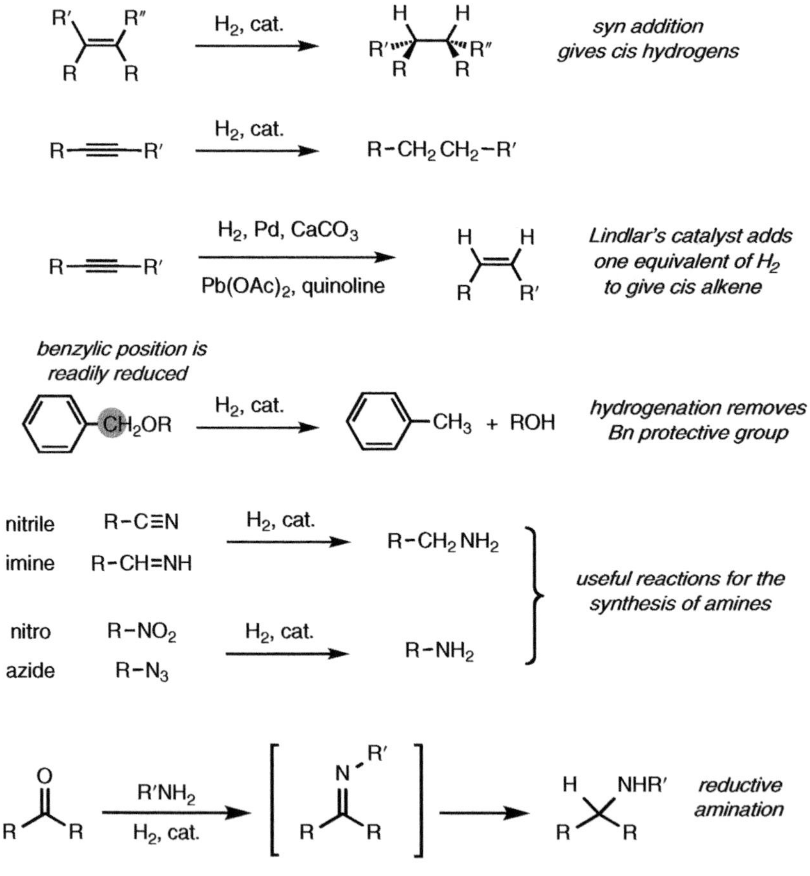

Hydrogenation of a Variety of Functional Groups

Under ordinary conditions, carbonyl groups of aldehydes and ketones are not readily reduced by catalytic hydrogenation, but if Raney nickel ($Ni-H_2$) is used as the catalyst, then a reduction takes place giving an alcohol product. If the ketone or aldehyde is first converted to a dithiane, then Raney Nickel reduction will give an alkane product (see Clemmenson and Wolff–Kishner reductions below as alternatives for this same transformation).

Hydrogenation/Reduction of Ketones and Aldehydes

Hydride Reagents

Hydride reagents (lithium aluminum hydride, $LiAlH_4$, and sodium borohydride, $NaBH_4$) are a source of nucleophilic hydride ("$H:^-$"). Reaction of hydride with a suitable carbon electrophile results in a reduction of that carbon (by increasing the number of C–H bonds). Reactions of carbonyl compounds with lithium aluminum hydride (LAH) generally give an alcohol product (after workup), with the exception of amides, which give amine products. Sodium borohydride is less reactive than LAH. It does not react with esters, amides, or carboxylic acids, so it is described as being "selective" for aldehydes and ketones. Sodium borohydride also fails to reduce nitroalkanes or alkyl halides, so LAH must be used in those reactions.

By varying the groups on aluminum, different reactivities can be achieved. One such useful reagent is diisobutylaluminum hydride (DIBAL-H). If one equivalent of DIBAL-H is used, this selective reagent only partially reduces acid chlorides, esters, and nitriles to give aldehyde products (addition of hydride to a C≡N triple bond gives a C=N double bond which is hydrolyzed to a C=O double bond upon workup).

Hydride Reductions

Metals as Reducing Agents

A variety of metals act as reducing agents, since they are capable of donating their valence electron(s). The previous section showed that aldehyde and ketone carbonyls can be reduced to alcohols with hydride reagents, but these carbonyls can also be completely reduced to a methylene (CH_2) by a metal-promoted method called the Clemmenson reduction (Zn/Hg, HCl). (The Wolff–Kishner reduction, using NH_2NH_2 and KOH, is a complimentary method with basic reaction conditions that is also commonly used to reduce a ketone or aldehyde to an alkane.)

Complete Reduction of Carbonyl

Sodium or lithium metal in ammonia causes a "dissolving metal reduction" of alkynes to give a trans alkene product. This reduction does not work for terminal alkynes, but other metals are available (such as Zn—Cu) to accomplish this transformation.

Dissolving Metal Reduction of Alkynes

These same conditions will reduce an aromatic ring to a 1,4-diene (called a Birch reduction). For a substituted benzene ring, the regiochemistry of the diene product depends on the nature of the substituent. Addition of an electron to the benzene ring results in a radical anion intermediate. Since this anion is stabilized by electron-withdrawing groups and destabilized by electron-donating groups, the pi bonds in the product are positioned according to the nature of the substituents present.

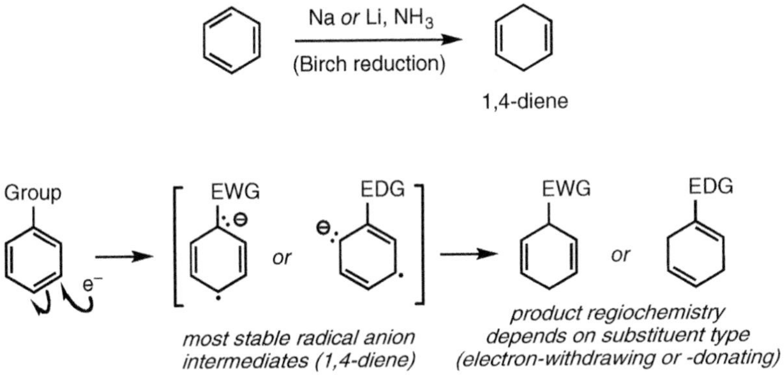
Na or Li, NH3
(Birch reduction)
1,4-diene
Group
e⁻
EWG
EDG
or
EWG
EDG
or
most stable radical anion intermediates (1,4-diene)
product regiochemistry depends on substituent type (electron-withdrawing or -donating)

Birch Reduction of Aromatic Rings

NUCLEOPHILES, ELECTROPHILES, AND REDOX

2-1. Anions are synthetically useful since they make good nucleophiles, and they are typically generated by deprotonation of a neutral compound. Which of the following compounds has an acidic proton? For each compound, predict the major product formed by its reaction with one equivalent of a strong base such as NaH. If no reaction is expected, write N.R.

A

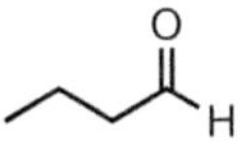

B

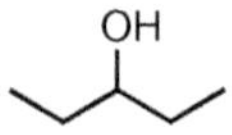

C

D

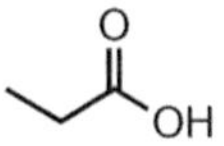

E

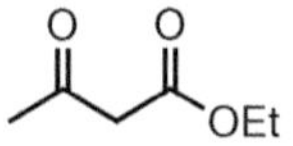

F

G

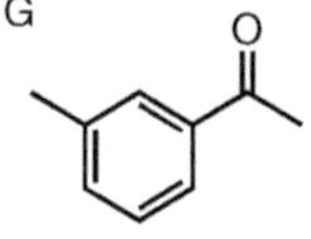

H

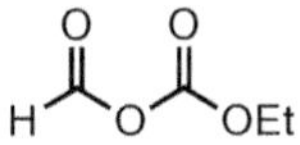

I

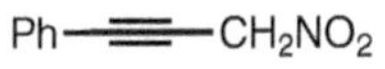

Ph—≡—CH$_2$NO$_2$

J

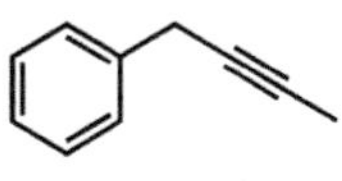

K

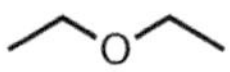

L

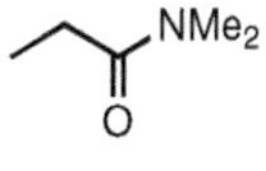

M

NC—CN

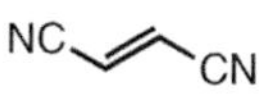

N

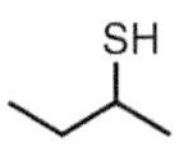

O

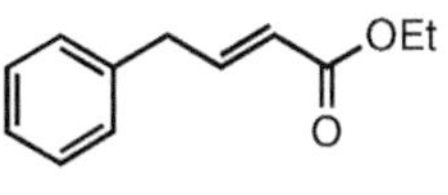

Introduction to Strategies for Organic Synthesis, Second Edition. Laurie S. Starkey.
© 2018 John Wiley & Sons, Inc. Published 2018 by John Wiley & Sons, Inc.

2-2. Electron-rich nucleophiles react with electron-poor electrophiles. Identify the electrophilic carbon(s) on each of the following compounds by placing a δ+ at the electron-deficient site(s) that are capable of reacting with nucleophiles. If no electrophilic carbon(s) exist, then no reaction with a nucleophile is expected, so write N.R.

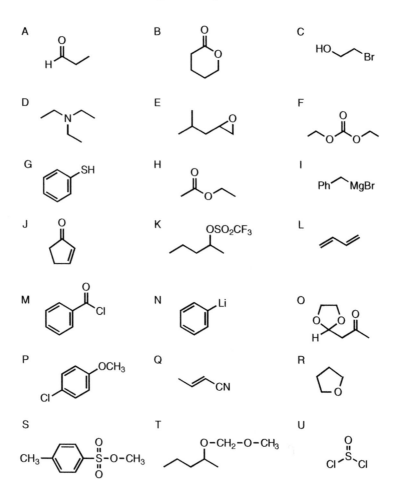

2-3. Identify the following reactions as either a reduction or an oxidation (or state that no reduction or oxidation has taken place, if that is the case) and provide the reagent(s) necessary to achieve the transformation.

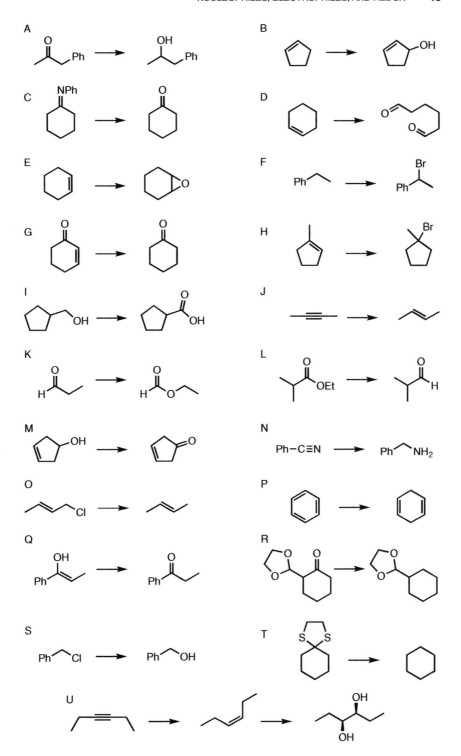

2-4. Provide the missing reagents for each of the following synthetic steps.

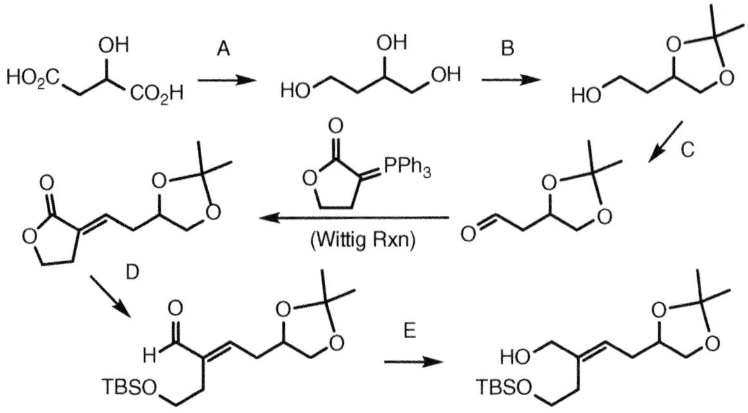

CHAPTER 3

SYNTHESIS OF MONOFUNCTIONAL TARGET MOLECULES (1-FG TMs)

The first synthetic targets to be explored are those containing a single functional group (e.g., a simple alcohol or alkene).

Introduction to Strategies for Organic Synthesis, Second Edition. Laurie S. Starkey.
© 2018 John Wiley & Sons, Inc. Published 2018 by John Wiley & Sons, Inc.

SYNTHESIS OF ALCOHOLS (ROH) AND PHENOLS (ArOH)

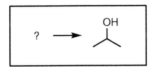

Alcohols are extremely versatile starting materials for organic synthesis, since they can be converted to a wide variety of functional groups. Many simple alcohols are commercially available, but this chapter will focus on the preparation of alcohols. These synthetic strategies will be useful not only for complex alcohol target molecules but also for alcohols that may be intermediates within a larger synthesis.

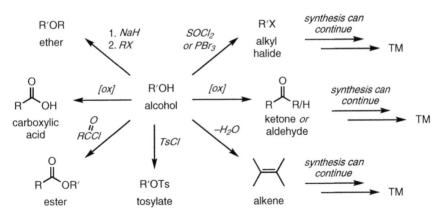

Alcohols are Versatile Starting Materials

Introduction to Strategies for Organic Synthesis, Second Edition. Laurie S. Starkey.
© 2018 John Wiley & Sons, Inc. Published 2018 by John Wiley & Sons, Inc.

3.1.1 SYNTHESIS OF ALCOHOLS BY FUNCTIONAL GROUP INTERCONVERSION (FGI)

Alcohols can be prepared from a variety of other functional groups, including by reduction of a carbonyl, hydration of an alkene, or substitution of a leaving group. Recall that regiochemistry is a concern when starting with an alkene, and water can be added with either Markovnikov (oxymercuration–demercuration) or anti-Markovnikov (hydroboration–oxidation) orientation.

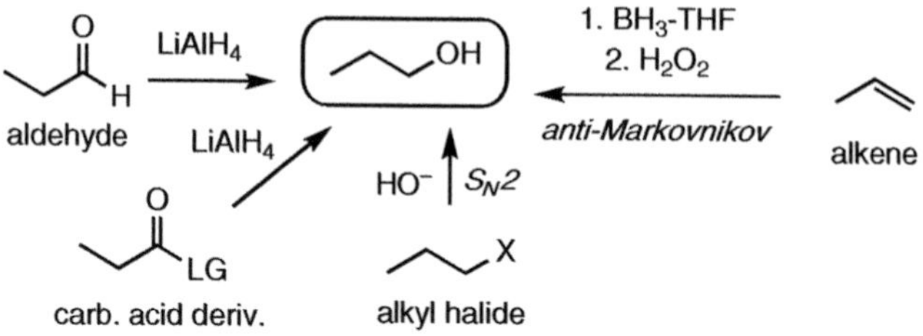

FGI to Make Primary Alcohols (1° ROH)

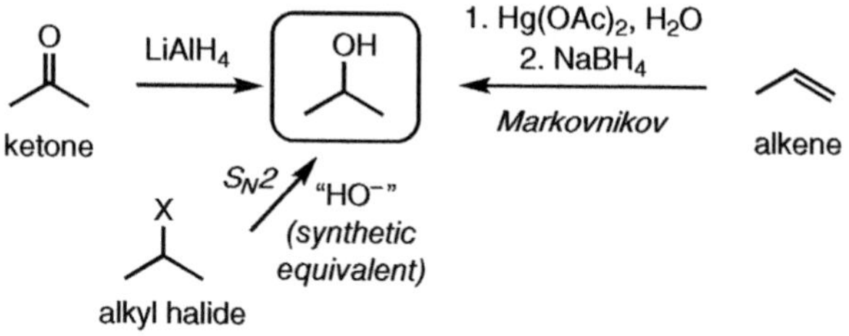

FGI to Make Secondary Alcohols (2° ROH)

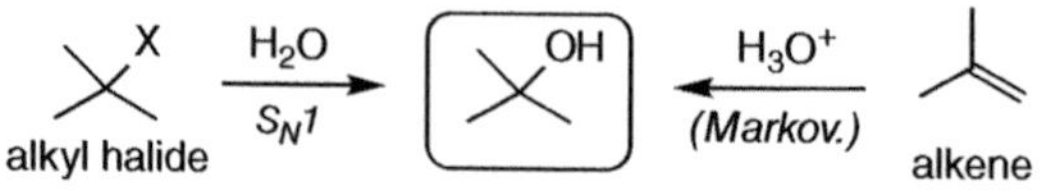

FGI to Make Tertiary Alcohols (3° ROH)

Also, recall that S_N2 substitution with strongly basic hydroxide (HO⁻) is reasonable only if there is an unhindered, 1° leaving group, or possibly with a leaving group on a 2° carbon that is allylic (i.e., next to a pi bond) or benzylic (i.e., next to a benzene ring). Otherwise, E2 elimination is the more favorable mechanism.

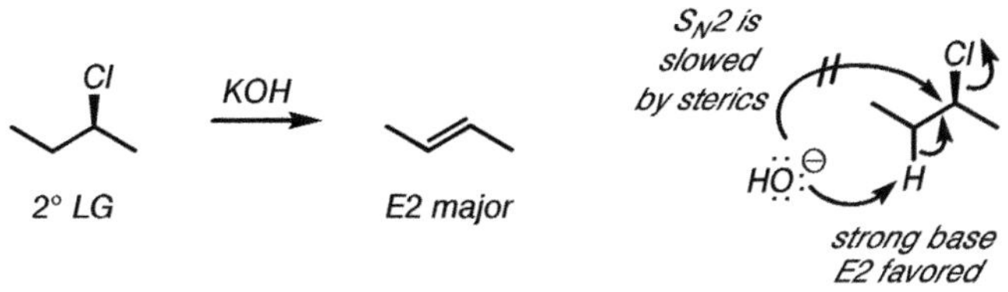

Reaction of Hydroxide with Typical 2° LG: E2 Favored

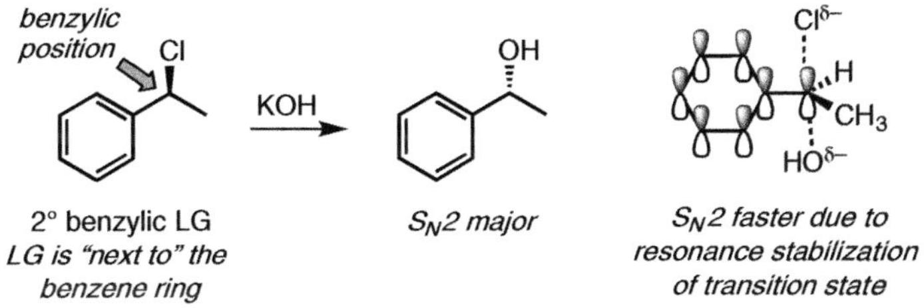

Reaction of Hydroxide with 2° *Allylic* LG: S$_N$2 Favored

Acetate as a Synthetic Equivalent of Hydroxide

In order to promote an S$_N$2 mechanism at a non-allylic 2° carbon, a different nucleophile is needed, such as acetate (AcO⁻). Since acetate is resonance-stabilized, it is a weaker base than hydroxide and the competing E2 elimination mechanism is minimized.

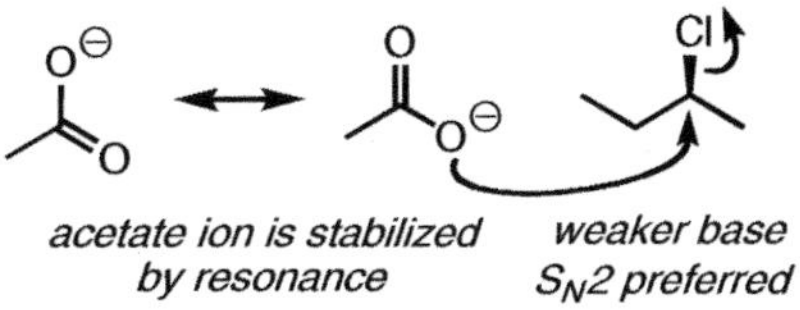

acetate ion is stabilized   weaker base
by resonance   S$_N$2 preferred

Reaction with Stabilized Nu: Favors S$_N$2

Substitution with acetate indeed replaces the leaving group with an oxygen atom, but it gives an *ester* product; hydrolysis of the ester affords the desired *alcohol* target molecule.

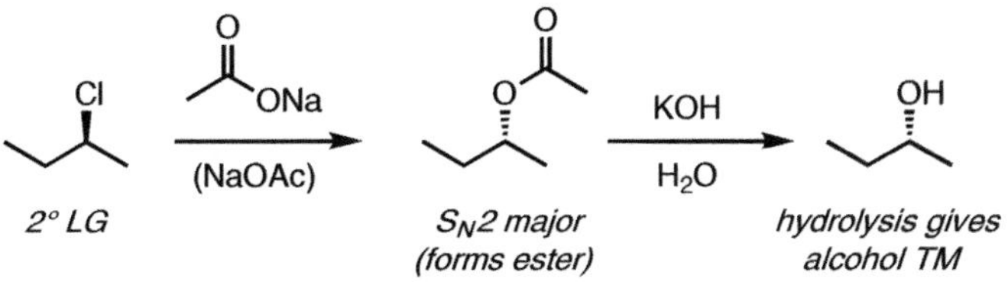

Ester Hydrolysis Gives Alcohol Product

This two-step synthesis (substitution followed by hydrolysis) is an effective strategy for replacing a 2° leaving group with a hydroxyl. Since acetate can be converted to a hydroxyl group, it is described as a "synthetic equivalent" of hydroxide; many such synthetic equivalents are employed in organic synthesis and will be described in this book.

acetate   hydroxide

Acetate (AcO⁻) Is Synthetic Equivalent of Hydroxide (HO⁻)

Retrosynthesis of Alcohols (FGI)

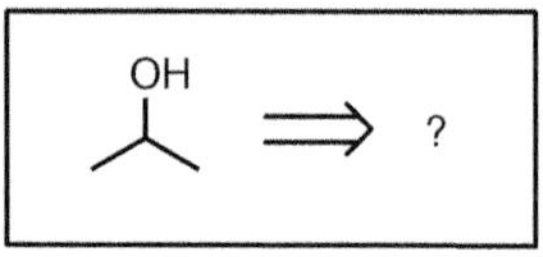

One possible retrosynthetic strategy for an alcohol target molecule is to simply convert the hydroxyl group to another functional group from which the hydroxyl can be derived, leaving the carbon chain intact. The possible starting materials include alkyl halides, alkenes, and carbonyl-containing compounds (ketone, aldehyde, carboxylic acid derivative).

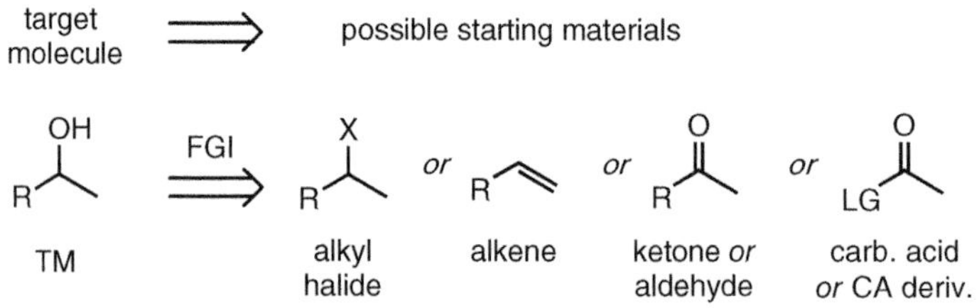

Alcohol Retrosynthesis via FGI

PRACTICE PROBLEM 3.1A: ALCOHOL SYNTHESIS BY FGI

3.1a Provide the reagents needed for each of the following transformations (more than one step may be required).

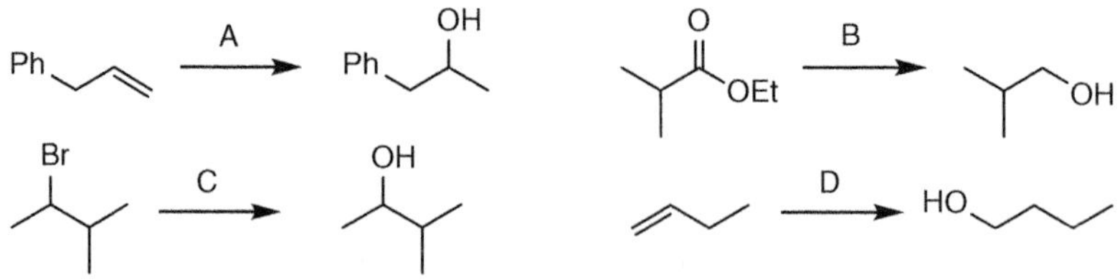

3.1.2 SYNTHESIS OF ALCOHOLS BY THE GRIGNARD REACTION

While it is important and very useful to know how to interconvert functional groups, the more significant goal of organic *synthesis* is the formation of *new carbon–carbon bonds*. Such reactions are the means to formation of new carbon skeletons and, thus, new organic molecules. Let us explore reactions that give alcohol products while creating new carbon–carbon bonds at the same time, such as the Grignard reaction. The Grignard reaction (which earned a Nobel Prize in Chemistry for Victor Grignard in 1912) is truly a classic reaction in organic synthesis.

Preparation of a Grignard Reagent (RMgX)

Grignard reagents (RMgX) are prepared by reacting the corresponding halide (RX; X = Br, Cl, or I) with magnesium metal in an aprotic solvent. Since Grignard reagents are strong bases, they cannot be formed or used in the presence of acidic protons; diethyl ether and tetrahydrofuran (THF) are commonly used aprotic solvents. The "R" group can be an ordinary sp^3 carbon chain (alkyl) or it can be an sp^2-hybridized carbon such as an alkene group (vinyl) or a benzene ring (aryl). Alkynyl (sp-hybridized) Grignard reagents can also be prepared ($RC \equiv CMgBr$); however, these latter reagents will be considered along with alkynyl anions ($RC \equiv CNa$).

$$R-X \xrightarrow{\text{Mg}} R-MgX \quad R = \textit{alkyl, vinyl, aryl}$$

Grignard Reagents are Prepared from Halides

Examples of Grignard Reagents

A Wide Variety of Grignard Reagents is Available

The Grignard reagent is a source of nucleophilic carbon (often written simply as "R:⁻") that will react with a variety of electrophilic carbons to create a new carbon–carbon bond. Since the development of Grignard's classic reagent, many experimental modifications have been made and a number of other metals can do similar reactions (e.g., Li, Cr, Ti, and Ce). For the sake of learning the basic strategies of synthesis, however, we will continue to use the Grignard reagent (RMgBr) as a great source of carbon nucleophiles.

Synthesis of Alcohols (Grignard)

Grignard reagents will react with aldehydes or ketones to furnish secondary or tertiary alcohol products, respectively, after a mildly acidic aqueous workup.

aldehyde *or*
ketone

2° *or* 3°
alcohol

Grignard Reaction with an Aldehyde or Ketone

Mechanism of the Grignard Reaction

The mechanism of the Grignard reaction begins with the nucleophilic ($\delta-$) Grignard carbon attacking the electrophilic ($\delta+$) carbonyl carbon, with the simultaneous breaking of the carbonyl pi bond. The resulting alkoxide intermediate is protonated upon workup to give an alcohol product.

*this "ionic" representation
of a Grignard reagent is
often used to simplify the
reaction mechanism*

Mechanism of the Grignard Reaction

This alcohol product contains a newly formed carbon–carbon bond. This is a key bond to be identified in an alcohol target molecule; a disconnection at this bond will lead to a possible retrosynthesis.

Retrosynthesis of an Alcohol (Grignard)

In order to form a new carbon–carbon bond, it is common that one of the carbons involved started out as a nucleophile (Nu:, electron-rich) and the other as an electrophile ($E+$, electron-deficient). One logical retrosynthesis of an alcohol TM involves making a disconnection of a carbon–carbon bond at the carbon bearing the OH group, breaking the molecule into two pieces. The carbon that ultimately ends up bearing an OH in the TM started out as a carbonyl carbon (C=O). Since a carbonyl is electrophilic, the new carbon group must have been introduced as a nucleophile (such as a Grignard) if it is expected to react with the carbonyl and form a new carbon–carbon bond.

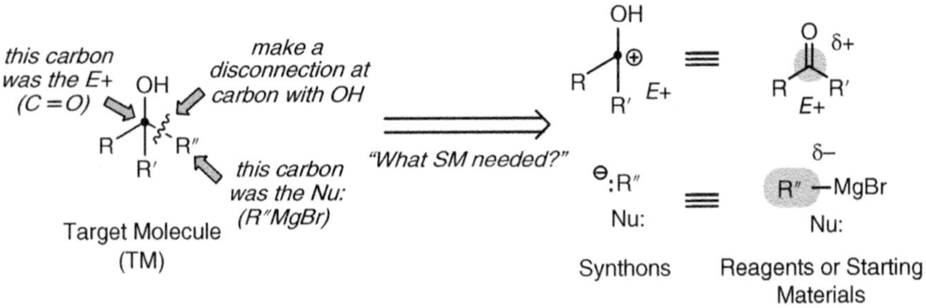

Retrosynthetic Strategy for Alcohol TMs

Using this strategy, we can devise a number of possible retrosyntheses for a given alcohol target molecule. The target molecule can be prepared by reacting a variety of aldehydes or ketones with a suitable Grignard reagent.

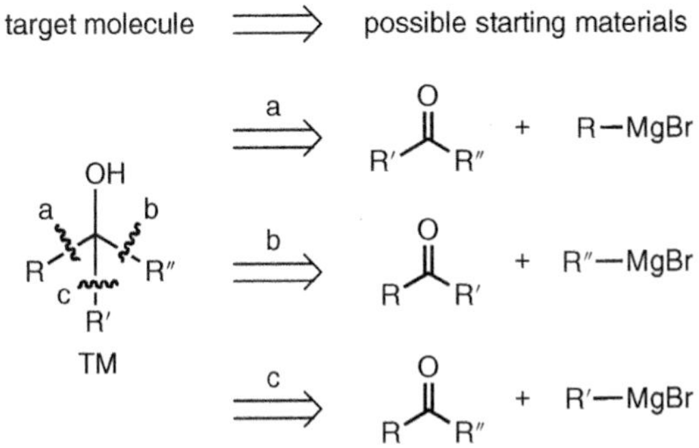

Alcohol Retrosynthesis via Grignard

Secondary and tertiary alcohols can be prepared by reacting a Grignard reagent with aldehydes and ketones, respectively, but these are not the only electrophiles that can be used with a Grignard. Grignard reagents also have useful reactions with esters and epoxides. These reactions also give alcohol products, but they lead to alternate retrosynthetic strategies for certain alcohol target molecules.

Reaction of a Grignard with an Ester

A Grignard reagent will react with an ester to give an alcohol product that contains two identical carbon groups attached to the carbon bearing the OH.

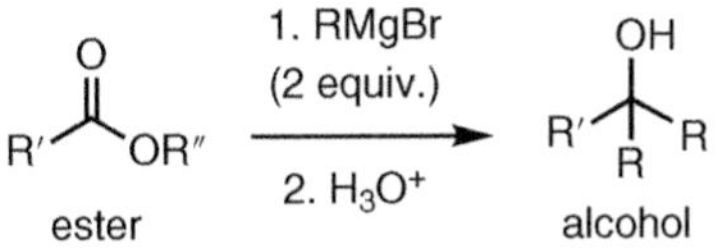

Grignard Reagents Add Twice to Esters

Mechanism of the Grignard Reaction with an Ester

Since esters contain a leaving group, they initially undergo *substitution* reactions at the carbonyl, in which a nucleophile replaces the leaving group. The so-called "addition–elimination" acyl substitution mechanism involves addition of a nucleophile to the carbonyl, followed by collapse of the resulting charged tetrahedral intermediate (CTI) to eject an alkoxide leaving group. When the nucleophile is a Grignard reagent, this acyl substitution initially results in a ketone product. However, since the ketone is a more reactive carbonyl electrophile than the ester starting material, it will rapidly react further with a second Grignard nucleophile. Ultimately, two equivalents of the Grignard are added to the ester to give a tertiary alcohol product.

Mechanism of the Grignard Reaction with an Ester

Retrosynthesis of an Alcohol Containing Two Identical Groups

Often, our analysis of a target molecule will extend beyond a simple inspection of functional groups. It is also important to be on the lookout for characteristic patterns, or perhaps certain combinations of functional groups, as you consider the possible disconnections. If the target molecule is an alcohol, we already know the Grignard reaction can be used to make it. However, if the alcohol TM has two identical groups attached at the carbon bearing the OH, then both of those carbon groups can be disconnected at the same time for one possible retrosynthesis. In this case, the synthesis would start with an ester, rather than a ketone or aldehyde, so the Grignard would be able to add twice.

Alcohol Retrosynthesis via Grignard + Ester

Reaction of a Grignard with an Epoxide

A Grignard reagent will also react with an epoxide to give an alcohol product, providing another potential disconnection to consider when synthesizing an alcohol.

epoxide   alcohol

Grignard Reagents Attack Less Hindered Carbon of Epoxides

Mechanism of an Epoxide Ring-Opening Reaction with a Grignard

Epoxides are strained molecules that readily undergo ring-opening reactions with good nucleophiles. Although this reaction involves the displacement of a strongly basic RO⁻ leaving group, it favors the more stable substitution product since a considerable amount of ring strain is released. As an S_N2 mechanism, the regiochemistry of the backside attack is governed by steric hindrance and the Grignard will attack the less crowded epoxide carbon.

Mechanism of an Epoxide Ring-Opening Reaction

Alternate Retrosynthesis of an Alcohol

The availability of epoxides as electrophilic reaction partners for Grignard reagents provides additional routes in the synthesis of alcohols. When a Grignard reagent reacts with a carbonyl group, the resulting new bond (and the corresponding disconnection) is at the carbon bearing the OH group. However, when a Grignard attacks an epoxide, the OH ends up on the carbon adjacent to the carbon that was attacked by the nucleophile. This disconnection of a TM (at carbon next to the carbon bearing the OH group) leads to a different synthon and a different starting material (the epoxide). This retrosynthetic approach is particularly evident when the target molecule contains a [—CH$_2$—CH$_2$—OH] unit, which can come from the simplest epoxide, ethylene oxide.

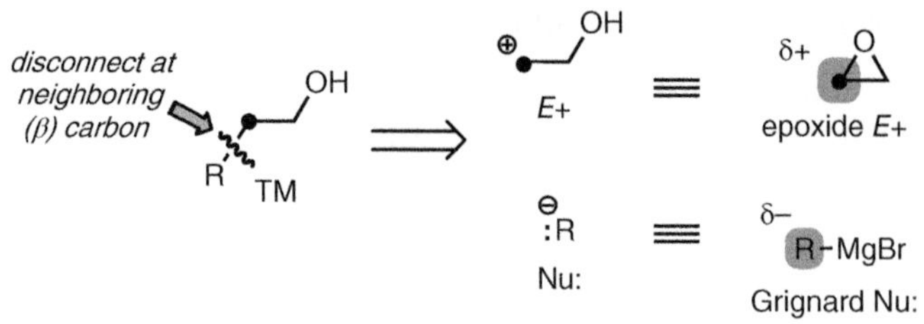

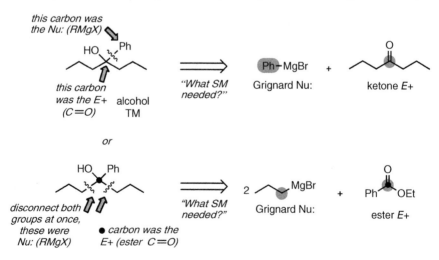

Alcohol Retrosynthesis via Grignard + Epoxide

EXAMPLE: ALCOHOL TM

Synthesize the following target molecule, using readily available starting materials and reagents. The synthesis must involve the formation of a new C—C bond.

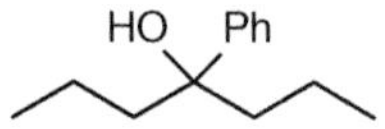

TM

At least two reasonable approaches can be taken for the retrosynthesis of this target molecule. A disconnection at the C—C bond between the phenyl group and the alcohol carbon leads to a simple phenyl Grignard and 4-heptanone. A different approach takes advantage of the two identical propyl groups attached to the alcohol carbon. Disconnection of both groups at once gives two equivalents of a propyl Grignard reagent reacting with some kind of a carboxylic acid derivative (the ethyl ester is typically commercially available and easy to handle, so ethyl benzoate is a good choice).

Retrosynthesis of Alcohol TM

The synthesis of the alcohol simply involves a Grignard reaction. Simple Grignard reagents are typically commercially available. Otherwise, they can be prepared from the corresponding halide.

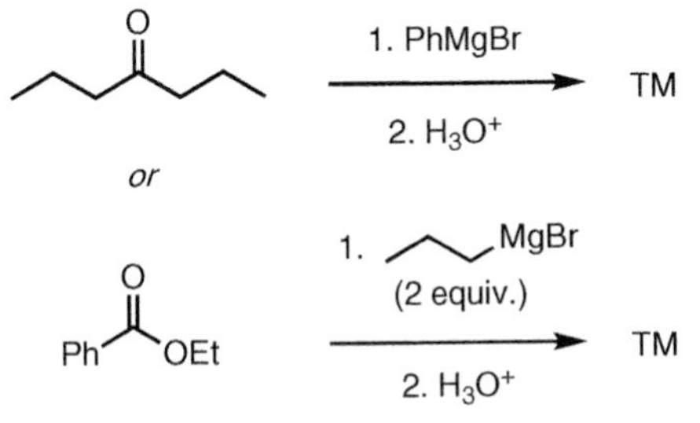

Possible Syntheses of Alcohol TM

3.1.3   SYNTHESIS OF PROPARGYLIC ALCOHOLS (RC≡CCH$_2$OH)

In a propargylic alcohol, the OH group is next to a triple bond. A logical disconnection of such an alcohol TM is made at the bond connecting the two functional groups. This disconnection would lead to an electrophilic carbonyl carbon and a nucleophilic alkynyl carbon. The sp-hybridized anionic synthon can be achieved with a Grignard reagent (RC≡CMgBr) as before, but it is also possible to use the acetylide anion instead (RC≡CNa).

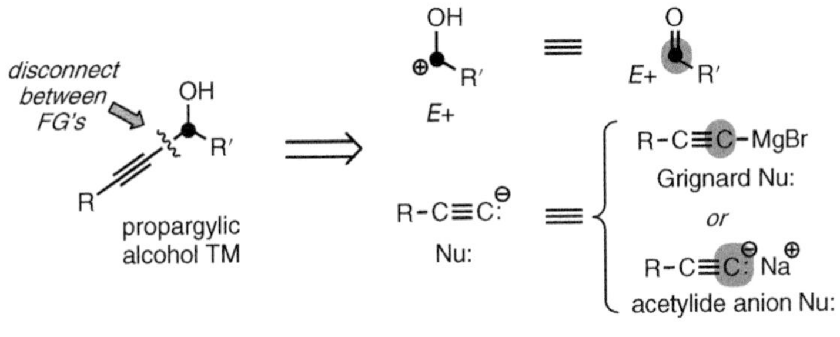

Retrosynthesis of a Propargylic Alcohol

Preparation of Alkynyl Nucleophiles (RC≡CNa)

Since sp-hybridized carbons are relatively acidic (p$K_a \approx 25$), strong bases can be used to deprotonate a terminal alkyne. If a Grignard reagent is used as the base, then the resulting product is drawn as a Grignard rather than an anion (this essentially sacrifices a more reactive, cheaper Grignard in the production of a new, more stable one…the desired alkynyl Grignard).

$$R-C{\equiv}C-H + CH_3CH_2-MgBr \longrightarrow R-C{\equiv}C-MgBr + CH_3CH_2-H$$

$pK_a \sim 25$ (left), $pK_a \sim 50$ (right)

Alkynyl Grignard Reagents are Prepared from a Grignard Base

Alternatively, a strong base such as sodium amide ($NaNH_2$) can be used to create a sodium salt. The resulting acetylide anion is stabilized by the sp-hybridization of the carbon; this is absolutely essential for the success of the deprotonation. Only terminal alkynes can be deprotonated in this fashion; alkenyl and alkyl anions/salts do not exist! Remember, we will use a Grignard reagent any time we have an anionic synthon (alkyl, alkenyl, or aryl), rather than attempt to deprotonate the corresponding non-acidic alkane, alkene, or benzene ring!

$$R-C{\equiv}C-H + NaNH_2 \longrightarrow R-C{\equiv}\overset{\ominus}{C}{:} \ \overset{\oplus}{Na} + NH_3$$

$pK_a \sim 25$ (left), $pK_a \sim 35$ (right)

Acetylide Anions are Prepared from a Strong Base such as $NaNH_2$

Because alkanes and alkenes contain no stabilizing functional groups, sp^3 and sp^2 C–H's of these compounds are NOT acidic enough to be deprotonated by $NaNH_2$. When carbons such as these are needed as nucleophiles, the corresponding Grignard reagents are employed.

R–H

Ph–H

$\overset{\ominus}{R}{:} \ \overset{\oplus}{Na}$ $\overset{\ominus}{Ph}{:} \ \overset{\oplus}{Na}$ anion $\overset{\ominus}{:} \overset{\oplus}{Na}$

NaNH₂ (base) → no reaction!

These anions are too unstable! Use Grignards as synthetic equivalent.

For Hydrocarbons, Only sp-Hybridized Carbons Can Be Deprotonated

PRACTICE PROBLEM 3.1B: ALCOHOL RETROSYNTHESIS

3.1b Each of the following TMs has a disconnection shown. Provide the starting materials needed to create the indicated bond.

A   B   C   D

(both)

3.1.4   SYNTHESIS OF PHENOL DERIVATIVES (ArOH)

Aromatic alcohols (phenols) can be prepared by treatment of a diazonium salt (ArN_2^+) with water, acid, and heat. In this substitution reaction, water acts as a nucleophile and replaces the excellent nitrogen leaving group (N_2) on the diazonium salt. The diazonium salt is prepared in three steps from benzene (or a substituted benzene derivative, but such compounds will be addressed in Chapter 5). First, nitration of benzene with HNO_3 and H_2SO_4 installs a nitro group $(-NO_2)$. The nitro is reduced to an amino group $(-NH_2)$ that is then treated with nitrous acid (HONO, prepared *in situ* with $NaNO_2$ and acid) to give the diazonium leaving group $(-N_2^+)$. The reaction of diazonium salts with nucleophiles other than water will be discussed in later sections including Part V, Synthesis of Aromatic Target Molecules.

Synthesis of Phenol

SYNTHESIS OF ALKYL (RX) AND ARYL HALIDES (ArX)

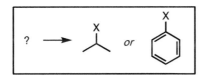

Like alcohols, alkyl halides are important starting materials for organic synthesis. The presence of a leaving group enables substitution and elimination mechanisms, and the electrophilic alkyl halide can be converted to a nucleophilic organometallic reagent such as a Grignard reagent (RMgX). Aryl halides are essential components in many widely used organometallic reactions, such as palladium-catalyzed coupling reactions.

Many simple alkyl halides are commercially available, and usually the chloroalkanes and bromoalkanes can be used interchangeably in a synthetic plan. (In the case of the methyl group, however, iodomethane, CH_3I, is the preferred choice since it is a liquid and the other methyl halides are gases at room temperature.) Alkyl halides can be prepared by functionalizing an alkane (via free-radical halogenation), or by functional group interconversions, including halogen substitution of an alcohol and addition of HX or X_2 to an alkene.

3.2.1 PREPARATION FROM ALKANES (RH → RX)

Free-radical halogenation of an alkane (Br_2 + light) will afford as the major product the bromide arising from the most stable radical intermediate. The rate of reaction varies by the ease of removal of the hydrogen atom

Introduction to Strategies for Organic Synthesis, Second Edition. Laurie S. Starkey.
© 2018 John Wiley & Sons, Inc. Published 2018 by John Wiley & Sons, Inc.

(and the stability of the resulting radical intermediate), with 3°, allylic and benzylic bromides being formed the fastest. *N*-Bromosuccinimide (NBS) is a convenient source of bromine atoms, and it is especially useful in allylic and propargylic brominations since using bromine as a reagent would also cause an addition reaction with the C—C double or triple bond. Free-radical chlorination (Cl$_2$ + light) is typically not synthetically useful since this reaction is less selective and results in a mixture of all possible monochlorination products.

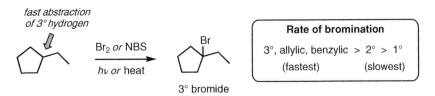

Synthesis of Alkyl Halides from Alkanes

3.2.2 PREPARATION FROM ALCOHOLS (ROH → RX)

Substitution reactions involving alcohols require the conversion of the hydroxyl group into a good leaving group. One option is the reaction of the alcohol with a strong acid such as HI, HBr, or HCl/ZnCl$_2$. After protonation of the alcohol, substitution of the water leaving group occurs via either S$_N$1 or S$_N$2 mechanisms, but carbocations are favored in these acidic conditions so carbocation rearrangements may be possible and racemization is likely.

Synthesis of Alkyl Halides from Alcohols Using HX

Alternative reagents such as thionyl chloride (SOCl$_2$), phosphorus tribromide (PBr$_3$), and phosphorus and iodine (P, I$_2$) are often preferred. These reagents also convert the hydroxyl into a good leaving group and provide a halide to do a substitution but avoid the strongly acid conditions of the haloacids and will not undergo rearrangements. Another advantage to these reagents is that they are stereospecific: phosphorus-based reagents result in inversion due to backside attack (S$_N$2) of the halide nucleophile, and thionyl chloride typically results in retention of stereochemistry due to internal delivery of the chloride nucleophile. An alternate strategy involves conversion of the alcohol into its corresponding tosylate (by reaction with TsCl/pyridine), followed by treatment with a halide nucleophile to displace the tosylate leaving group.

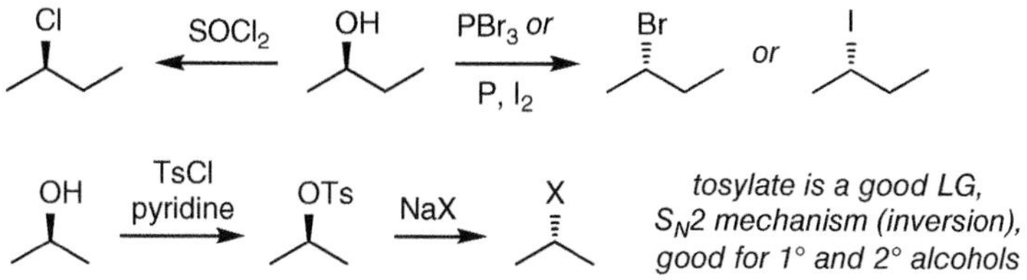

Synthesis of Alkyl Halides from Alcohols—More Options

3.2.3 PREPARATION FROM ALKENES (C=C → RX)

Electrophilic addition of HCl or HBr to an alkene normally occurs with Markovnikov regiochemistry, resulting in an alkyl halide with the halogen located at the more substituted carbon of the starting alkene. Since the presence of peroxides (ROOR) initiates a radical mechanism, HBr can also be added to an alkene with anti-Markovnikov orientation. Anti addition of bromine or chlorine to an alkene gives a trans-dibromo or dichloro product.

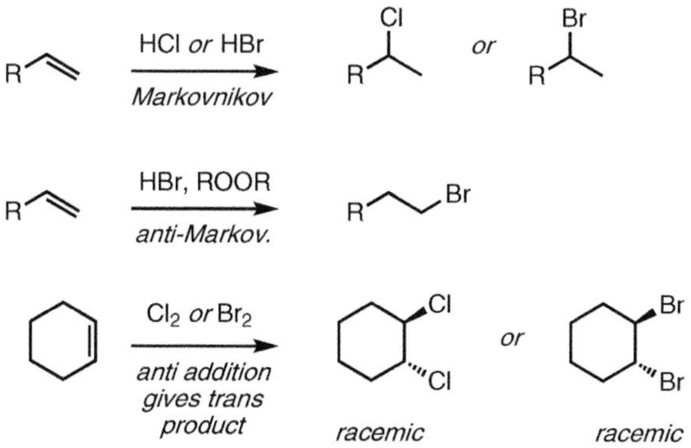

Synthesis of Alkyl Halides from Alkenes

3.2.4 RETROSYNTHESIS OF ALKYL HALIDES

The retrosynthesis of an alkyl halide target molecule involves a functional group interconversion to a suitable alkane, alcohol, or alkene starting material.

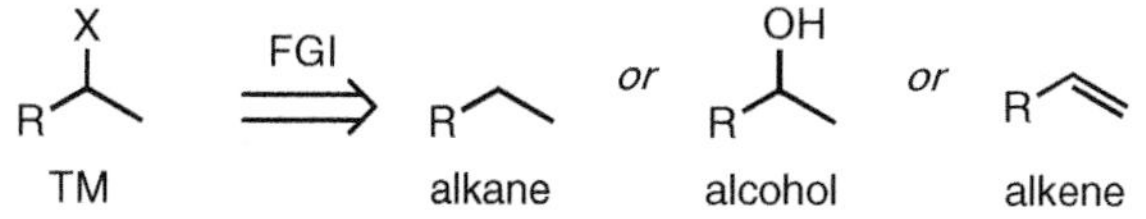

Retrosynthesis of Alkyl Halides

3.2.5 SYNTHESIS OF ARYL HALIDES (ArX)

The methods described earlier for alkyl halides (on sp^3-hybridized carbons) cannot be applied to the synthesis of aryl halides (on sp^2-hybridized carbons). The introduction of a halogen to benzene (or a benzene derivative) can be accomplished either by an electrophilic aromatic substitution reaction ($Cl_2/FeCl_3$ or $Br_2/FeBr_3$) or by treatment of a diazonium salt (ArN_2^+) with a halide nucleophile (CuI, $CuBr$, HBF_4, or KI).

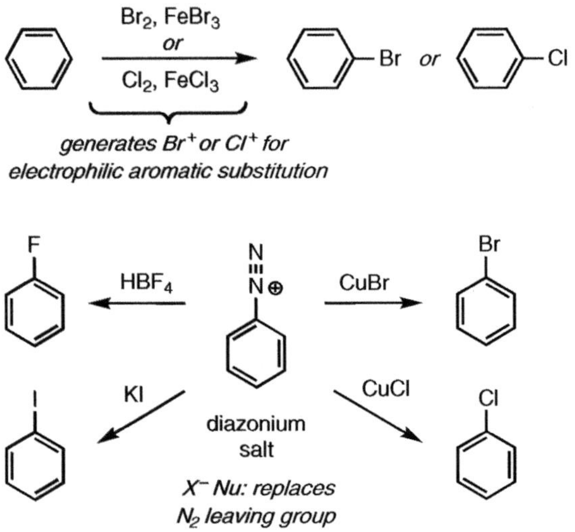

Synthesis of Aryl Halides

Since electrophilic aromatic substitution begins with benzene (or a benzene derivative), the retrosynthesis of an aryl chloride or bromide simply involves the replacement of the halide with a hydrogen atom. Another possible retrosynthesis that can be applied to any aryl halide is a functional group interconversion that replaces the halide with a nitro group, since the nitro group can be converted to any halide via the diazonium salt.

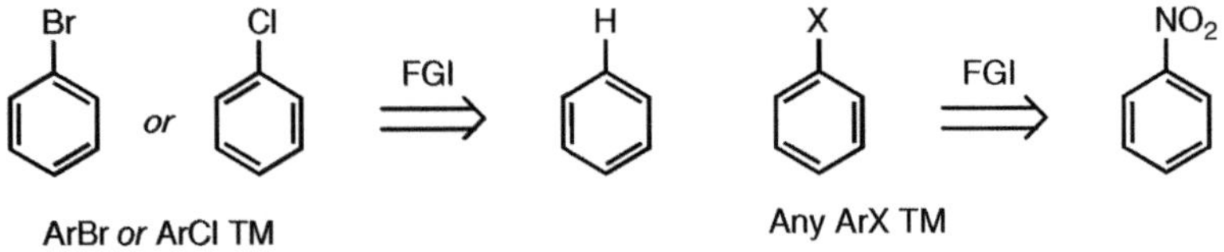

Aryl Halide Retrosynthesis

PRACTICE PROBLEM 3.2: ALKYL HALIDE SYNTHESIS

3.2 For each transformation, provide the missing starting material or reagent.

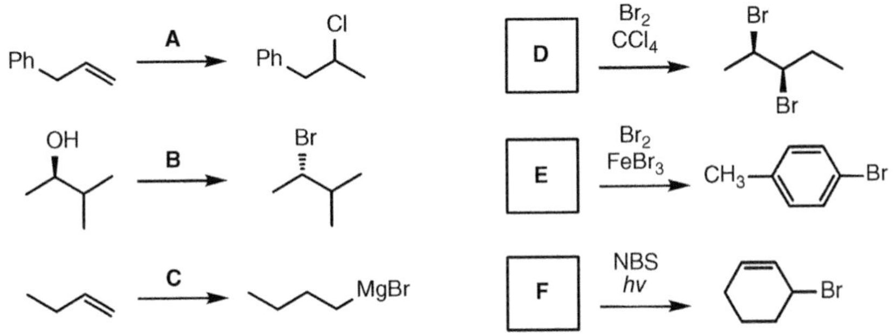

EXAMPLE: ALKYL HALIDE TM

Provide the reagents necessary to transform the given starting material into the desired product. More than one step may be required.

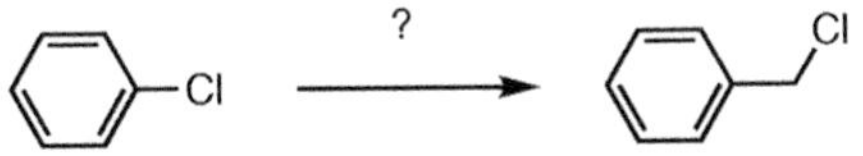

A "transform" problem should begin like any synthesis problem: with a retrosynthetic analysis of the product. The given starting material will guide that retrosynthesis, by pointing to the disconnections that are required. In this case, the product has an additional carbon on the aromatic ring. Retrosynthesis of the halide TM leads to an alcohol structure. Disconnection at the C–C bond between the benzene ring and the carbon bearing the hydroxyl is a typical one for an alcohol, affording a Grignard nucleophile and a formaldehyde electrophile. The magnesium chloride version of the Grignard is chosen since the given starting material is chlorobenzene.

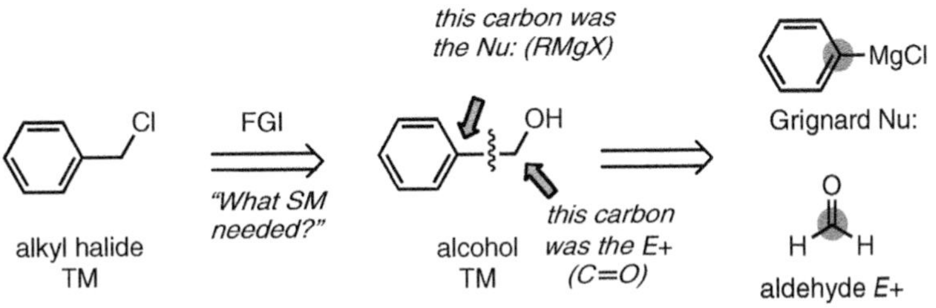

Retrosynthesis of Alkyl Halide TM

After planning the synthesis, the reagents required for the transform problem can be provided: magnesium metal to make the Grignard reagent, formaldehyde to give an alcohol product with one additional carbon, and $SOCl_2$ to convert benzyl alcohol to benzyl chloride.

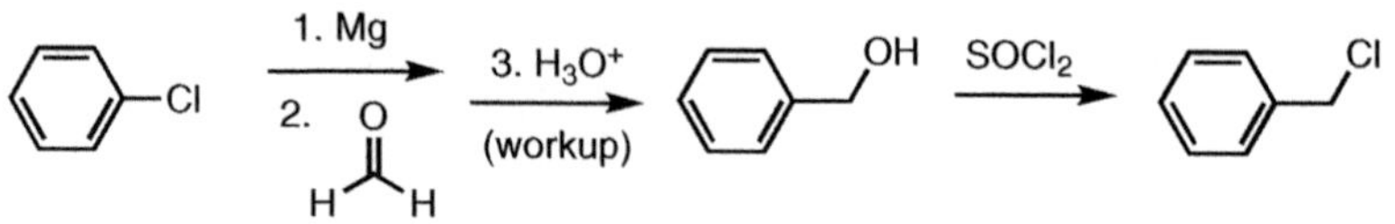

Synthesis of Alkyl Halide TM

SYNTHESIS OF ETHERS (ROR′)

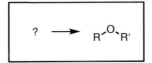

3.3.1 WILLIAMSON ETHER SYNTHESIS (RX + R′O⁻ → ROR′)

Ethers are frequently prepared via the Williamson ether synthesis, which involves the reaction of an alkyl halide electrophile (RX) with an alkoxide nucleophile (RO⁻). As usual, the S_N2 backside attack is sensitive to sterics, and the E2 elimination reaction is expected to compete here since alkoxides are strong bases. The S_N2 substitution can be expected to give good yields of the ether if the halide is on a methyl, primary, allylic, or benzylic carbon. Simple alkoxides may be commercially available; otherwise, the alkoxide can be prepared from the corresponding alcohol by treatment with a strong base (NaH) or a Group I metal (Na or K).

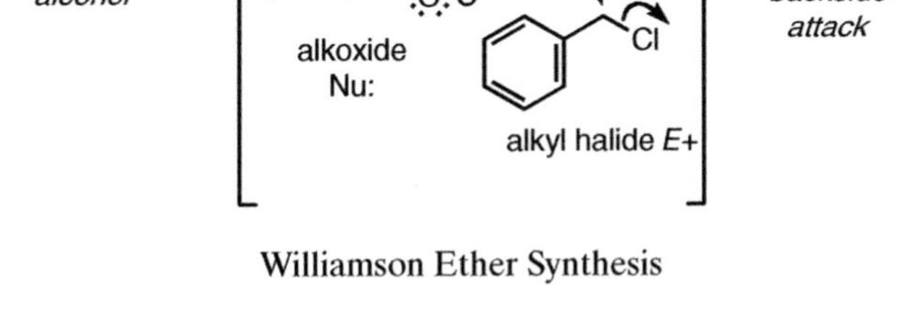

Williamson Ether Synthesis

Introduction to Strategies for Organic Synthesis, Second Edition. Laurie S. Starkey.
© 2018 John Wiley & Sons, Inc. Published 2018 by John Wiley & Sons, Inc.

Treatment of properly substituted difunctional haloalcohols with base (NaOH) initiates deprotonation, followed by an intramolecular backside attack, to afford 3-, 5-, or 6-membered cyclic ether products. Since 5- and 6-membered rings have little to no ring strain, these reactions have favorable transition states and proceed smoothly. Three-membered rings form quickly in spite of the ring strain involved, because the nucleophile and the electrophile are so closely located. However, the greater distance between reactive centers in a four-membered ring, combined with the ring strain in the transition state, cause these rings to form slowly. Reactions to form 7-membered or larger rings require too much organization to bring the nucleophile and the electrophile together, so this high cost in entropy slows these reactions.

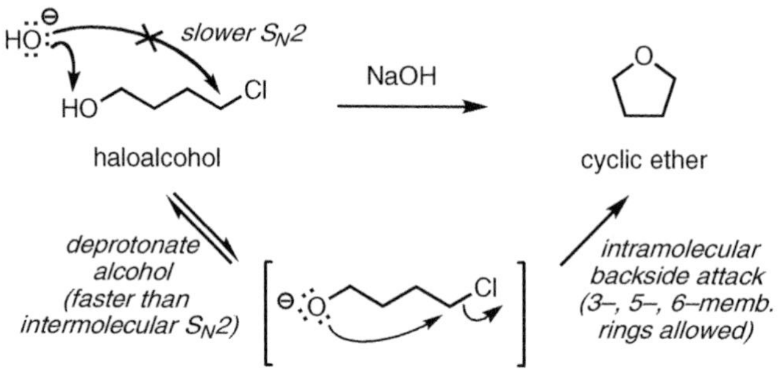

Intramolecular Williamson Produces Cyclic Ethers

3.3.2 ALTERNATE ETHER PREPARATIONS

The Williamson ether synthesis is not the only ether preparation available, and it is not suitable for sterically crowded ethers. Ethers can also be synthesized by addition of an alcohol to an alkene. Just as hydration of an alkene gives an alcohol product (water as a nucleophile installs an OH group), addition reactions in the presence of an alcohol gives an ether product (alcohol as a nucleophile installs an OR group). Two examples are shown below; the alkene can react initially with either a strong acid or a mercury cation (alkoxymercuration–demercuration).

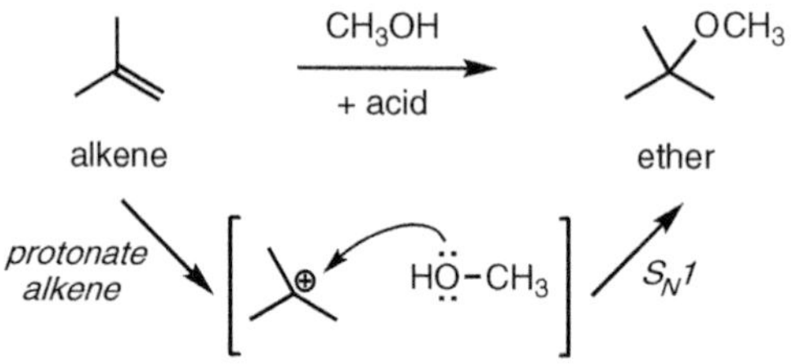

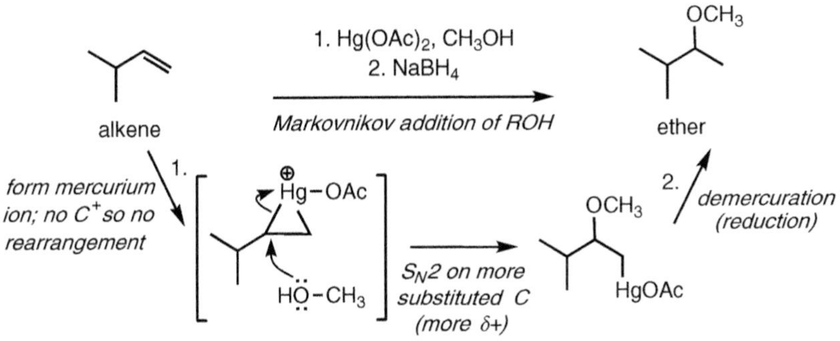

Ether Synthesis in Acidic Conditions

Symmetrical ethers can be prepared by treatment of an alcohol with a strong acid. After protonation of the alcohol, substitution of the water leaving group occurs via either S_N1 or S_N2 mechanisms. By using a non-nucleophilic acid, such as H_2SO_4, the only nucleophile present to replace the water leaving group is a second molecule of the alcohol, resulting in an ether product.

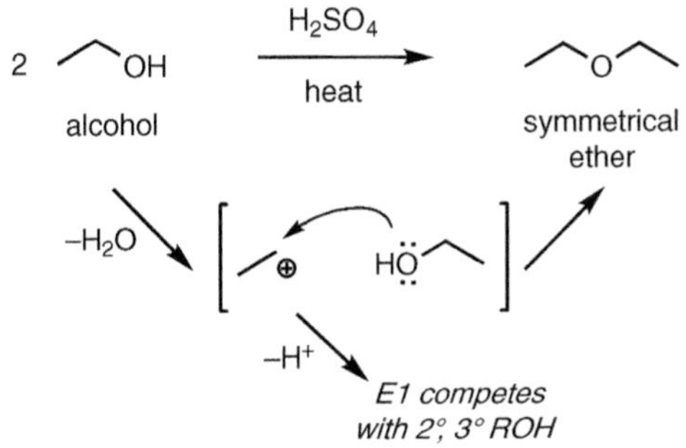

Synthesis of Symmetrical Ethers via Alcohol Dehydration

3.3.3 SYNTHESIS OF EPOXIDES

Epoxides are unique ethers that can be synthesized by oxidation of an alkene with a peroxyacid (RCO_3H, such as mCPBA) or by an intramolecular S_N2 on a vicinal haloalcohol, called a halohydrin (this is an intramolecular version of the Williamson ether synthesis). Since halohydrins are prepared by reaction of an alkene with a halogen and water (such as Br_2 and H_2O), both approaches to an epoxide target molecule begin with an alkene starting material.

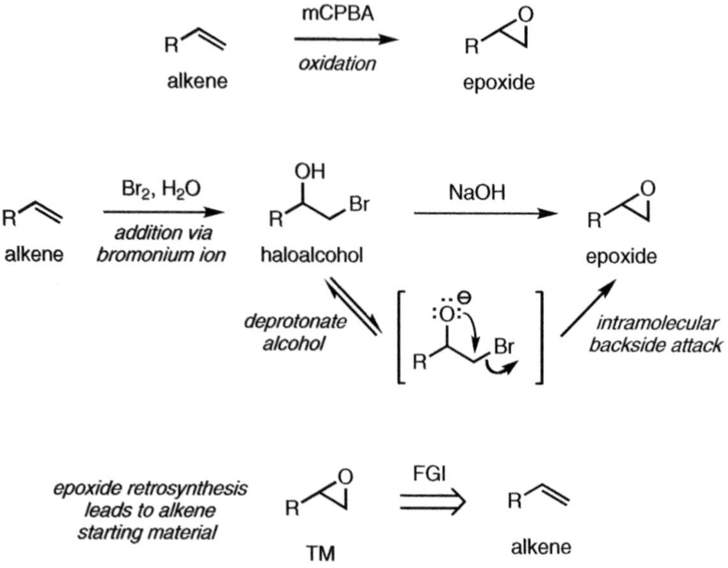

Epoxide Synthesis and Retrosynthesis

3.3.4   RETROSYNTHESIS OF ETHERS

Disconnection of an ether target molecule occurs at either C—O bond, leading to an alkoxide nucleophile and alkyl halide electrophile. This Williamson ether synthesis involves an S_N2 mechanism, so minimization of steric hindrance is the main consideration when determining which disconnection to make in the retrosynthesis.

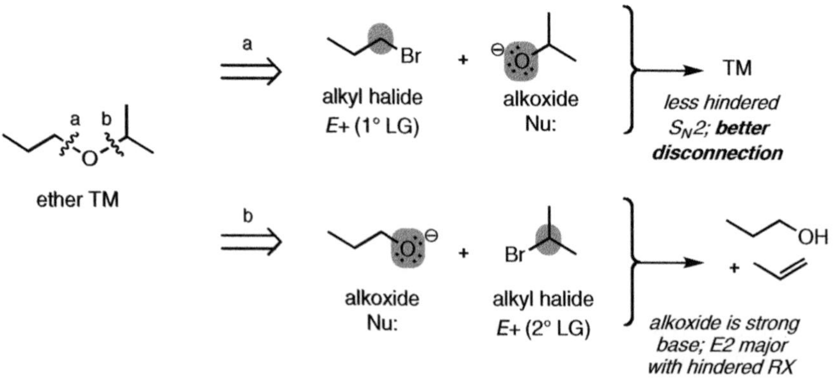

Retrosynthesis of Ethers via Williamson Ether Synthesis

EXAMPLE: ETHER TM

Synthesize the following target molecule, using only benzene and alcohol starting materials as sources of carbon (along with any necessary reagents that are commercially available).

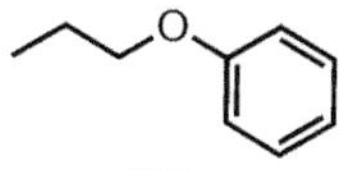

TM

The retrosynthesis begins by disconnecting the C—O bond that can be formed more easily, typically via an S_N2 mechanism. Since formation of the C—O bond at the aromatic ring would be more challenging, the disconnection shown is made instead to give a primary alkyl halide electrophile and a phenoxide nucleophile. Each of these compounds can be prepared from the corresponding alcohols, as required, and phenol can be synthesized from benzene via the phenyl diazonium salt.

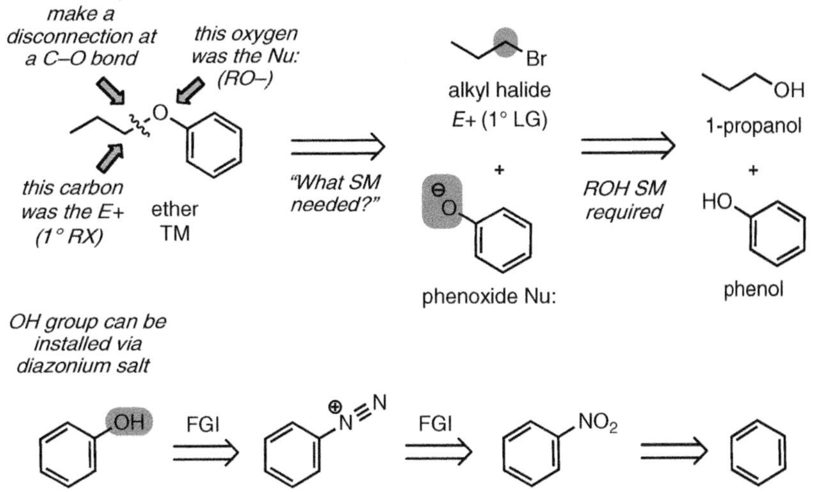

Retrosynthesis of Ether TM

Treatment of 1-propanol with PBr_3 gives 1-bromopropane with no risk of rearrangement. Preparation of the diazonium salt begins with nitration of benzene and reduction of the nitro group to give aniline. Treatment of aniline with nitrous acid (formed *in situ* from $NaNO_2$ and HCl) generates a diazonium salt that can be converted to phenol by reaction with water. Since phenol is more acidic than an ordinary alcohol,

it can be deprotonated with NaOH. Reaction of the resulting phenoxide
with 1-bromopropane produces the desired target molecule.

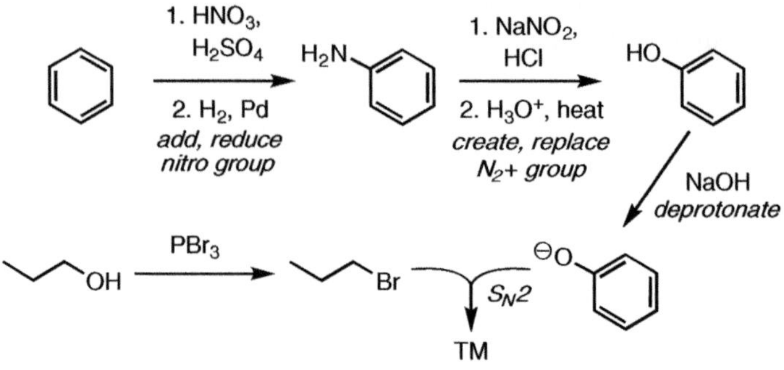

Synthesis of Ether TM

PRACTICE PROBLEM 3.3: ETHER SYNTHESIS

3.3 Provide the reagents needed for each of the following transformations
(more than one step may be required).

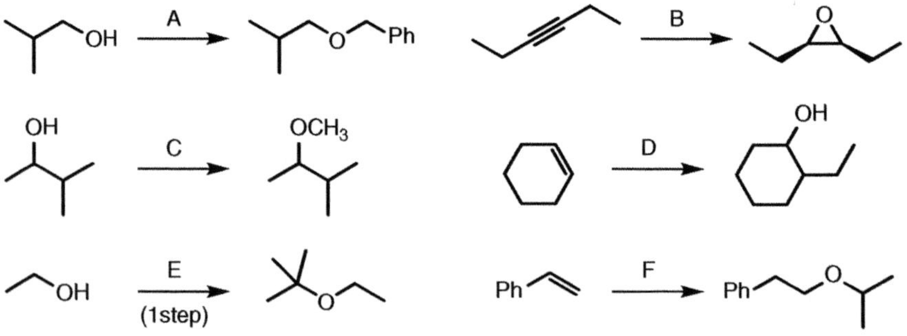

SYNTHESIS OF THIOLS (RSH) AND THIOETHERS (RSR′)

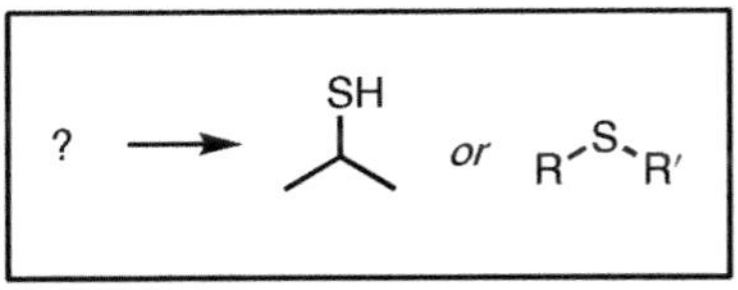

Thiols and thioethers are typically prepared by substitution of a leaving group with a sulfur group, so the synthesis will begin with an alkyl halide and a suitably substituted sulfur anion. This strategy is analogous to the ones already presented for the preparation of alcohols and ethers. Because sulfur is larger than oxygen, a negative charge on sulfur (HS⁻or RS⁻) is better delocalized and more stable, making it a weaker base (compared to HO⁻or RO⁻). In fact, even the S^{2-} anion is commercially available in the form of sodium sulfide (Na_2S). The larger sulfur anion is also very polarizable, making it an excellent nucleophile. As a result, E2 elimination mechanisms do not compete much with S_N2 substitution mechanisms involving sulfur anions such as sodium hydrosulfide (NaSH). The same general strategy of the Williamson ether synthesis can be applied to the synthesis of thioethers (also called sulfides). Since thiols are more acidic than alcohols, a weaker base such as NaOH can be used for the deprotonation of the thiol.

Introduction to Strategies for Organic Synthesis, Second Edition. Laurie S. Starkey.
© 2018 John Wiley & Sons, Inc. Published 2018 by John Wiley & Sons, Inc.

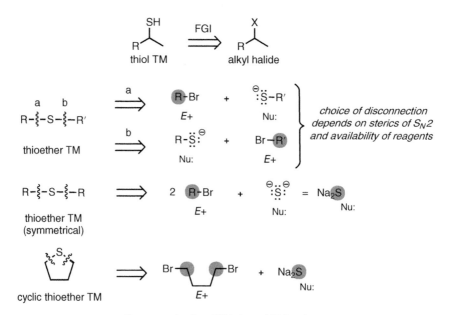

Synthesis of Thiols and Thioethers via S$_N$2

While the retrosynthesis of a thioether should still take into consideration the more favorable, less hindered S$_N$2 mechanism, it is possible to do a backside attack of RS⁻ on a secondary alkyl halide to synthesize a sterically crowded thioether. A symmetrical thioether or cyclic ether can be prepared via sequential S$_N$2 displacements of two halides with sodium sulfide (Na$_2$S).

Retrosynthesis of Thiols and Thioethers

EXAMPLE: THIOETHER TM

Provide the reagents necessary to transform the given starting material into the desired product. More than one step may be required.

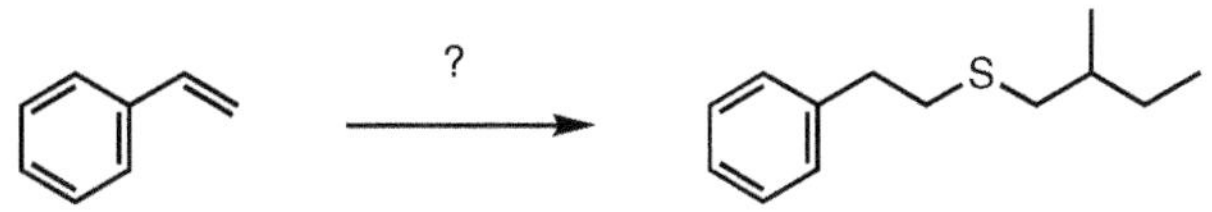

   A retrosynthetic analysis of the product involves the eventual disconnection of both C—S bonds. Disconnection at the branched alkyl group gives a thiolate nucleophile and a complex (i.e., expensive) alkyl halide that should be prepared from the readily available alcohol starting material 2-methyl-1-butanol. Further disconnection of the thiol leads to the original carbon framework with a leaving group at the end carbon.

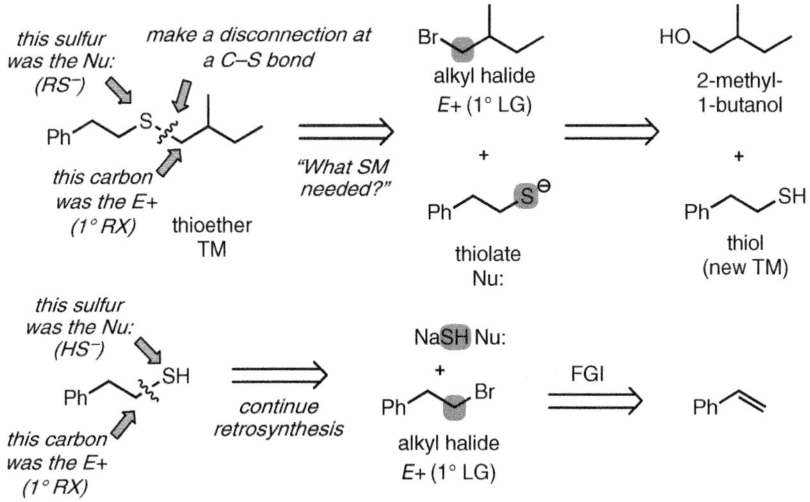

Retrosynthesis of Thioether TM

Installation of the required leaving group requires an anti-Markovnikov addition to the alkene double bond of the given starting material. This can be achieved directly via radical addition of HBr (HBr with peroxides). Another route involves a hydroboration–oxidation to give the primary alcohol that can be converted to a suitable leaving group (halide or tosylate by treatment with TsCl). Displacement of the leaving group with NaSH gives a thiol that can be deprotonated before a second S_N2 reaction installs the second alkyl group.

Synthesis of Thioether TM

SYNTHESIS OF AMINES (RNH₂) AND ANILINES (ArNH₂)

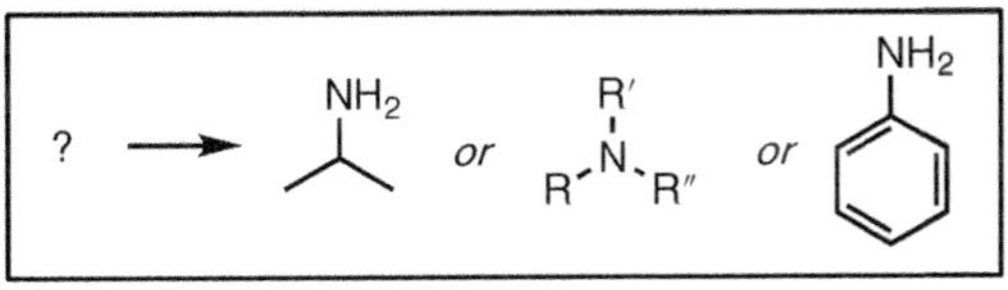

Utilizing a strategy analogous to thiols and thioethers, one might imagine simply disconnecting an amine target molecule at a C—N bond and introducing the nitrogen via S_N2 displacement of the leaving group of an alkyl halide with ammonia (NH_3) or an amine nucleophile. Ammonia and all neutral amines (including primary RNH_2, secondary R_2NH, and tertiary R_3N) are excellent nucleophiles and readily undergo such S_N2 substitutions. However, this reaction is difficult to control because the addition of an electron-donating alkyl group makes the nitrogen in the amine product an even better nucleophile, so overalkylation is common. In fact, reaction of an amine with an excess of alkyl halide results in multiple S_N2 reactions to give a quaternary ammonium salt product.

Introduction to Strategies for Organic Synthesis, Second Edition. Laurie S. Starkey.
© 2018 John Wiley & Sons, Inc. Published 2018 by John Wiley & Sons, Inc.

Possible Retrosynthesis of Amine TM

a logical disconnection
leads to reasonable
starting materials...

amine TM

Nu:
synthons

Nu:
starting materials

Attempted Synthesis of Amine TM

H$_3$N:
great Nu:

S_N2
$-H^+$

also a great Nu:
so it is difficult
to stop here

S_N2

...but reaction of an amine
with an alkyl halide can
lead to overalkylation

Simple S$_N$2 with Amine Nu: is Not So Simple

While using an excess of the amine can minimize the amount of di- and tri-alkylated side products, this is not always a practical option. A wide variety of strategies have been developed to avoid the problem of overalkylation, including the use of NH$_3$ equivalents and the preparation of other nitrogen-containing functional groups that can be converted to amines.

3.5.1   SYNTHETIC EQUIVALENTS OF NH$_3$ (RX → RNH$_2$)

The Gabriel Synthesis of Amines

Much like using acetate (AcO$^-$) as a synthetic equivalent of hydroxide (HO$^-$), the anion of phthalimide is used as a synthetic equivalent of ammonia. Phthalimide itself is non-nucleophilic since the nitrogen lone pair electrons are delocalized by resonance with the carbonyls. Upon deprotonation, however, the anion can do an S$_N$2 substitution with an unhindered (1°) alkyl halide. This introduces nitrogen in place of the halogen leaving group, and no further substitution can take place with the neutral imide product. Since the imide is a carboxylic acid derivative, hydrolysis releases the amino group (an acyl substitution mechanism: addition of hydroxide nucleophile, elimination of leaving group) and furnishes an amine product. Alternatively, acyl substitution is accomplished using hydrazine (NH$_2$ NH$_2$) as a nucleophile that can react with both carbonyls of the imide. This overall sequence is known as the Gabriel synthesis.

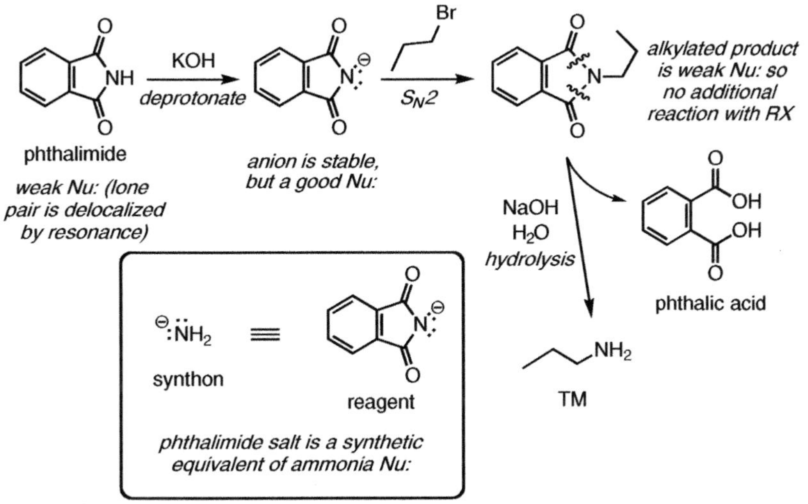

Gabriel Synthesis of Amines

Amine Synthesis via Azides (RX → RN$_3$ → RNH$_2$)

The azide anion (N$_3^-$) can also be used as a synthetic equivalent of ammonia. Sodium azide (NaN$_3$) is commercially available and the azide anion can do an S$_N$2 substitution with an alkyl halide. Like using phthalimide, this introduces nitrogen in place of the halogen-leaving group. Reduction of the resulting azide by catalytic hydrogenation provides an amine product.

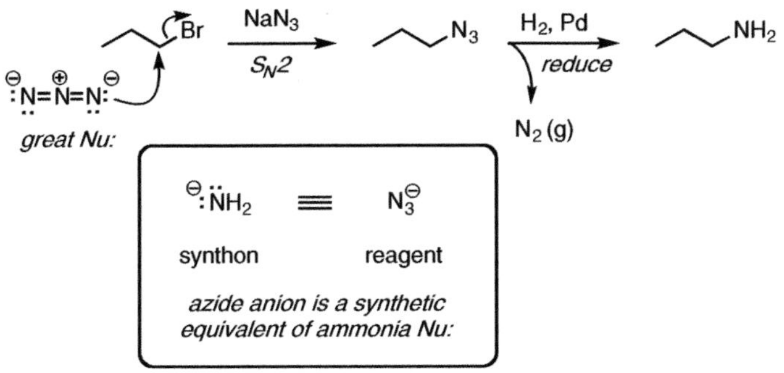

Amine Synthesis via Azide Nucleophile

The availability of these ammonia synthetic equivalents provides two good options when considering the retrosynthesis of an amine target molecule. The −NH$_2$ group can be converted to either the phthalimide group or an −N$_3$ group. After this FGI, disconnection of the resulting structure at the C−N bond leads to reasonable starting materials.

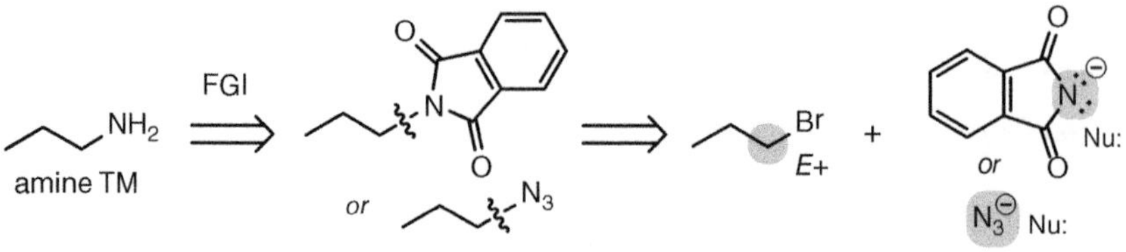

Retrosynthesis of Amines via NH$_3$ Equivalent

3.5.2 SYNTHESIS OF AMINES VIA REDUCTION REACTIONS

Amine Synthesis via Nitriles (RX → RC≡N → RCH$_2$NH$_2$)

The C≡N triple bond of nitriles can be reduced with LiAlH$_4$ or by catalytic hydrogenation to give amine products. Since the cyano group can be introduced via an S$_N$2 on an alkyl halide, this offers a useful route to amine target molecules.

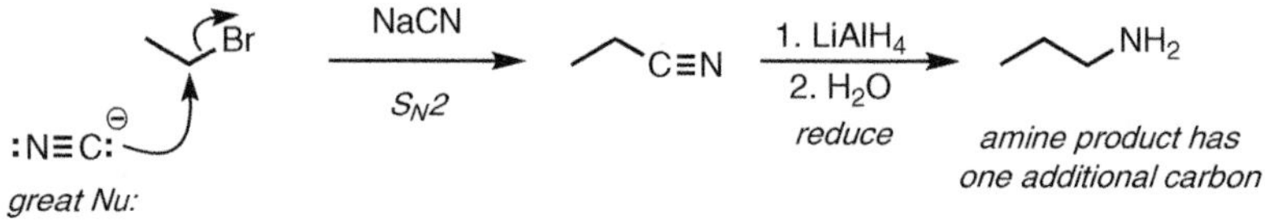

Amine Synthesis via a Nitrile

Note that the cyanide nucleophile introduces both nitrogen and a new carbon. This extra carbon must be accounted for in the retrosynthetic planning.

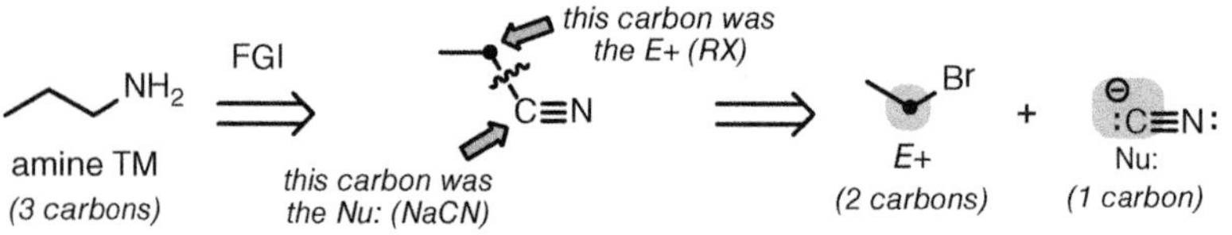

Amine Retrosynthesis via a Nitrile

Amine Synthesis via Amides (RCO$_2$H → RCONHR′ → RCH$_2$NHR′)

Reduction of amides with LiAlH$_4$ adds two equivalents of hydride to afford amine products. Reaction of an acid chloride with an amine (ammonia NH$_3$, primary RNH$_2$, or secondary R$_2$NH) generates the desired amide product.

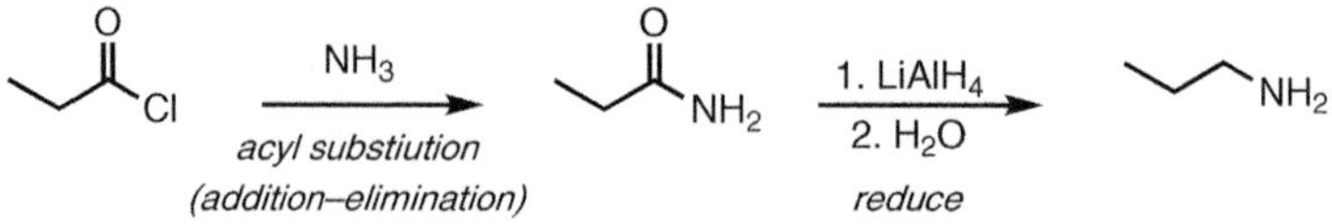

Amine Synthesis via an Amide

A possible retrosynthesis of an amine TM begins with an FGI to an amide by inserting a carbonyl on one of the alkyl groups. Disconnection of the amide at the C—N bond leads to the required acid chloride and amino starting materials.

Amine Retrosynthesis via an Amide

Reductive Amination of Ketones ($R_2C=O \rightarrow [R_2C=NR'] \rightarrow R_2CHNHR'$)

Amines react with aldehydes or ketones to give imine products. If this reaction is conducted in the presence of a reducing agent, then the imine C=N double bond will be reduced *in situ* to give an amine product. Since reduction of the carbonyl C=O double bond is to be avoided, reduction of the imine is achieved either by catalytic hydrogenation (H_2/Pd) or with sodium cyanoborohydride ($NaBH_3CN$). This reaction, called a reductive amination, can be performed on either an aldehyde or a ketone, using ammonia (NH_3), a primary amine (RNH_2) or a secondary amine (R_2NH), to give a 1°, 2°, or 3° amine product, respectively.

Amine Synthesis via Reductive Amination

A possible retrosynthesis of an amine TM begins with an FGI to an imine by inserting a double bond to nitrogen. Disconnection of the imine at the C=N double bond leads to the required aldehyde/ketone and amino starting materials.

Amine Retrosynthesis via Reductive Amination

PRACTICE PROBLEM 3.5: AMINE SYNTHESIS

3.5 Provide the reagents needed for each of the following transformations (more than one step may be required).

3.5.3 RETROSYNTHESIS OF AMINES

Almost every amine synthesis involves an eventual disconnection at the C—N bond (since the cyanide nucleophile delivers a carbon with a nitrogen attached, that retrosynthesis involves a disconnection of the C—C bond next to the C—N bond). However, when planning an amine synthesis, the first step is a functional group interconversion (FGI). What starting material functional groups can be converted to an amine? There are several possibilities, including a phthalimide, an azide, a nitrile, an amide, or an imine. The Gabriel synthesis as well as the azide and nitrile approaches afford a primary amine (RNH₂) and require an S_N2 substitution on an unhindered alkyl halide (methyl CH_3I and primary RCH_2X halides are ideal candidates). Secondary, tertiary, and branched amine target molecules are better prepared via an amide or by reductive amination of an aldehyde or ketone. It is likely that there is more than one reasonable synthesis for a given amine TM. Also, it is worth noting that both NaN_3 and NaCN are toxic and that certain azides including NaN_3 may explode when heated (involving a rapid evolution of N_2), so it is good to know that such a wide variety of amine synthetic strategies are available.

3.5.4 SYNTHESIS OF ANILINE DERIVATIVES (ArNH₂)

Aromatic amines can be prepared by reduction of a nitro group ($-NO_2$). Nitration of an aromatic ring with HNO_3 and H_2SO_4 affords the necessary nitro substituent. Aromatic target molecules containing multiple substituents are the focus of Chapter 5.

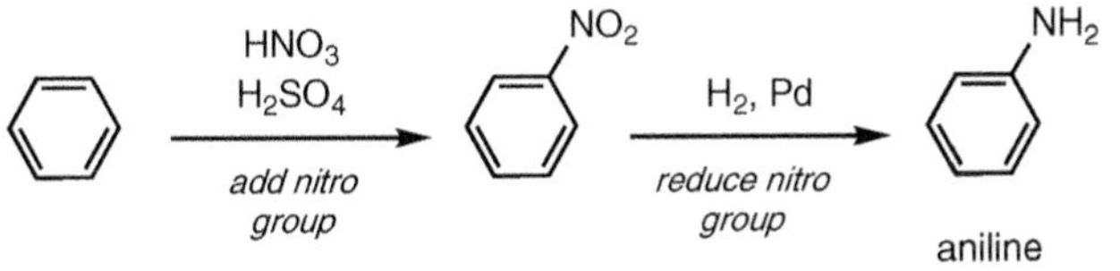

Aniline Synthesis

EXAMPLE: AMINE TM

Synthesize the following target molecule, using readily available starting materials and reagents.

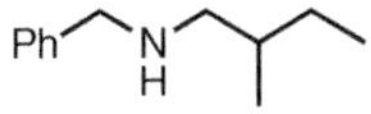

TM

An FGI on the secondary amine TM leads to either an amide or an imine. Further disconnection of either the amide or the imine results in a commercially available electrophile (benzoyl chloride or benzaldehyde, respectively) and an amine nucleophile (2-methylbutan-1-amine). Presumably, this amine structure is complex enough (i.e., expensive enough) that it too should be synthesized, so the retrosynthesis continues with another FGI. The use of an ammonia synthetic equivalent, such as an azide, does not help to simplify the carbon structure, but a retrosynthesis involving a nitrile includes an eventual disconnection of a C–C bond.

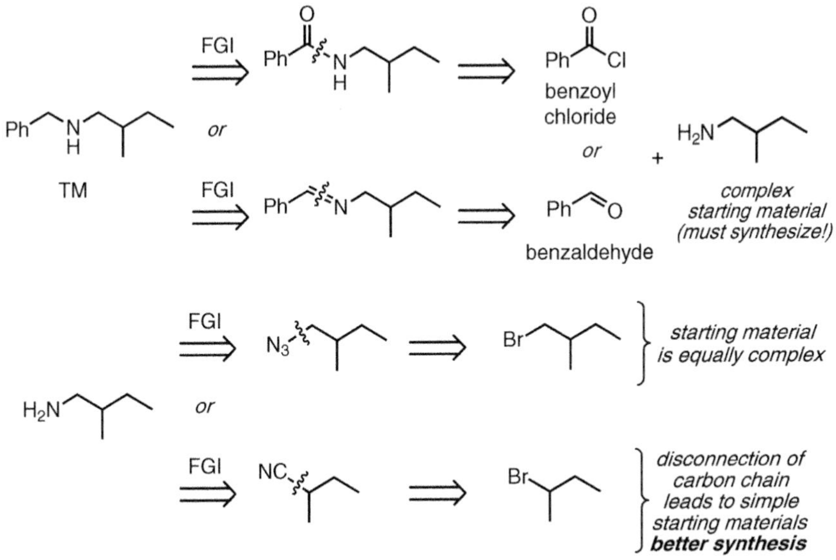

Retrosynthesis of Amine TM

One possible synthesis of the target molecule begins with the reaction of 2-bromobutane with NaCN. Hydride reduction of the resulting nitrile affords 2-methylbutan-1-amine. Addition of benzoyl chloride and reduction of the resulting amide product installs the second alkyl group on the secondary amine TM.

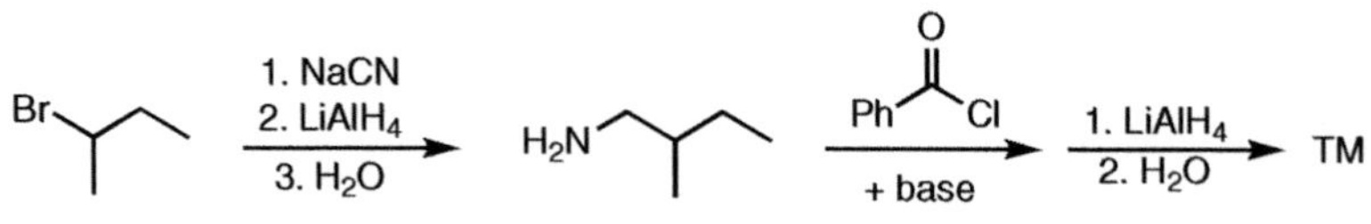

Possible Synthesis of Amine TM

SYNTHESIS OF ALKENES (R$_2$C=CR$_2$)

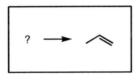

3.6.1   SYNTHESIS OF ALKENES VIA FGI

Alkenes via E2 Elimination (RX Starting Material)

Dehydrohalogenation (loss of HX) of an alkyl halide is promoted by treatment with a strong base (HO$^-$, RO$^-$, or R$_2$N$^-$) and involves an E2 mechanism to give an alkene product. The use of a sterically crowded base (such as *t*-BuOK), as well as the addition of heat, encourages E2 elimination over S$_N$2 substitution. The one-step mechanism is typically both stereospecific (*anti*-elimination) and regioselective to give the more stable (more substituted) double bond as the major product, following Zaitsev's (or Saytzeff's) Rule. However, if a bulky base is used on a hindered alkyl halide, the less sterically hindered proton will be preferentially attacked, leading to the *less substituted* alkene product (described as following Hofmann's Rule).

Introduction to Strategies for Organic Synthesis, Second Edition. Laurie S. Starkey.
© 2018 John Wiley & Sons, Inc. Published 2018 by John Wiley & Sons, Inc.

Mechanism and Regiochemistry of an E2 Elimination

Alkenes via E1 Elimination (ROH Starting Material)

Dehydration (loss of H_2O) of an alcohol is promoted by treatment with a strong, non-nucleophilic acid (such as H_2SO_4 or H_3PO_4) and heat. Alcohol dehydration involves an E1 mechanism with a carbocation intermediate. Since carbocations can rearrange, the double bond can end up anywhere on the carbon chain and the most stable, most highly substituted alkene is expected as the major product (follows Zaitsev's Rule, with possible rearrangement of the carbon skeleton). The planar carbocation intermediate also results in a loss of stereochemistry, so the more stable stereoisomer will be produced as the major product (trans or E isomer). The dehydration reaction is most useful when only a single product is possible; otherwise, a mixture of alkene products is likely to be obtained.

Mechanism of Alcohol Dehydration

Retrosynthesis of Alkenes (Elimination)

If an elimination reaction is planned, the retrosynthesis of an alkene involves a functional group interconversion that adds either HX or water to the alkene to give an alkyl halide or an alcohol starting material, respectively. The stability of the alkene target molecule will help determine whether an E2 or E1 elimination would be more suitable. For example, an unstable, terminal alkene cannot be prepared via dehydration because a rearrangement of the carbocation intermediate would be expected. Care must be taken such that the alkyl halide or alcohol starting material would give only the desired target molecule as the major product.

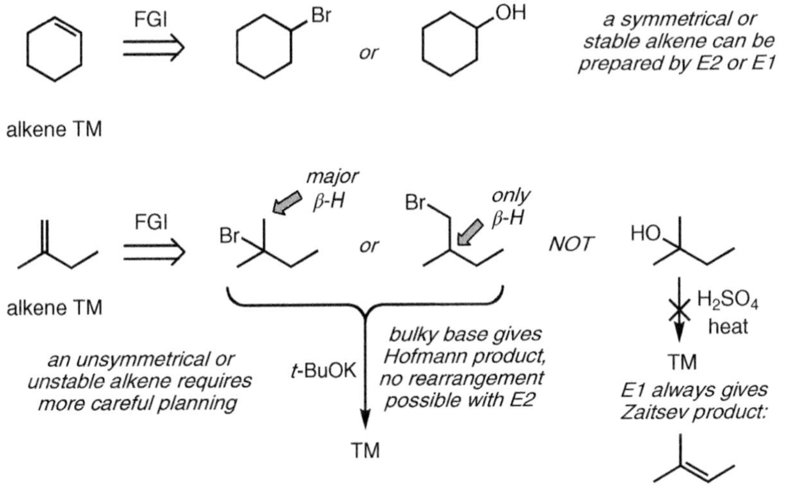

Alkene Retrosynthesis via E1 or E2

Alkenes via Reduction of Alkynes (RC≡CR → RCH=CHR)

Alkynes can be reduced to either a cis or trans alkene. Catalytic hydrogenation of an alkyne using a "poisoned" catalyst (H$_2$, Lindlar's catalyst) results in the syn addition of one equivalent of H$_2$ to give a cis alkene product. Dissolving metal reduction (Na, NH$_3$) of an alkyne produces the corresponding trans alkene. This strategy is well suited for synthesizing monosubstituted alkenes and disubstituted alkenes with a specific stereochemistry.

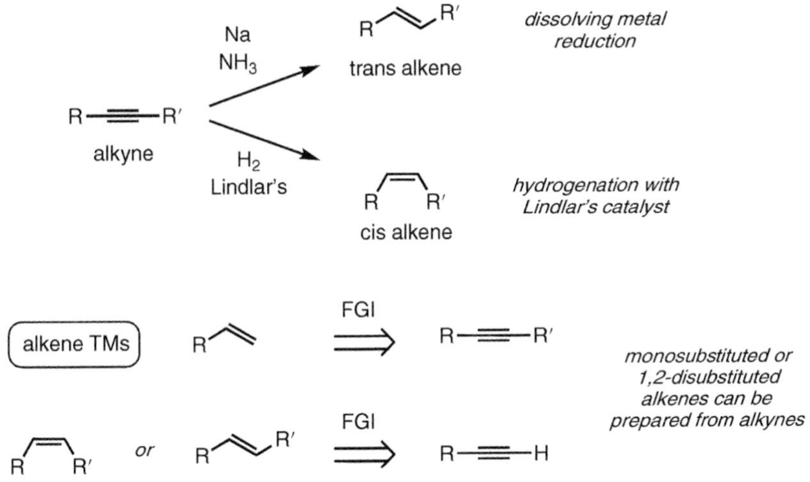

Synthesis and Retrosynthesis of Alkenes via Alkynes

3.6.2 SYNTHESIS OF ALKENES VIA THE WITTIG REACTION

The Wittig reaction (which earned a Nobel Prize in Chemistry for Georg Wittig in 1979) is a powerful method for synthesizing alkenes from aldehydes or ketones.

Preparation of a Wittig Reagent (R_2C=PPh_3)

Wittig reagents (R_2C=PPh_3) are prepared in two steps from alkyl halides. A methyl, 1° or 2° alkyl halide (R_2CHX; X = Br, Cl, or I) is first reacted with triphenylphosphine (PPh_3), causing an S_N2 displacement of the halide leaving group. Deprotonation of the resulting phosphonium salt with a strong base (such as BuLi) affords the Wittig reagent, a resonance-stabilized ylide.

Wittig Reagents are Prepared from Alkyl Halides

Synthesis of Alkenes (Wittig)

Nucleophilic Wittig reagents react with aldehyde or ketone electrophiles to furnish alkene products. To properly predict the product of a Wittig reaction, the carbonyl oxygen is replaced by the carbon group of the Wittig reagent, converting the C=O double bond into a C=C double bond. The reaction is regiospecific since the alkene always replaces the carbonyl. It is typically stereoselective to give cis (or Z) alkenes, but this can vary according to the Wittig reagent and carbonyl compound involved, as well as the specific reaction conditions (e.g., choice of base and solvent).

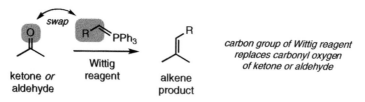

Wittig Reactions Produce Alkenes

Mechanism of the Wittig Reaction

The mechanism involves addition of the carbanion nucleophile to the electrophilic carbon of the carbonyl, followed by a breakdown of the resulting oxaphosphetane intermediate. A molecule of triphenylphosphine oxide ($Ph_3P=O$) is generated along with the alkene product.

ketone *or* aldehyde E+	Wittig reagent Nu:	oxaphosphetane intermediate is formed and then cleaves to produce alkene and $Ph_3P=O$		alkene product

Mechanism of the Wittig Reaction

Retrosynthesis of an Alkene (Wittig)

The product of this reaction is an alkene that contains a newly formed C=C double bond. This is the key bond to be identified in an alkene target molecule; a disconnection at this bond will lead to a possible retrosynthesis. One carbon of the alkene used to be a carbonyl carbon (C=O), which is an electrophilic carbon. The other carbon group must have been introduced as a nucleophile (Wittig reagent).

Retrosynthesis of Alkene via Wittig: Two Reasonable Options

Often, both combinations of possible starting materials are reasonable, and the choice made will depend on factors such as availability and ease of handling. However, a Wittig reaction involving an aldehyde electrophile is faster and may be preferred over a Wittig reaction with a ketone starting material.

PRACTICE PROBLEM 3.6A: ALKENE SYNTHESIS BY FGI AND WITTIG

3.6a For each transformation, provide the missing starting material and name the reaction type.

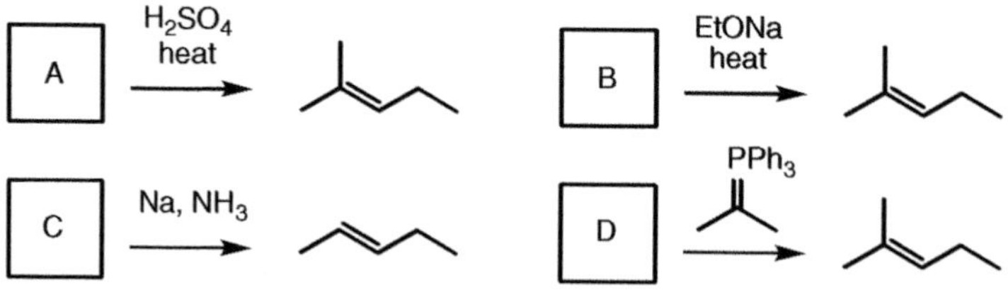

PRACTICE PROBLEM 3.6B: ALKENE SYNTHESIS

3.6b Provide the reagents needed for each of the following transformations (more than one step may be required).

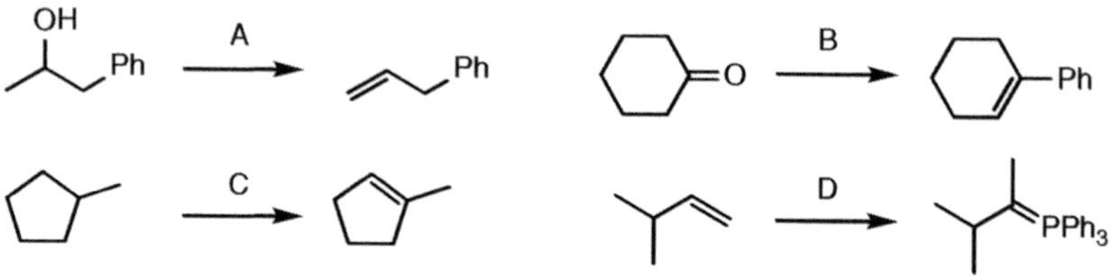

EXAMPLE: ALKENE TM

Provide the reagents necessary to transform the given starting material into the desired product. More than one step may be required.

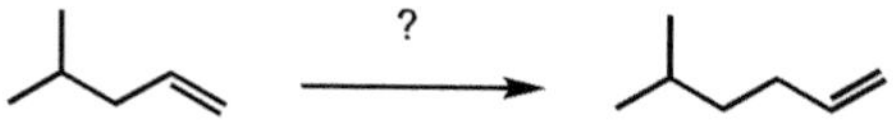

A close look at this transformation reveals that the target molecule is an alkene with one extra carbon. Since it is a relatively unstable, terminal alkene, the Wittig reagent offers the best option for its synthesis. As usual, there are two possible paths for the disconnection of the C=C double bond, giving two possible combinations of nucleophile and electrophile. Both aldehyde starting materials are expected to be good substrates for the Wittig reaction, but formaldehyde is more difficult to handle so the first disconnection (Wittig "a") may be the preferred one. The retrosynthesis of the aldehyde in "a" is an FGI that leads to an alcohol that can be prepared from the given starting material. The retrosyntheses of the aldehyde in "a" and

Wittig reagent in "b" lead to an alcohol and a bromide, respectively, and both of these can be prepared from the given starting material.

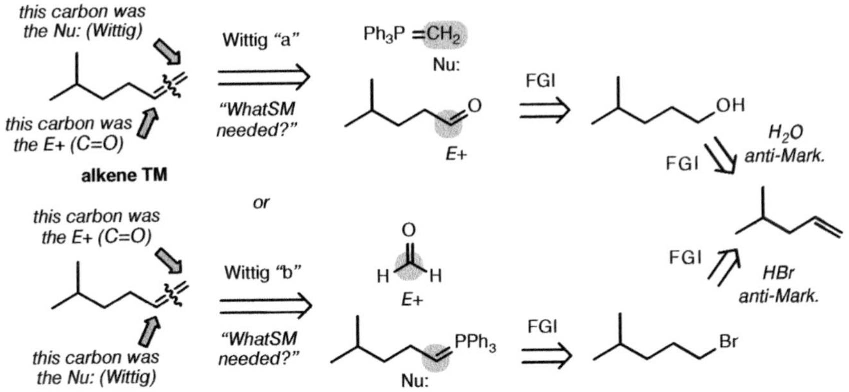

Retrosynthesis of Alkene TM

Transformation "a" begins with a hydroboration–oxidation reaction to give a primary alcohol that can be oxidized with DMP to give an aldehyde. Reaction of this aldehyde with the Wittig reagent shown gives the desired product. This Wittig reagent ($Ph_3P=CH_2$) may be commercially available, or it can be prepared from CH_3I. Transformation "b" involves the anti-Markovnikov, radical addition of HBr to give a primary alkyl halide. This bromide is converted in two steps to a Wittig reagent that will react with formaldehyde to give the desired TM.

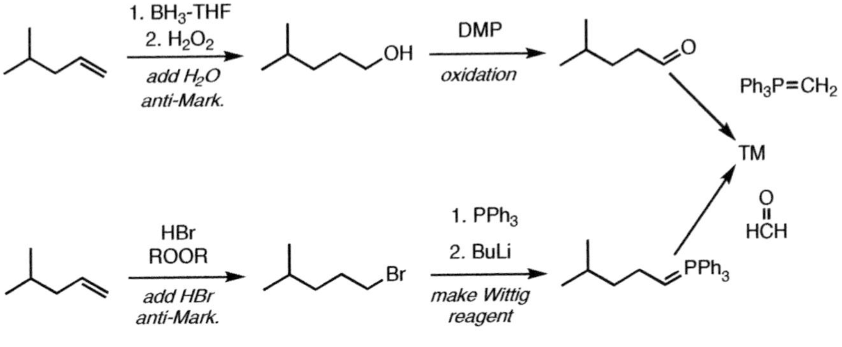

Synthesis of Alkene TM

SYNTHESIS OF ALKYNES (RC≡CR′)

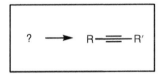

3.7.1 SYNTHESIS OF ALKYNES VIA FGI

Alkynes via E2 Elimination

When treated with a strong base (such as NaNH$_2$), a vicinal dihalide will undergo a double E2 dehydrohalogenation (loss of HX x2) to give an alkyne product. This dihalide can be prepared by addition of either bromine or chlorine to an alkene, so this two-step strategy can be used to convert an alkene into an alkyne.

Conversion of an Alkene to an Alkyne via Halogenation, Double E2

Introduction to Strategies for Organic Synthesis, Second Edition. Laurie S. Starkey.
© 2018 John Wiley & Sons, Inc. Published 2018 by John Wiley & Sons, Inc.

3.7.2 SYNTHESIS OF ALKYNES FROM OTHER ALKYNES (RC≡CH → RC≡CR′)

In the synthesis of propargylic alcohols, we saw the reaction of an alkynyl nucleophile (either the anion RC≡CNa or the Grignard RC≡CMgBr, both prepared from the alkyne RC≡CH) with a carbonyl electrophile to give an alcohol product. Such acetylide-type nucleophiles will also undergo S_N2 reactions (backside attack) with unhindered alkyl halides. With this two-step sequence (deprotonation followed by alkylation), acetylene can be converted to a terminal alkyne, and a terminal alkyne can be converted to an internal alkyne. Because acetylide anions are strong bases, the alkyl halide used must be methyl or 1°; otherwise, the E2 elimination is favored over the S_N2 substitution mechanism.

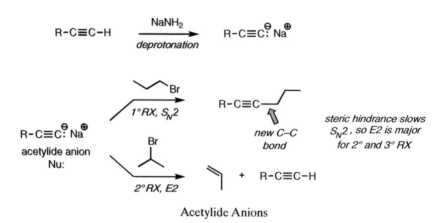

Acetylide Anions

Retrosynthesis of Alkynes (Alkylation)

The retrosynthesis of an alkyne involves a disconnection on either side of the C≡C triple bond, with the less hindered electrophile (alkyl halide) being preferred as the more successful S_N2.

Retrosynthesis of Alkynes via Alkylation

EXAMPLE 1: ALKYNE TM

Provide the reagents necessary to transform the given starting material into the desired product. More than one step may be required.

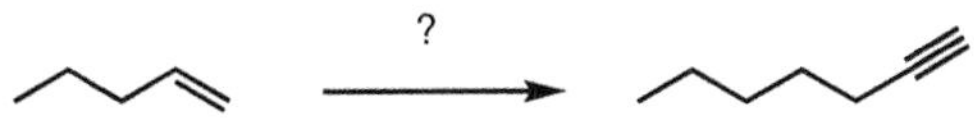

At first, this transformation might seem challenging, since no reaction exists that adds an alkynyl group to an alkene. However, a systematic approach to these problems always begins with the ending: a retrosynthesis of the desired product. Once the alkyne target molecule is disconnected to give an acetylide nucleophile and a five-carbon electrophile (1-bromopentane), the solution becomes clear.

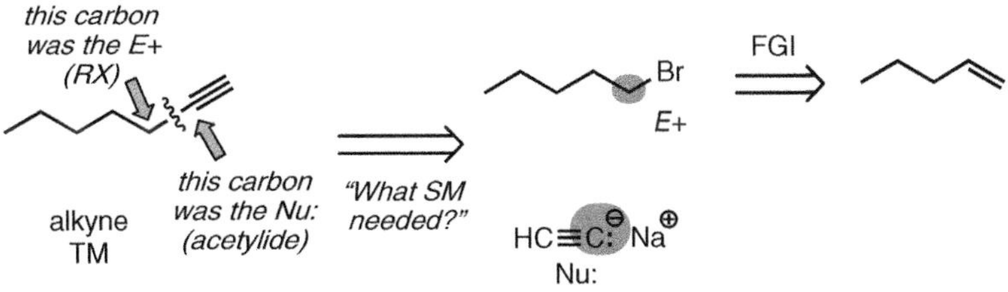

Retrosynthesis of Alkyne TM

This transformation begins with the anti-Markovnikov, radical addition of HBr to give a primary alkyl halide. This bromide undergoes an S_N2 displacement with acetylide to give the desired TM (the acetylide anion is commercially available, or it can be prepared from acetylene).

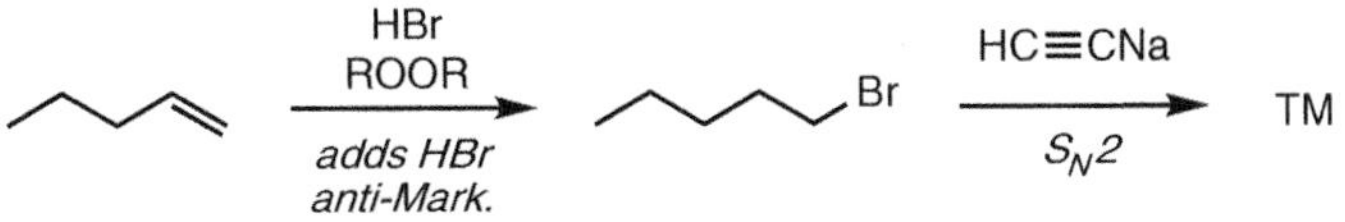

Synthesis of Alkyne TM

PRACTICE PROBLEM 3.7: ALKYNE SYNTHESIS

3.7 Provide the reagents needed for each of the following transformations (more than one step may be required).

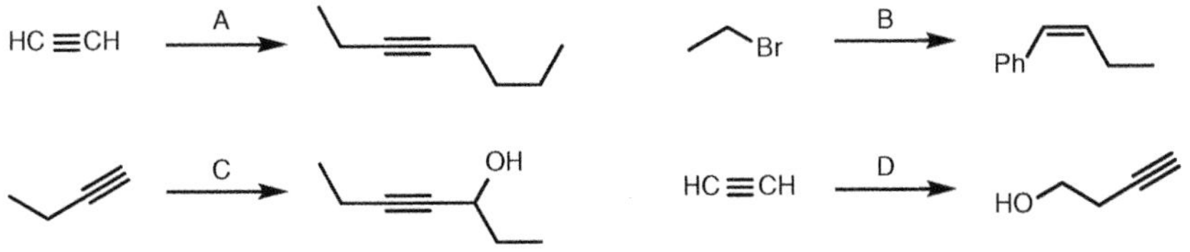

EXAMPLE 2: ALKYNE TM

Provide the reagents necessary to transform the given starting material into the desired product. More than one step may be required.

The key to this transformation is to first identify which carbons in the target molecule came from the starting material. By numbering the carbon chain, the four original carbons are located in the product, and the new C—C bond is identified. This transformation can be considered in two parts: first converting the alcohol starting material into an alkyne and then alkylating the alkyne with a benzyl group to give the desired product. To install the triple bond, an FGI retrosynthesis leads to a dibromide, which can come from an alkene.

Retrosynthesis of Alkyne TM

Conversion of the given alcohol to a terminal alkene cannot be achieved by dehydration, since that would give the more stable, internal alkene product. Instead, an E2 elimination is employed. A halide could be prepared, or treatment of the alcohol with tosyl chloride gives a tosylate that will undergo elimination when treated with a strong, bulky base (*tert*-butoxide or diisopropylamine). Addition of Br_2 installs two leaving groups, and treatment with strong base (KOH) and heat gives an alkyne product after a double dehydrohalogenation. This terminal alkyne is deprotonated with strong base ($NaNH_2$) and reacted with benzyl bromide to produce the target molecule.

Synthesis of Alkyne TM

SYNTHESIS OF ALKANES (RH)

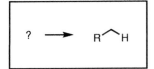

3.8.1 SYNTHESIS OF ALKANES VIA FGI

Alkanes can be prepared via functional group interconversion starting with a variety of functional groups, including alkenes, alkynes, alkyl halides, alcohols, ketones, and aldehydes. In each case, the functional group is being replaced by simple C—H bonds, so these transformations are all described as reduction reactions.

Alkane Synthesis via Substitution (RLG → RH)

A substitution reaction using a hydride nucleophile (LiAlH$_4$) to replace a halogen or tosylate leaving group with an S$_N$2 mechanism will give an alkane product. Another way to replace a halogen with hydrogen is to convert the alkyl halide into a Grignard reagent and then treat it with a protic solvent. Normally, quenching the Grignard reagent is something one tries to avoid (by using aprotic solvents, dry glassware, etc.), but it can be employed as a useful reaction to remove a halogen, or to introduce a deuterium (^{2}H isotope) onto a carbon chain at a specific location by reacting the Grignard reagent with a deuterated protic solvent such as D$_2$O.

Introduction to Strategies for Organic Synthesis, Second Edition. Laurie S. Starkey.
© 2018 John Wiley & Sons, Inc. Published 2018 by John Wiley & Sons, Inc.

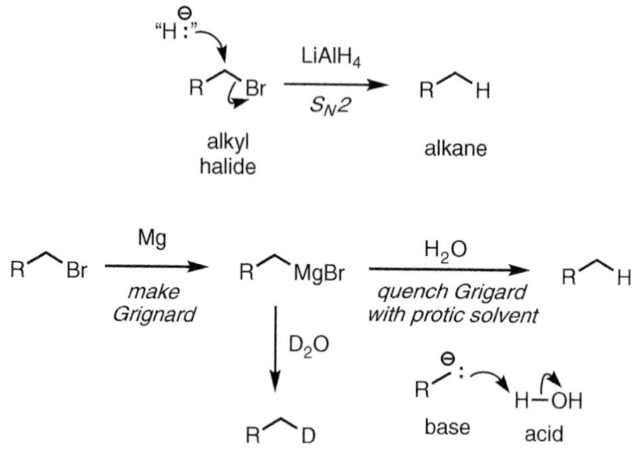

Alkane Synthesis from Alkyl Halides (Hydride/Grignard)

Alkane Synthesis via Reduction (C=C, C≡C, C=O → Alkane)

Catalytic hydrogenation (H_2, Pd) of an alkene or alkyne affords an alkane product. Aldehyde and ketone carbonyls can be completely reduced to a methylene group (CH_2) by a variety of methods, including the acidic Clemmenson reduction (Zn/Hg, HCl) and the basic Wolff–Kishner reduction (NH_2NH_2, KOH). An alternate strategy is to convert the carbonyl to a thioacetal that can then be reduced to an alkane with Raney Nickel ($Ni-H_2$).

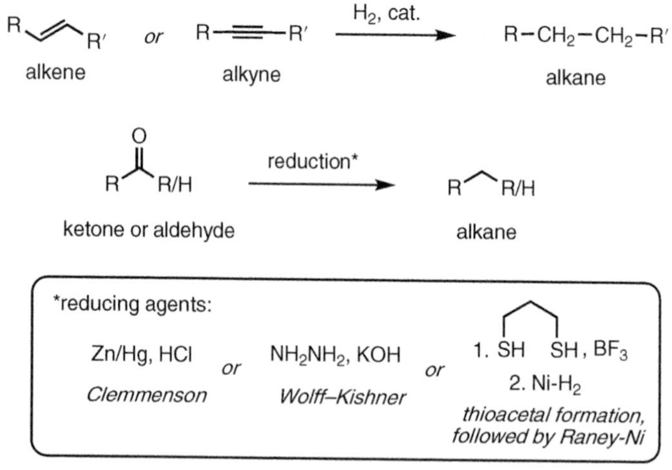

Alkanes via Reduction Reactions

PRACTICE PROBLEM 3.8A: ALKANE SYNTHESIS VIA FGI

3.8A Provide the missing reagents.

3.8.2 SYNTHESIS OF ALKANES VIA C—C BOND FORMATION

Alkanes via Metal Coupling Reactions (RM + R′X → R—R′)

The ever-increasing use of organometallic reagents in organic synthesis is both wide and varied. For example, we will see in Chapter 8 how palladium-catalyzed coupling reactions can be used to form C—C bonds between various combinations of alkyl, vinyl, alkynyl, and aryl groups. As a simple introduction to this synthetic strategy in this chapter, only a single example will be presented for the preparation of alkanes, using a reagent that should be familiar to most undergraduate students. Organocuprates (R_2CuLi) are a source of nucleophiles ("R:⁻") with reactivities that are different from Grignard (RMgX) and organolithium (RLi) reagents. Three significant differences include the organocuprate's ability to undergo coupling reactions with alkyl halides, its preference for conjugate (1,4-) additions with α,β-unsaturated carbonyls, and its reaction with acid halides to give ketone products. Organocuprates are prepared by treatment of the corresponding organolithium reagent with copper(I) iodide (CuI). Reaction of the cuprate with an alkyl halide produces an alkane product.

Synthesis of Alkanes via Organocuprate Coupling

Synthesis of Aromatic Alkanes (Friedel–Crafts Reaction)

Reaction of an aromatic compound with a carbocation (called the Friedel–Crafts alkylation) affixes an alkyl group onto the aromatic ring. In this electrophilic aromatic substitution reaction, there are several methods for generating the carbocation, including from alkyl halides (RX with FeX_3 or $AlCl_3$), alcohols (ROH with a strong acid or BF_3), and alkenes (alkene plus acid).

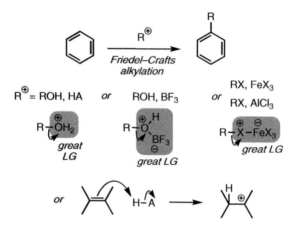

Friedel–Crafts Alkylation Begins with Carbocation Formation

Since carbocations are prone to rearrangement, the Friedel–Crafts alkylation is only suitable for the addition of stable carbocations (or ones that cannot rearrange to more stable carbocations).

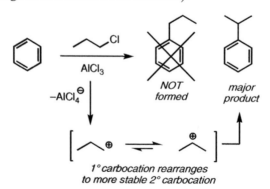

Friedel–Crafts Alkylation Involves most Stable Carbocation

To synthesize a target molecule containing a 1° alkyl group attached to a benzene ring, a Friedel–Crafts *acylation* reaction is employed instead. Reaction of an acid chloride with a strong Lewis acid ($AlCl_3$) forms an acylium ion that reacts with benzene to install an acyl group. Reduction of the resulting aromatic ketone (by hydrogenation, Clemmenson, or Wolff–Kishner) provides the desired alkyl group. Friedel–Crafts reactions involving substituted benzenes will be discussed in Chapter 5.

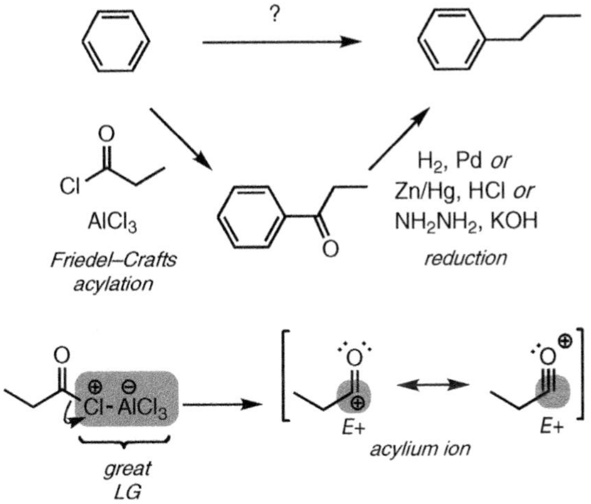

Alkylbenzenes can be Prepared via Friedel–Crafts Acylation

Retrosynthesis of Alkanes

The retrosynthesis of an alkane target molecule generally begins with an FGI that *adds* a functional group of your choice (an alkyl halide, alkene, alkyne, aldehyde, or ketone) and continues with a disconnection consistent with that functional group. Addition of the functional group at a branch point will likely lead to a good disconnection, but its precise location is not critical since it will ultimately be removed.

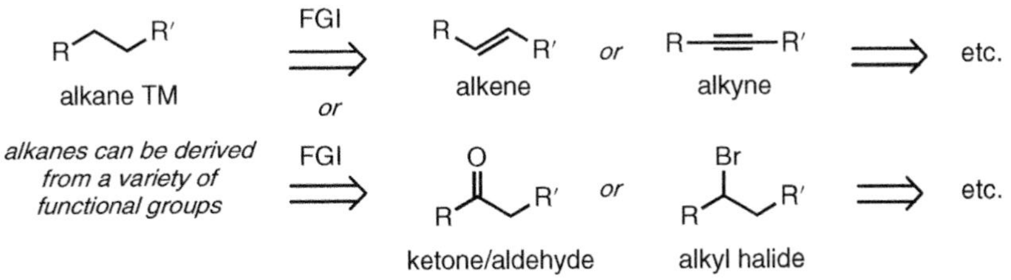

Alkane Retrosynthesis via FGI

It is also possible to directly disconnect the alkane at any position that would lead to simple carbon chains (one that would start as an organocuprate and the other as an alkyl halide).

one carbon
was the E+
(RX)

R'

R

disconnect any
C—C bond

one carbon was
the Nu: (R₂CuLi)

alkane TM

$R^{\oplus}$ ≡ R—Br
E+   E+

$^{\ominus}R'$ ≡ $\left(R'\right)_2$CuLi
Nu: Nu:

Alkane Retrosynthesis via Cuprate Coupling

The retrosynthesis of aryl alkanes can occur at the C–C bond connecting the alkyl group to the benzene ring. Retrosynthetic analysis places a hydrogen atom on the benzene ring, since the ring acts as a nucleophile in the reaction and the H⁺ will ultimately be replaced by the E+. The disconnected alkyl group requires either a halide or a hydroxyl at the point of disconnection, since either of these will result in a carbocation when treated with a Lewis acid. In the case of a target molecule containing a 1° alkyl group attached to the aromatic ring (Ar–CH₂R), then the retrosynthesis begins with an FGI that adds a carbonyl to the benzylic position. The resulting aromatic ketone is disconnected as shown (ketone synthesis will be discussed further in Chapter 3.9).

this carbon
was the E+
(carbocation)

this carbon was
the Nu: (PhH)

disconnect
between alkyl
group and
aromatic ring

alkylbenzene
TM

$R^{\oplus}$ ≡ R—OH, BF₃
E+   R—X, AlCl₃
         E+

Nu: H
Nu:

Retrosynthesis of Alkylbenzene TMs

PRACTICE PROBLEM 3.8B: ALKANE SYNTHESIS, TRANSFORM PROBLEMS

3.8B Provide the reagents needed for each of the following transformations (more than one step may be required).

EXAMPLE: ALKANE TM

Provide at least TWO synthetic routes to the following target molecule, using readily available starting materials and reagents. Each synthesis must involve the formation of a new C—C bond.

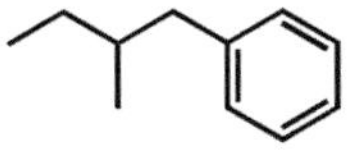

TM

There are several possible syntheses of this compound. One possible FGI on the branched alkane TM leads to an alkene. This alkene can be prepared in a number of ways, including by dehydration of an alcohol or with a Wittig reaction.

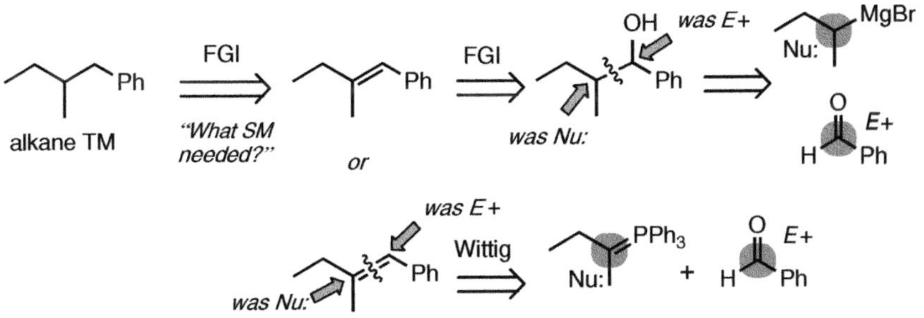

Possible Retrosyntheses of Alkane TM: FGI to Alkene

Another option is to consider disconnecting at the aromatic ring and utilizing a Friedel–Crafts reaction. Since the alkyl group would require an unstable 1° carbocation, synthesis via Friedel–Crafts alkylation is impossible. Instead, an FGI that installs a carbonyl at the benzylic position gives an aromatic ketone that is suitable for a Friedel–Crafts acylation disconnection.

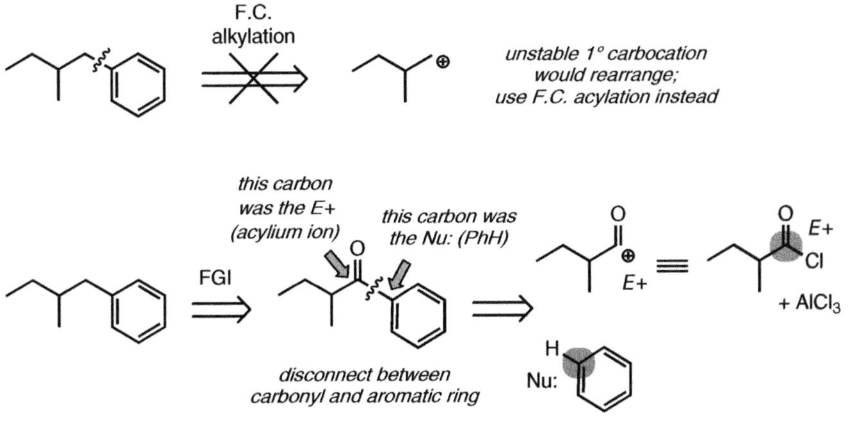

Possible Retrosynthesis of Alkane TM: Aromatic Disconnection

Three possible syntheses are presented for the given target molecule, each of which forms a new C—C bond as required (via Grignard, Wittig, or Friedel–Crafts reactions). The acid chloride required for the Friedel–Crafts acylation can be prepared from the corresponding carboxylic acid (see Section 3.11).

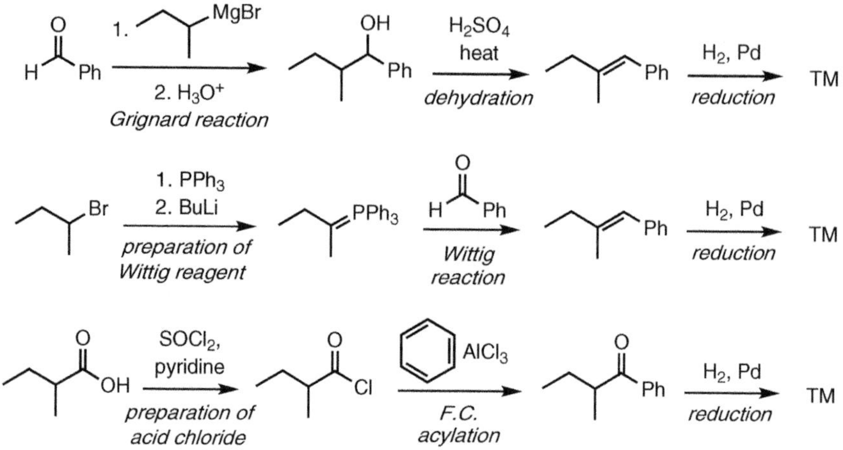

Possible Syntheses of Alkane TM

SYNTHESIS OF ALDEHYDES AND KETONES (RCHO, R₂C═O)

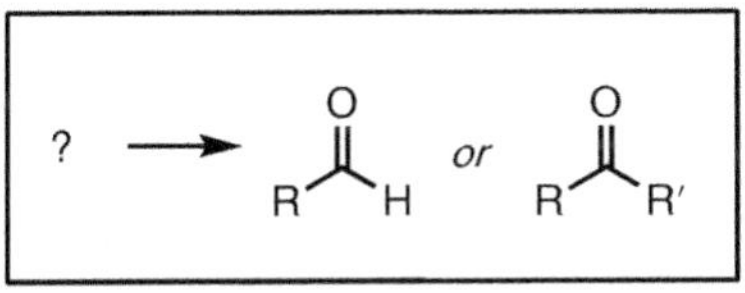

Aldehyde and ketone carbonyls can be created via functional group interconversions (FGI), and compounds already containing a carbonyl can be alkylated by reaction at either the electrophilic carbonyl carbon (e.g., acyl substitution reactions) or the nucleophilic alpha carbon. A wide variety of strategies can be used to produce aldehyde and ketone target molecules.

3.9.1 SYNTHESIS OF ALDEHYDES/KETONES VIA FGI

Aldehydes and ketones can be prepared by oxidation of alcohols or alkenes, by the partial reduction of carboxylic acid derivatives (aldehydes only), or by hydration of alkynes.

Aldehydes/Ketones via Oxidation or Reduction Reactions

Oxidation of a 2° alcohol produces a ketone while partial oxidation of a 1° alcohol (DMP or Swern) gives aldehyde products. Ozonolysis of an alkene, followed by a reductive workup, gives ketone and/or aldehyde products, depending on the alkene starting material. Esters, acid chlorides, and nitriles can be partially reduced with diisobutylaluminum hydride (DIBAL-H) to produce aldehydes as well.

Introduction to Strategies for Organic Synthesis, Second Edition. Laurie S. Starkey.
© 2018 John Wiley & Sons, Inc. Published 2018 by John Wiley & Sons, Inc.

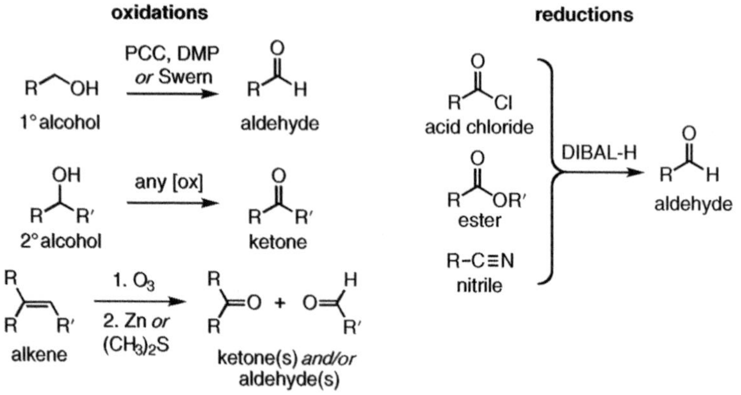

oxidations   **reductions**

Synthesis of Aldehydes and Ketones via Oxidation or Reduction Reactions

Aldehydes/Ketones via Alkyne Hydration (RC≡CR → [enol] → ket/ald)

Addition of the components of water (—H and —OH) to an alkyne gives an enol intermediate that spontaneously tautomerizes to give a ketone or aldehyde product. As in the hydration of alkenes, the regiochemistry of this reaction with a terminal alkyne can be controlled to give either Markovnikov or anti-Markovnikov orientation (via H$_3$O$^+$/HgSO$_4$ or hydroboration–oxidation, respectively). When water is added to a terminal alkyne with Markovnikov regiochemistry, the hydrogen goes to the terminal carbon and a methyl ketone is produced. Anti-Markovnikov regioselectivity on a terminal alkyne is improved when using a bulky boron reagent, such as disiamylborane (Sia$_2$BH), and results in an aldehyde product, since the hydroxyl group is added to the terminal carbon. Hydration of an internal alkyne is synthetically useful only if it is symmetrically substituted, such that only a single ketone product is possible.

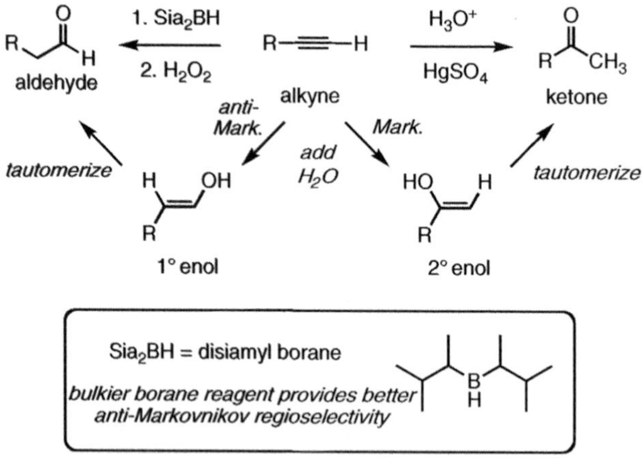

Synthesis of Aldehydes or Ketones from Terminal Alkynes

PRACTICE PROBLEM 3.9A—ALDEHYDE/KETONE SYNTHESIS BY FGI

3.9A Provide the missing reagents.

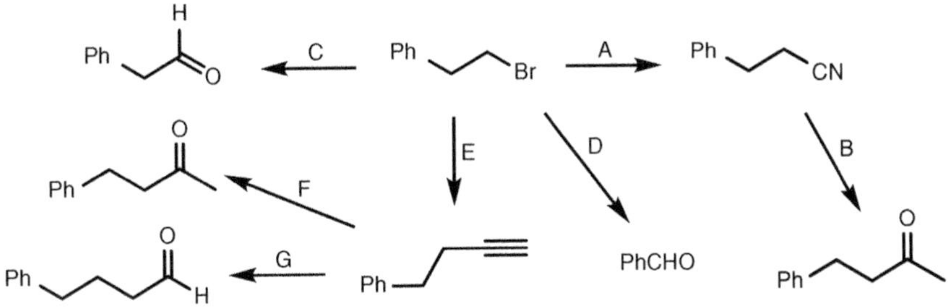

3.9.2   SYNTHESIS OF ALDEHYDES/KETONES VIA ACYL SUBSTITUTIONS

Synthesis of Ketones via Organometallic Reagents

Reaction of an organometallic reagent, such as a Grignard (R′MgX), with a carboxylic acid derivative, such as an ester (RCO_2R), usually results in the addition of two equivalents of the organometallic groups. However, in certain situations it is possible to control the reaction such that only a single equivalent of the organometallic nucleophile is added. The result is an acyl substitution reaction to give a ketone product. Two examples of such syntheses include the reaction of an acid halide with an organocuprate and the reaction of a nitrile with a Grignard reagent (followed by hydrolysis of the resulting imine product).

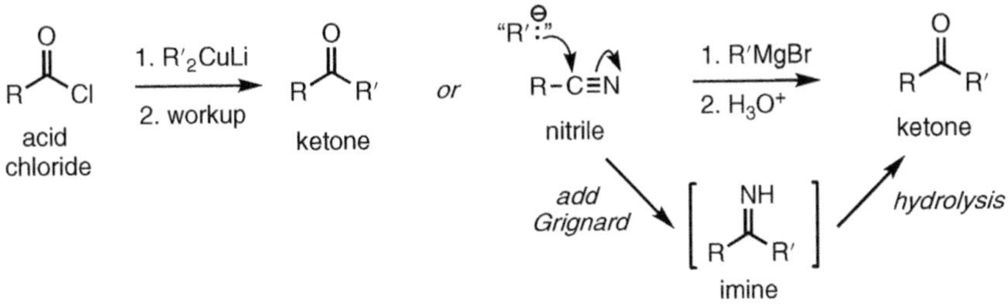

Synthesis of Ketones via Organometallic Reactions

Synthesis of Aromatic Ketones (Friedel–Crafts Acylation)

Friedel–Crafts acylation, the reaction of an aromatic compound with an acid chloride and a Lewis acid (such as $AlCl_3$), adds an acyl group to the aromatic ring to give an aromatic ketone product. Like the previously discussed alkylation reaction, it involves a strongly electrophilic carbocation (called an acylium ion), but this carbocation is not subject to rearrangement since it is stabilized by resonance.

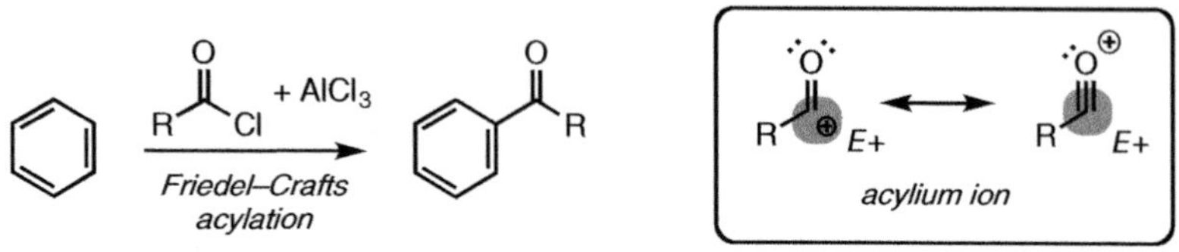

Friedel–Crafts Acylation Reaction

Synthesis of Aromatic Aldehydes (Formylation Reactions)

In order to synthesize benzaldehyde using the Friedel–Crafts acylation, formyl chloride would be required. Since this acid chloride is too reactive to be prepared, several synthetic equivalents have been developed to accomplish this formylation reaction.

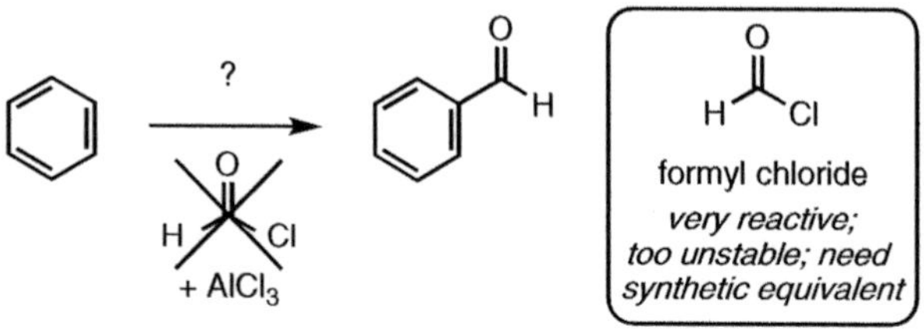

Friedel–Crafts Reaction Fails with Formyl Chloride

The Gattermann–Koch reaction provides one such formylation method. In this reaction, benzene (or an alkylbenzene) is treated with carbon monoxide (CO), HCl, and AlCl$_3$. The CO and HCl combine to produce formyl chloride *in situ*. The usual reaction of the acid chloride with AlCl$_3$ enables the Friedel–Crafts acylation to proceed, affording benzaldehyde.

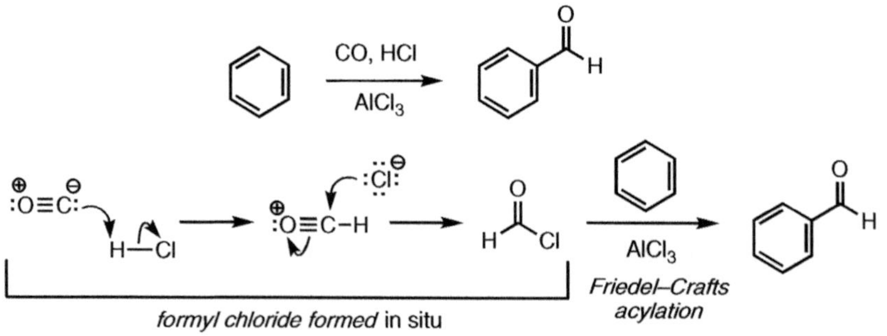

The Gattermann–Koch Reaction

In the Vilsmeier–Haack reaction, a reagent such as *N,N*-dimethylformamide (DMF) is the source of the formyl group. Reaction of DMF with phosphorus oxychloride (POCl$_3$) creates an electrophilic iminium ion *in situ* that reacts with activated aromatic compounds. Hydrolysis of the resulting iminium ion results in a benzaldehyde derivative.

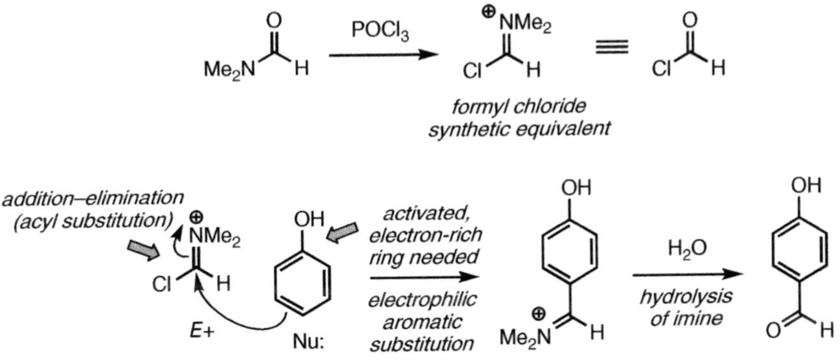

formyl chloride
synthetic equivalent

The Vilsmeier–Haack Reaction

Retrosynthesis of Ketones (Acyl Substitution)

One possible retrosynthesis of a ketone target molecule is by disconnection at the carbonyl carbon. The carbonyl carbon used to be the electrophile (acid chloride or nitrile) and the other carbon was the nucleophile (organometallic reagent, or a benzene ring if it is an aromatic ketone TM).

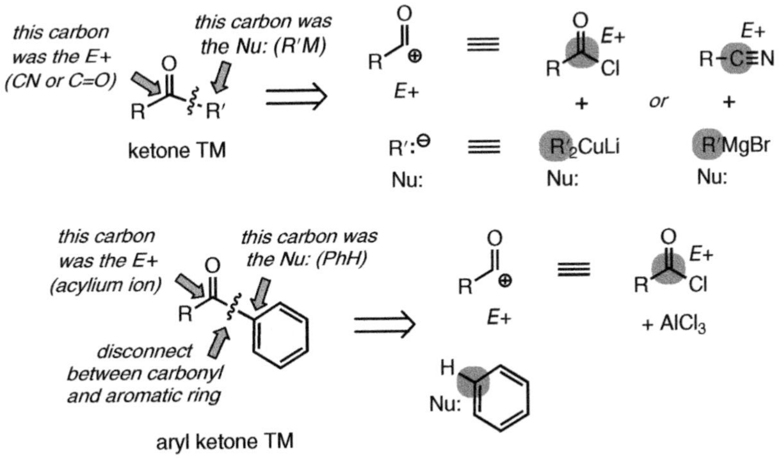

Retrosynthesis of a Ketone TM via Acyl Disconnection

PRACTICE PROBLEM 3.9B: ALDEHYDE/KETONE SYNTHESIS I

3.9B   Provide the reagents needed for each of the following transformations (more than one step may be required).

3.9.3   SYNTHESIS OF KETONES VIA α-ALKYLATION

Formation and Reactivity of Enolates

Aldehydes and ketones are acidic at the alpha (α) position because deprotonation results in a resonance-stabilized conjugate base (called an enolate). The enolate ion is nucleophilic at the alpha carbon. Enolates prepared from aldehydes are difficult to control, since aldehydes are also very good electrophiles and a dimerization reaction often occurs (self-aldol condensation). However, the enolate of a ketone is a versatile synthetic tool since it can react with a wide variety of electrophiles. For example, when treated with an unhindered alkyl halide (RX), an enolate will act as a nucleophile in an S_N2 mechanism that adds an alkyl group to the alpha carbon. This two-step "α-alkylation" process begins by deprotonation of a ketone with a strong base, such lithium diisopropylamide (LDA) at −78°C, followed by the addition of an alkyl halide. Since the enolate nucleophile is also strongly basic, the alkyl halide must be unhindered to avoid the competing E2 elimination (ideal RX for S_N2=methyl, 1°, allylic, benzylic).

α-Alkylation of a Ketone Using LDA

Kinetic versus Thermodynamic Regiocontrol of Enolate Formation

When starting with an unsymmetrical ketone such as 2-butanone, deprotonation at either alpha carbon will result in two possible enolates. Varying the reaction conditions can control the regioselectivity of enolate formation. When LDA is used at a low temperature, the irreversible deprotonation is controlled by kinetics and the least sterically hindered proton is removed to form the so-called "kinetic" enolate.

pK_a ~20   *lithium diisopropyl amide* pK_a ~40
                   (LDA)—very strong base

Deprotonation by LDA is Essentially Irreversible

Deprotonation at the more substituted alpha carbon is slower since it is more crowded, but the resulting "thermodynamic" enolate is more stable since it has a more substituted double bond. The thermodynamic enolate is favored when the reaction is allowed to equilibrate; using higher temperatures and either an excess of ketone or a weaker base allows the reverse reaction to occur. In another approach, an enolate is trapped with trimethylsilyl chloride (TMSCl) to give the thermodynamic silyl enol ether. The reversible mechanism of the silyl enol ether formation, along with the warmer reaction conditions, promotes equilibration and, therefore, favors the more stable product.

Choice of Base/Reaction Conditions Determines Enolate Regiochemistry

Treatment of the silyl enol ether with fluoride ion (tetrabutylammonium fluoride, TBAF) regenerates the enolate, making alkylation at the more substituted alpha carbon possible.

α-Alkylation via Silyl Enol Ether

The Acetoacetic Ester Synthesis

Another way to control the regiochemistry of enolate formation is by temporarily adding an ester group to the alpha carbon that is to be deprotonated. Addition of an electron-withdrawing group (EWG) dramatically increases the acidity of the alpha carbon, making enolate formation highly selective. Treatment with a weaker base (such as K$_2$CO$_3$, HO$^-$, or RO$^-$) causes complete deprotonation since the enolate product has extra resonance stabilization. These less reactive conditions tend to allow greater selectivity, and a stabilized enolate can react with a variety of electrophiles, such as reaction with an alkyl halide to alkylate the alpha carbon. Finally, the ester group can be removed by hydrolysis followed by decarboxylation of the resulting β-ketoacid. Since ethyl acetoacetate serves as the simplest starting material, this strategy is known as the acetoacetic ester synthesis.

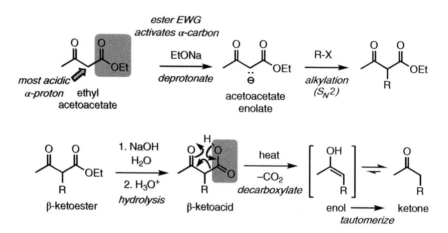

Synthesis Using Ethyl Acetoacetate

Retrosynthesis of a Ketone (α-Alkylation)

One possible retrosynthesis of a ketone is disconnection at the alpha carbon. The alpha carbon was the nucleophile (enolate) and the other carbon was the electrophile (an alkyl halide). If the target molecule is a derivative of acetone, then this disconnection leads to the enolate of acetone. Rather than using the unstable enolate of acetone in a synthesis, the use of the ethyl acetoacetate enolate as a synthetic equivalent is preferred.

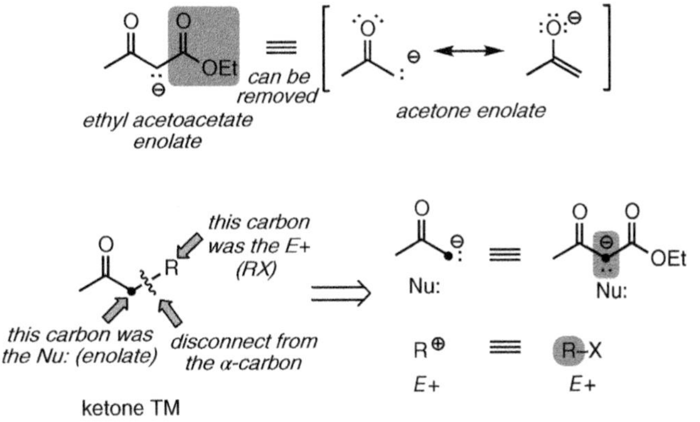

ethyl acetoacetate enolate

acetone enolate

ketone TM

Ethyl Acetoacetate as a Synthetic Equivalent of Acetone

Alkylation of Dienolates

If ethyl acetoacetate is treated with two equivalents of LDA, a dianion is formed. This dienolate is more reactive at the end carbon, since that negative charge is delocalized over just one EWG; the other negative charge is alpha to two EWGs, making that position less reactive. Treatment of the dienolate with an alkyl halide will alkylate the terminal alpha carbon.

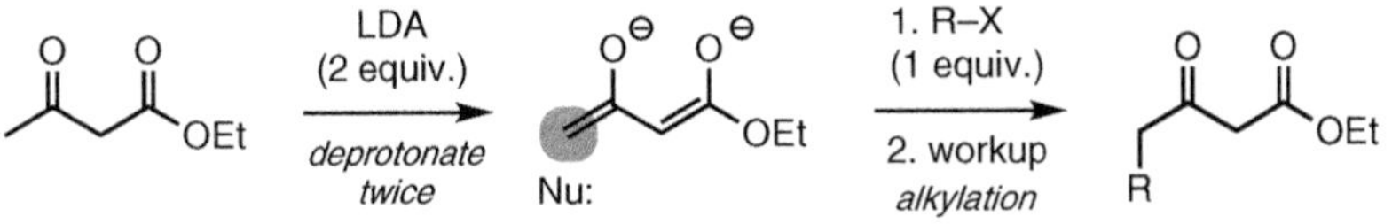

Dienolate Formation and Alkylation

This strategy offers a useful way to functionalize the commercially available ethyl acetoacetate. If it is followed with another α-alkylation and decarboxylation (acetoacetic ester synthesis), the final product will be an acetone molecule with alkyl groups added to both alpha carbons.

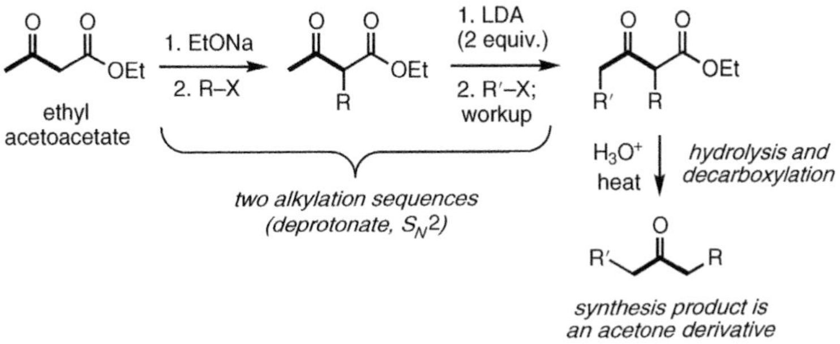

Ethyl Acetoacetate as a Versatile Starting Material

PRACTICE PROBLEM 3.9C: ALDEHYDE/KETONE SYNTHESIS II

3.9C Provide the reagents needed for each of the following transformations (more than one step may be required).

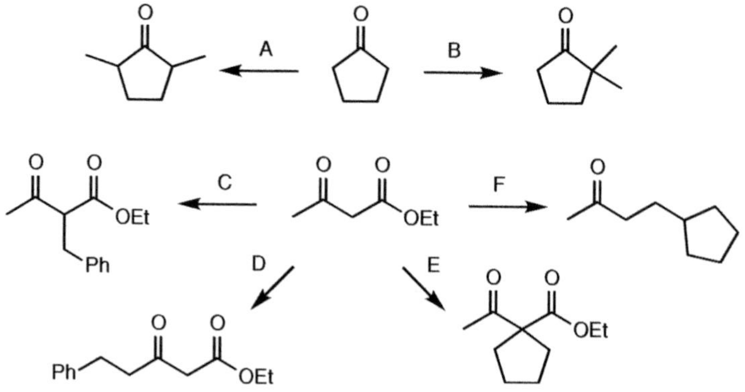

EXAMPLE: KETONE TM

Synthesize the following target molecule, using readily available starting materials and reagents. The synthesis must involve the formation of a new C—C bond.

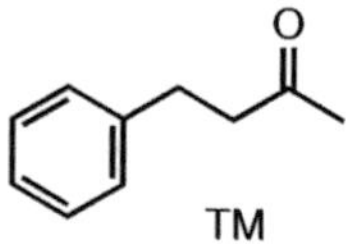

TM

Once again, there are several possible syntheses of this compound. One
possibility is an FGI to an alcohol. This alcohol can then be disconnected at two
different places and prepared by a Grignard reaction with either an aldehyde or
an epoxide. While the epoxide disconnection may have a slight advantage by
breaking the target molecule into more nearly equal pieces, both routes involve
readily available starting materials and represent acceptable syntheses.

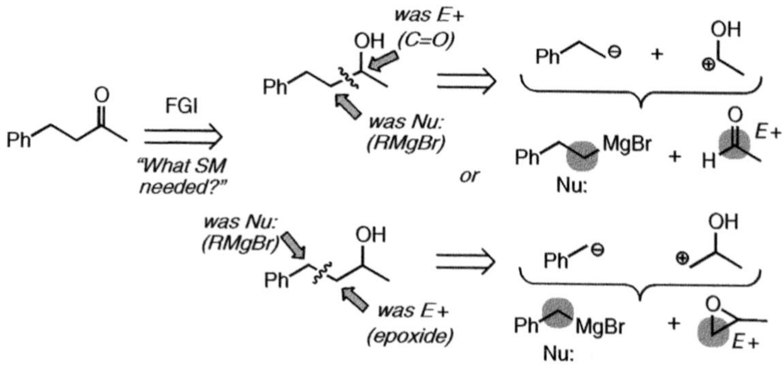

Possible Retrosynthesis of Ketone TM (FGI)

Another good option is to consider disconnecting at the nucleophilic alpha carbon. The resulting acetone enolate would be replaced with its ethyl acetoacetate synthetic equivalent.

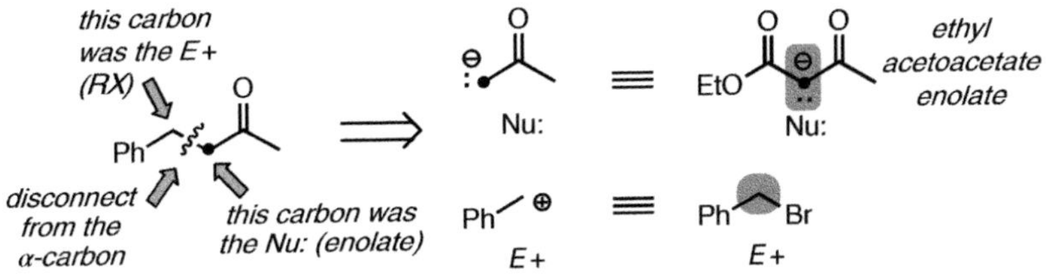

Possible Retrosynthesis of Ketone TM (Acetoacetate Ester)

Three possible syntheses are presented for the given target molecule, each of which forms a new C—C bond as required, via Grignard or enolate reactions.

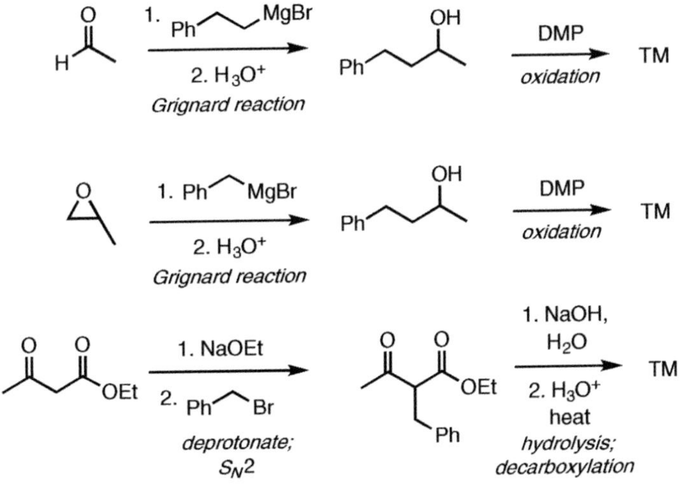

Possible Syntheses of Ketone TM

SYNTHESIS OF CARBOXYLIC ACIDS (RCO$_2$H)

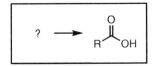

Carboxylic acids can be created via functional group interconversions (FGI), including oxidations and acyl substitutions. A variety of carbon–carbon bond-forming reactions can also be used to synthesize carboxylic acids.

3.10.1   SYNTHESIS OF CARBOXYLIC ACIDS VIA FGI

Carboxylic acids are formed by oxidation of aldehydes or 1° alcohols with a strong oxidizing agent such as chromic acid (Na$_2$Cr$_2$O$_7$/H$_2$SO$_4$, Jones) or permanganate (KMnO$_4$). These same conditions can be used to oxidize benzylic carbons, to give benzoic acid derivatives. Alkynes and certain alkenes (those not tetrasubstituted) will also give carboxylic acid products when treated with vigorous KMnO$_4$ conditions (warm, concentrated) or by ozonolysis (with an oxidative workup for alkenes).

Introduction to Strategies for Organic Synthesis, Second Edition. Laurie S. Starkey.
© 2018 John Wiley & Sons, Inc. Published 2018 by John Wiley & Sons, Inc.

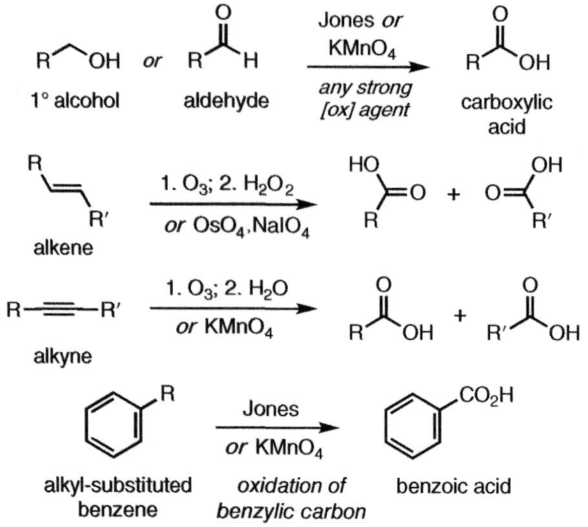

Synthesis of Carboxylic Acids by Oxidation

Another functional group interconversion that gives carboxylic acid products is the hydrolysis of carboxylic acid derivatives such as esters and nitriles. Such hydrolysis reactions can either be acid-catalyzed (H_3O^+) or base-promoted (1. NaOH, H_2O; 2. H_3O^+) and involve an acyl substitution mechanism (addition–elimination) that replaces any acyl leaving group with a hydroxyl group. The synthesis of carboxylic acids via nitriles is especially noteworthy since the introduction of the cyano group via S_N2 with CN⁻ involves the formation of a new C—C bond (adds one new carbon to the alkyl halide carbon chain).

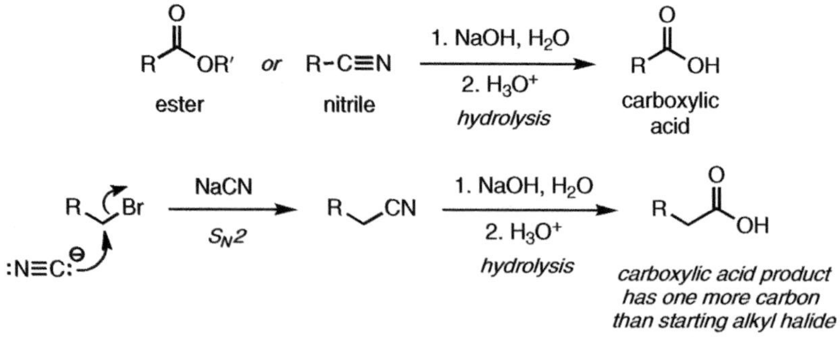

Synthesis of Carboxylic Acids via Ester or Nitrile Hydrolysis

3.10.2 SYNTHESIS OF CARBOXYLIC ACIDS VIA GRIGNARD ($RMgBr + CO_2 \rightarrow RCO_2H$)

Reaction of a Grignard reagent with carbon dioxide produces a carboxylic acid product upon workup. This synthesis converts an alkyl (or aryl) halide to a carboxylic acid, while also extending the carbon chain by one carbon.

Synthesis of Carboxylic Acids by Reaction of Grignard Reagent with CO_2

3.10.3 RETROSYNTHESIS OF CARBOXYLIC ACIDS (DISCONNECT AT CARBONYL)

The retrosynthesis of a carboxylic acid target molecule can involve a disconnection between the alpha carbon and the carbonyl carbon. This bond can either be formed via a Grignard reaction (alpha carbon was the nucleophile, reacting with CO_2) or via S_N2 with cyanide (alpha carbon was the electrophile as an alkyl halide, reacting with NaCN, followed by conversion of $-C\equiv N$ to $-CO_2H$). Not all TMs can be effectively prepared by either method. An S_N2 mechanism requires an unhindered alkyl halide (methyl, 1°, or 2°), so the nitrile approach is unsuitable for a highly substituted target molecule or for a benzoic acid derivative since there cannot be an S_N2 mechanism on an sp^2-hybridized carbon.

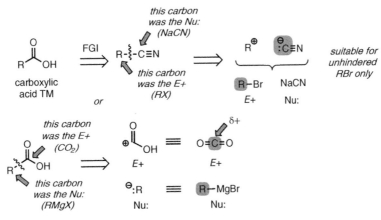

Retrosynthesis of Carboxylic Acids

3.10.4 CARBOXYLIC ACIDS VIA α-ALKYLATION: MALONIC ESTER SYNTHESIS

Deprotonation of the alpha carbon of acetic acid with LDA would be impossible since the carboxylic acid proton is more acidic than the alpha proton. Protection as an ester overcomes this problem, but the resulting ester enolate is not particularly stable and its reactions can be low yielding. Addition of a second ester to the alpha carbon serves as an activating group and allows the formation of a stabilized enolate. As seen with the aceto-acetic ester synthesis, this ester group can be eventually removed by a decarboxylation reaction. The malonic ester synthesis starts with commercially available diethyl malonate. Deprotonation, alkylation of the resulting enolate with an alkyl halide, and hydrolysis followed by decarboxylation furnishes a carboxylic acid product.

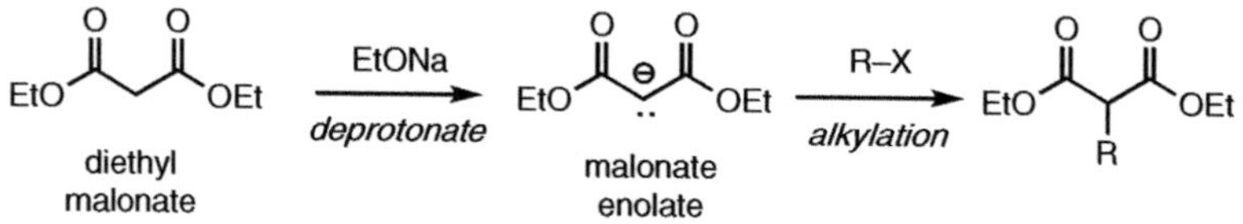

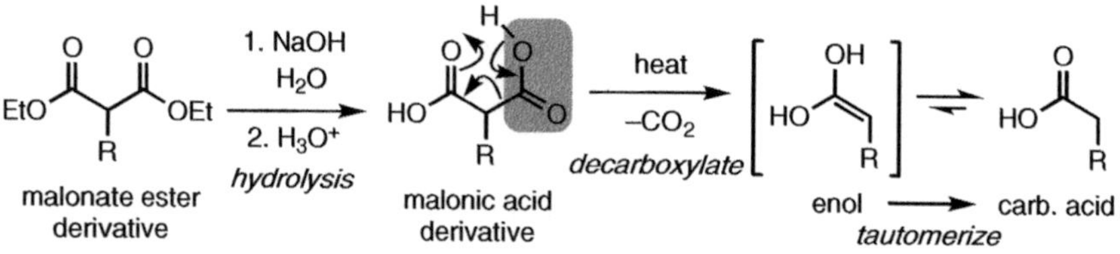

Synthesis Using Diethyl Malonate

If desired, the two-step alkylation process can be repeated before hydrolysis to give a wide variety of acetic acid derivatives after decarboxylation.

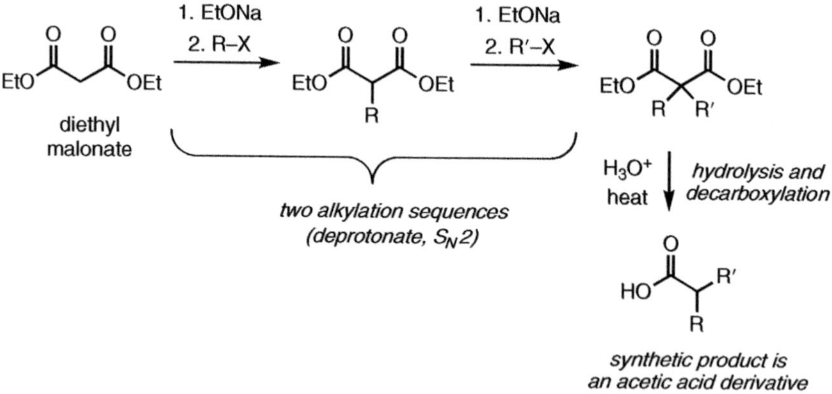

Malonic Ester Synthesis of Acetic Acid Derivatives

3.10.5   RETROSYNTHESIS OF CARBOXYLIC ACIDS (DISCONNECT AT α-CARBON)

The retrosynthesis of a carboxylic acid target molecule can involve a disconnection between the alpha and the beta carbons. The alpha carbon was the nucleophile (enolate) and the beta carbon was the electrophile (alkyl halide). If there are two groups attached to the alpha carbon, it is common practice to disconnect both alkyl groups and work backward to the two-carbon acetic acid enolate (since diethyl malonate can be used as its synthetic equivalent).

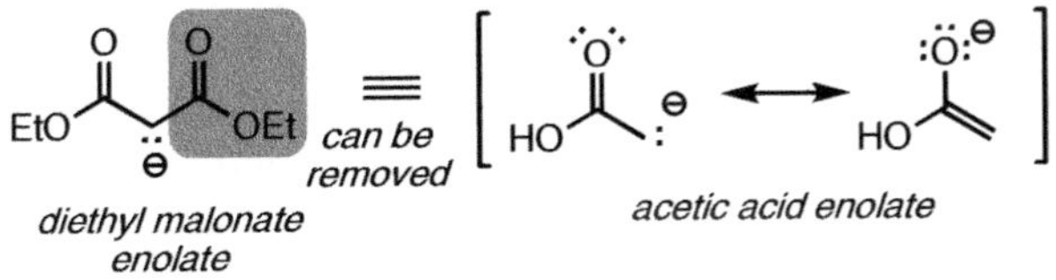

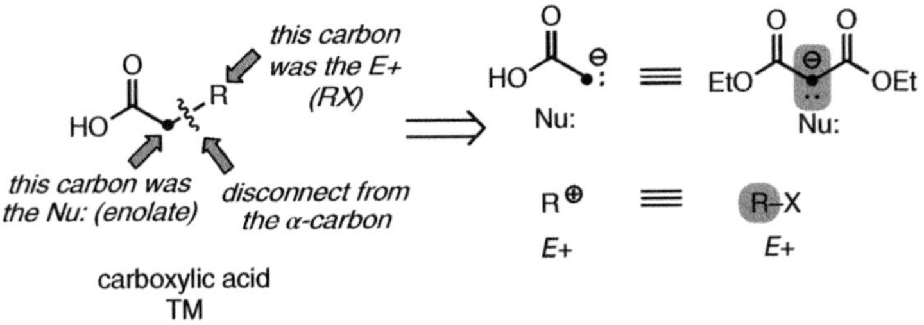

Diethyl Malonate as a Synthetic Equivalent of Acetic Acid

PRACTICE PROBLEM 3.10: CARBOXYLIC ACID SYNTHESIS

3.10   For each problem, provide a synthesis of the TM, using the given retrosynthesis.

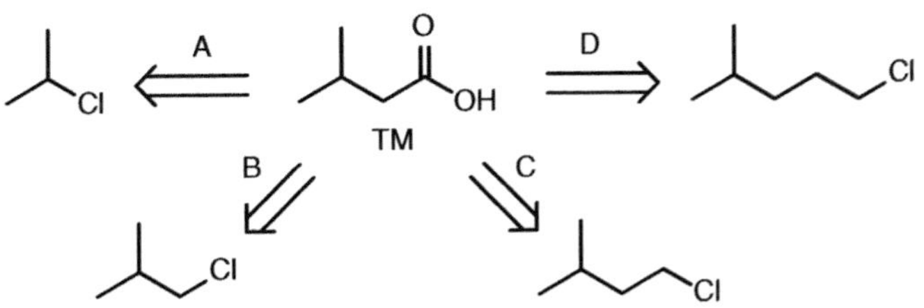

EXAMPLE: CARBOXYLIC ACID TM

Synthesize the following target molecule from readily available starting materials that contain no carbon chains longer than four carbons.

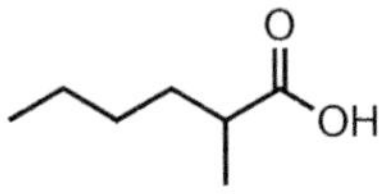

TM

The instructions for this problem point to a disconnection at the nucleophilic alpha carbon, thereby creating a four-carbon electrophile (RX) and a three-carbon nucleophile (enolate). However, the resulting enolate is not stable. Addition of an ester electron-withdrawing group leads to a starting material that can be further simplified, and a second disconnection at the alpha carbon results in the familiar diethyl malonate enolate. While diethyl 2-methylmalonate (ethyl methylmalonate) is commercially available, it is not inexpensive. It can be as much as 10 times the price of diethyl malonate! Recognizing the target molecule as a derivative of acetic acid can help you recognize the double disconnection that is possible.

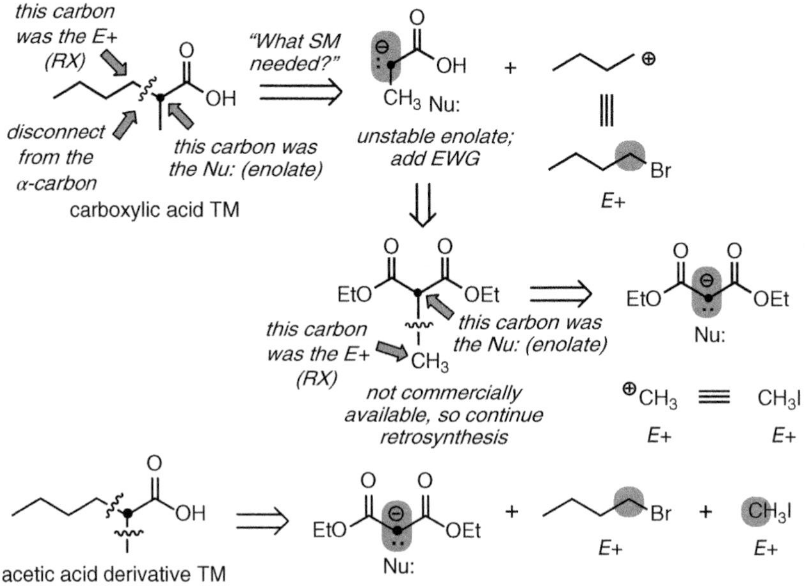

Retrosynthesis of Carboxylic Acid TM

The synthesis of this carboxylic acid begins with the deprotonation of diethyl malonate with NaOEt and alkylation with methyl iodide. A second deprotonation and alkylation with *n*-butyl bromide installs the necessary carbon chain. Hydrolysis followed by heat causes the decarboxylation that removes the extra carboxyl group and affords the desired target molecule.

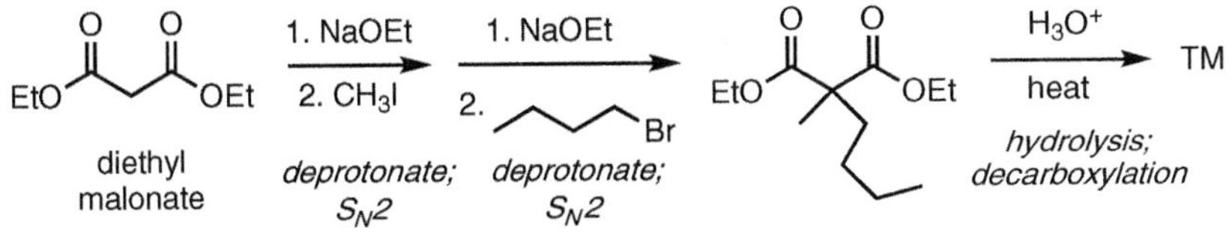

Synthesis of Carboxylic Acid TM

SYNTHESIS OF CARBOXYLIC ACID DERIVATIVES

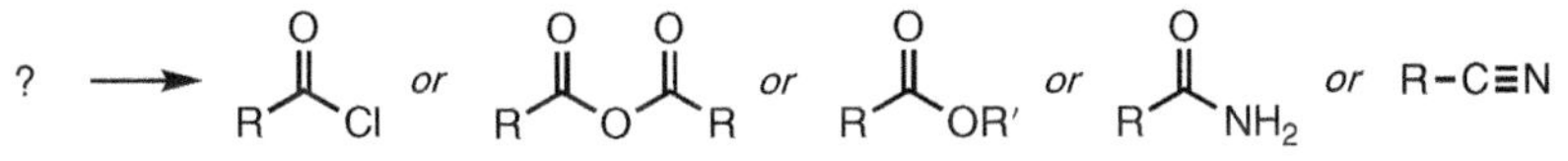

$$? \longrightarrow \quad \underset{R}{\overset{O}{\underset{Cl}{\bigwedge}}} \quad or \quad \underset{R}{\overset{O}{\underset{O}{\bigwedge}}} \underset{R}{\overset{O}{\bigwedge}} \quad or \quad \underset{R}{\overset{O}{\underset{OR'}{\bigwedge}}} \quad or \quad \underset{R}{\overset{O}{\underset{NH_2}{\bigwedge}}} \quad or \quad R-C\equiv N$$

Carboxylic acid derivatives can be created via functional group interconversions (FGI), beginning with either a carboxylic acid or from another carboxylic acid derivative. Since it is possible to make enolates from esters, tertiary amides, and nitriles, these compounds can also be prepared by alkylation at the alpha carbon.

3.11.1 RELATIVE REACTIVITIES OF CARBOXYLIC ACID DERIVATIVES (RCOLG)

The interconversion of carboxylic acid derivatives relies on the ability to replace one leaving group with another (addition–elimination mechanism involving collapse of a charged tetrahedral intermediate, CTI).

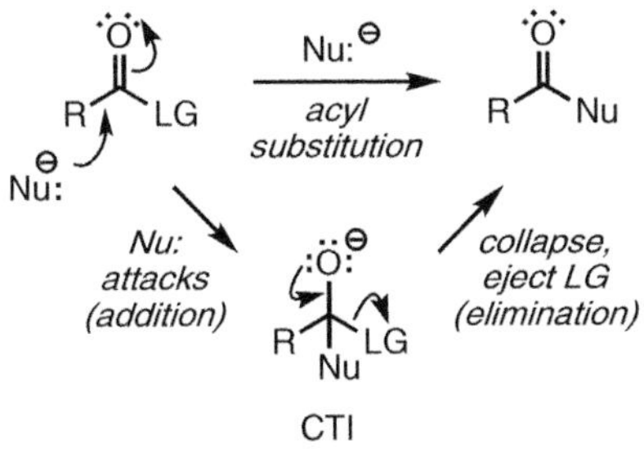

Synthesis of Carboxylic Acid Derivatives via Acyl Substitution

Introduction to Strategies for Organic Synthesis, Second Edition. Laurie S. Starkey.
© 2018 John Wiley & Sons, Inc. Published 2018 by John Wiley & Sons, Inc.

As generally observed with all substitution reactions, collapse of a CTI proceeds by ejecting the better leaving group. The best leaving group in the case of carboxylic acid derivatives is Cl^-, making the acid chloride the most reactive of the carboxylic acid derivatives and the ideal starting material to prepare all other derivatives. The next best leaving group is the resonance-stabilized carboxylate (RCO_2^-), so anhydrides are also very useful reagents (acid chlorides and anhydrides are often used interchangeably). When comparing the leaving groups on esters and amides, one must consider the difference in electronegativities of oxygen and nitrogen. Since oxygen is more electronegative, it better handles a negative charge and makes a better leaving group (RO^- and HO^- are both better leaving groups than the unstable and strongly basic NH_2^-). Amides are very unreactive at the carbonyl, both because the nitrogen group is such a poor leaving group (NH_2^-) and because it is so good at donating its lone pair of electrons to the carbonyl by resonance (making the carbonyl more electron-rich and, therefore, a poor electrophile). As summarized below, the overall order of leaving group ability is parallel to the reactivity (electrophilicity) of the various carboxylic acid derivatives.

Relative Reactivities (as Electrophiles) of Carboxylic Acid Derivatives

One must keep these relative reactivities in mind when planning a synthesis, since the better leaving group will always be displaced in an acyl substitution reaction.

Favorable Acyl Substitutions Involve Displacement of Good LG

When attempting a substitution in which the two groups have similar leaving group abilities (e.g., a transesterification reaction that replaces one —OR group with a different —OR′ group), Le Châtelier's principle can be employed to push the equilibrium in the desired direction, by the addition of an excess of one starting material and/or the removal of a product. In the case of hydrolysis, base-promoted reactions are preferred since deprotonation of the carboxylic acid product drives the reaction forward, even to favor the loss of a poor leaving group such as in the hydrolysis of an amide.

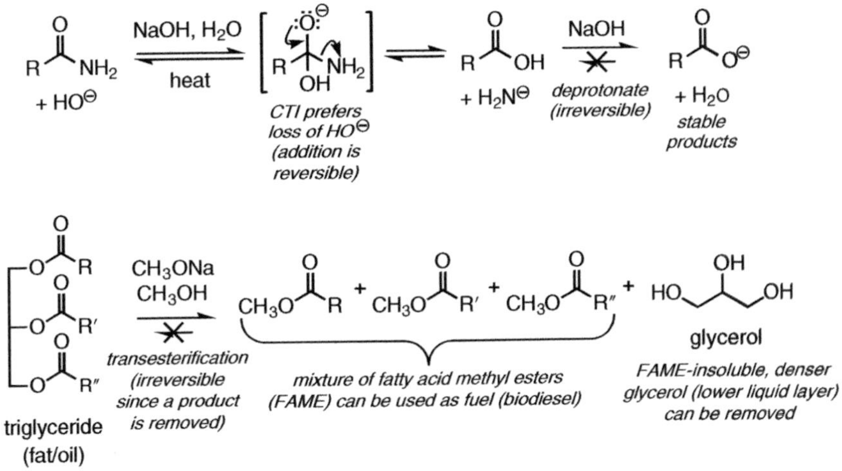

Driving an Equilibrium in One Direction

3.11.2 SYNTHESIS OF ACID CHLORIDES (RCOCl)

Formation of an acid chloride is a challenge since it involves a substitution reaction that introduces an excellent leaving group. Reaction of a carboxylic acid with a reagent such as thionyl chloride ($SOCl_2$) overcomes this problem by first converting the OH group into an excellent leaving group that can then be replaced with Cl.

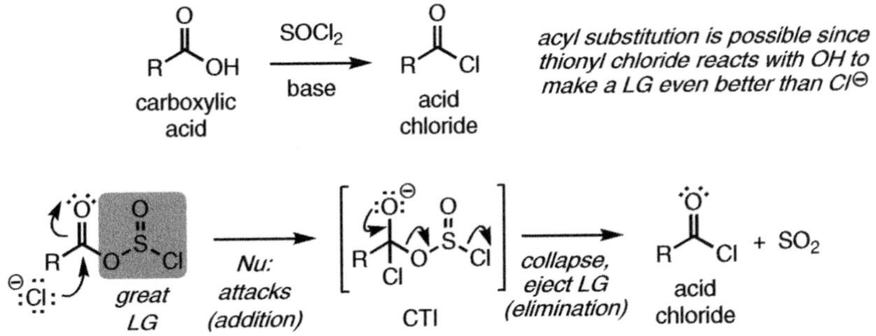

Synthesis of Acid Chlorides

3.11.3 SYNTHESIS OF ANHYDRIDES (RCO₂COR)

Anhydrides contain a highly electrophilic center bearing a good leaving group, so they are reactive compounds that can serve as useful intermediates in organic synthesis. As the name implies, anhydrides can be formed by dehydration of a carboxylic acid. The OH oxygen of one carboxylic acid molecule acts as a nucleophile, attacking the carbonyl carbon of another molecule. The acyl substitution (addition–elimination mechanism) results in a loss of water. Treatment of a carboxylic acid with heat, typically along with a drying agent (something that reacts with the water being formed, such as acetic anhydride, Ac₂O, or phosphorus pentoxide, P₂O₅) results in a symmetrical anhydride.

Synthesis of Symmetrical Anhydrides via Dehydration

Mixed anhydrides can be formed by reaction of the acid chloride of one carboxylic acid with a different carboxylic acid in the presence of base; the mechanism involves the addition of the carboxylate nucleophile and elimination of the chloride leaving group from the resulting charged tetrahedral intermediate (CTI).

Synthesis of Mixed Anhydrides

3.11.4 SYNTHESIS OF ESTERS (RCO₂R)

Esters can be created via functional group interconversions (FGI), including oxidation and substitutions (acyl or S$_N$2). The acidity of the alpha protons can also be exploited to functionalize esters at the alpha carbon.

Synthesis of Esters via FGI

The most common way to synthesize an ester is by a substitution reaction. Synthesis of an ester (RCO$_2$R′) from a carboxylic acid (RCO$_2$H) involves an acyl substitution in which the −OH leaving group is replaced with an −OR′ group. This substitution can be initiated by first forming the acid chloride, followed by treatment with the desired alcohol nucleophile in the presence of base. Alternatively, the Fischer esterification involves the acid-catalyzed reaction of a carboxylic acid with an alcohol. Since both −OH and −OR′ have similar leaving group abilities, the equilibrium must be driven forward (by having an excess of alcohol, by removal of water, or by removal of the ester product as it is formed, such as by distillation).

Synthesis of Esters via Acyl Substitution

   Reaction of a carboxylate nucleophile (RCO$_2^-$) with an alkyl halide (R′X) would also create an ester functional group (RCO$_2$R′). Rather than having a substitution at the acyl carbon, this reaction involves a backside attack of the oxygen nucleophile on an alkyl halide, analogous to the Williamson ether synthesis. As usual, this S$_N$2 mechanism requires a good leaving group (halide or tosylate) on an unhindered carbon. This reaction can be especially slow if there is any steric hindrance since the resonance-stabilized carboxylate is not strongly nucleophilic, but since the carboxylate is also not strongly basic, S$_N$2 reactions on 2° alkyl halides are possible with minimal E2 competition.

Synthesis of Esters via S$_N$2

Methyl esters can be prepared with a reagent called diazomethane (CH_2N_2). When reacted with a carboxylic acid, diazomethane becomes protonated and serves as an extremely reactive electrophile since it contains an extremely good leaving group (N_2). A rapid and irreversible S_N2 displacement by the carboxylate nucleophile gives the methyl ester product.

Synthesis of Methyl Esters Using Diazomethane

Esters can also be formed upon oxidation of ketones with a peroxyacid oxidizing agent (such as mCPBA or CF_3CO_3H). This Baeyer–Villiger oxidation is especially useful in the preparation of cyclic esters (called lactones) from cyclic ketones.

Synthesis of Esters via Baeyer–Villiger Oxidation of Ketones

Retrosynthesis of Esters (FGI)

The retrosynthesis of an ester target molecule typically begins with a discon-nection at either C—O bond. Formation of this bond always involves a nucleophilic oxygen reacting with an electrophilic carbon that is either acyl (a carboxylic acid derivative such as acid chloride) or alkyl (RX), causing a substitution to take place. If the alkoxy group is highly substituted (or is a phenoxy derivative), the S_N2 approach is unsuitable so the target molecule must be prepared by acyl substitution or by peroxyacid oxidation of the corresponding ketone.

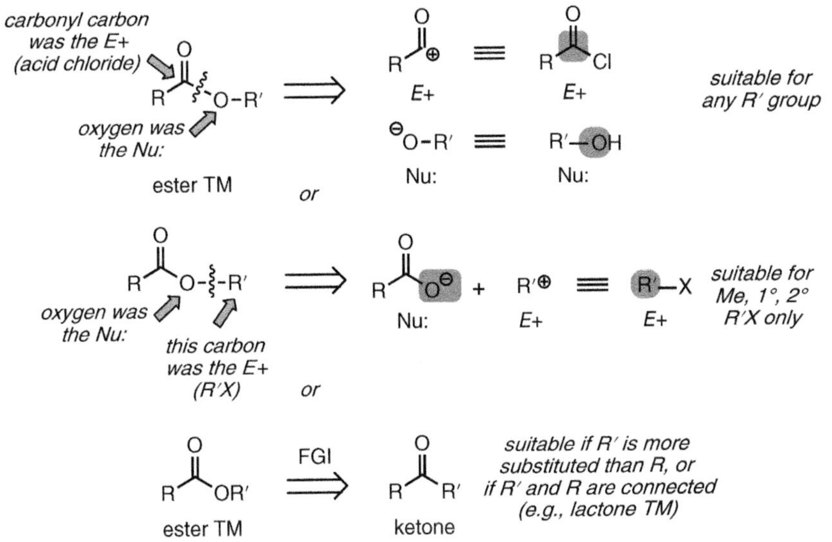

carbonyl carbon
was the E+
(acid chloride)

ester TM

or

oxygen was
the Nu:

$E+$ ≡ $E+$

$^{\ominus}$O-R' ≡ R'—OH
Nu: Nu:

*suitable for
any R' group*

oxygen was
the Nu:

this carbon
was the E+
(R'X)

or

R $O^{\ominus}$ + $R'^{\oplus}$ ≡ R'—X
Nu: E+ E+

*suitable for
Me, 1°, 2°
R'X only*

ester TM

FGI

ketone

*suitable if R' is more
substituted than R, or
if R' and R are connected
(e.g., lactone TM)*

Retrosynthesis of Esters via FGI

Retrosynthesis of Lactones

Lactones can be prepared by oxidation or possibly by a cyclization reaction, such as an intramolecular Fischer esterification, if the lactone is a five- or six-membered ring (called γ-"gamma" and δ- "delta" lactones, respectively).

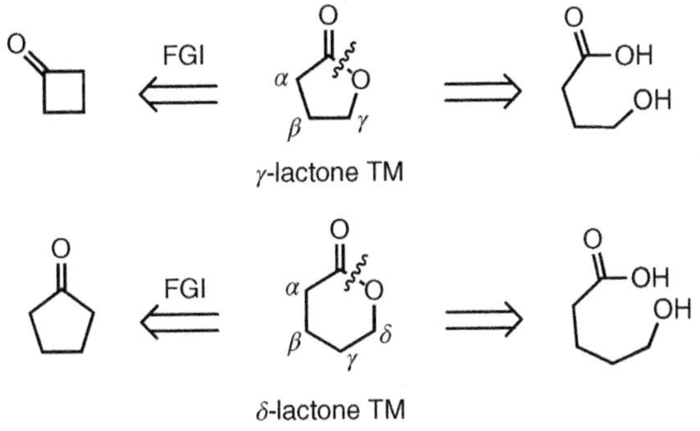

FGI   α β γ   ⟹   OH OH

γ-lactone TM

FGI   α β γ δ   ⟹   OH OH

δ-lactone TM

Retrosynthesis of Lactones

Esters via α-Alkylation

As with an aldehyde or ketone, deprotonation of an ester's alpha carbon with a strong base (such as LDA) gives a nucleophilic enolate that can be alkylated with an unhindered alkyl halide.

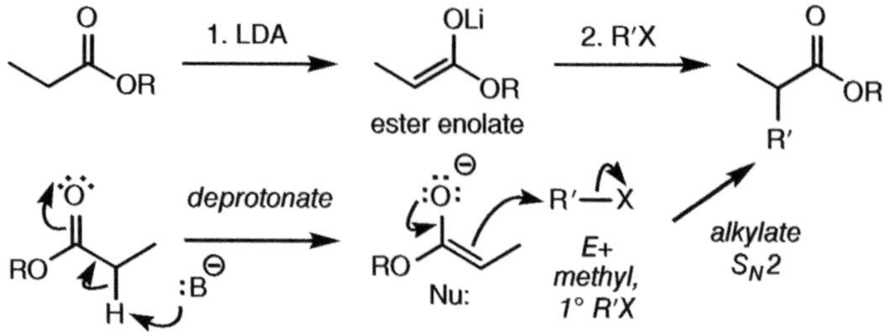

α-Alkylation of Esters

Retrosynthesis of Esters (α-Alkylation)

As seen with other target molecules containing a carbonyl, a possible retrosynthesis of an ester can begin with the disconnection of an alkyl group from the alpha carbon. As usual, the alpha carbon was the nucleophile (enolate) and the other carbon was the electrophile (R'X). Since the reaction between these two starting materials is an S_N2 backside attack, the appropriate disconnection is the one that results in an unhindered alkyl halide starting material to avoid competition from the E2 elimination. An alternative choice uses the malonic ester synthesis (i.e., deprotonation, alkylation, hydrolysis, decarboxylation), followed by esterification of the resulting carboxylic acid product. Although this would clearly be a longer synthesis, an S_N2 with a *stabilized* enolate (such as the one derived from diethyl malonate) is often higher yielding, so this may be the preferred route in practice.

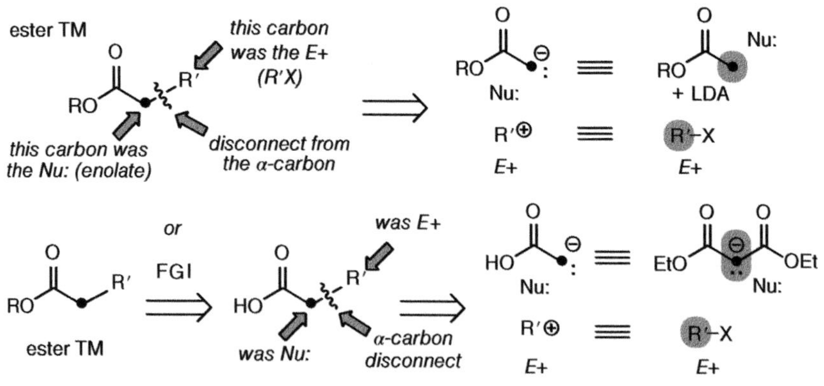

Retrosynthesis of an Ester via α-Alkylation

PRACTICE PROBLEM 3.11A: ESTER SYNTHESIS

3.11A Provide four possible syntheses of the given TM, using the disconnections shown.

(two ways)

3.11.5 SYNTHESIS OF AMIDES (RCONH₂)

While tertiary (*N,N*-disubstituted) amides can also undergo α-alkylation
using LDA, the most common way to synthesize an amide is by an acyl
substitution reaction on a carboxylic acid derivative (R′COLG) in which
the leaving group is replaced with a nucleophilic amino group. The nitrogen
nucleophile can be ammonia (NH_3), a primary amine (RNH_2), or a secondary
amine (R_2NH) to give a primary amide ($R'CONH_2$), an *N*-substituted
secondary amide ($R'CONHR$), or a tertiary amide ($R'CONR_2$), respectively. Since all other acyl leaving groups are better than the nitrogen group,
amides can be formed from any other carboxylic acid derivative including
acid chlorides, anhydrides, and esters. However, amides cannot be easily
prepared directly from a carboxylic acid, since an acid–base reaction with
the basic amine occurs first. Addition of extreme heat will enable the
substitution reaction to ultimately occur, but this approach precludes other
sensitive functionality in the molecule. An easier route is to first convert the
carboxylic acid to its acid chloride before reacting it with the amine.

Synthesis of Amides via Acyl Substitution

Amides without any *N*-alkyl groups ($RCONH_2$) can also be prepared
by partial hydration of a nitrile ($RC{\equiv}N$). Addition of one equivalent of
water across the nitrile triple bond results in an enol-like intermediate

that tautomerizes to give a carbonyl. This mechanism is analogous to the hydration of an alkyne to give a ketone product. Since amides can also be hydrolyzed to give carboxylic acids, the nitrile partial hydrolysis reaction conditions are kept relatively mild when the amide product is desired.

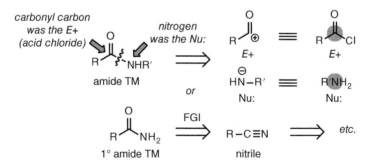

Synthesis of Amides via Hydrolysis of Nitriles

Retrosynthesis of Amides

The retrosynthesis of an amide target molecule typically begins with a disconnection at the bond between the carbonyl and the nitrogen. Formation of this bond involves a nucleophilic nitrogen reacting with an electrophilic acyl carbon bearing a leaving group (can be any carbonyl-containing carboxylic acid derivative, but it is typically an acid chloride or anhydride). If the target molecule is a primary amide (i.e., it does not have any N-alkyl groups, RCONH$_2$), then another retrosynthesis option is to do a functional group interconversion (FGI) to a nitrile.

Retrosynthesis of Amides

Retrosynthesis of Lactams

Five-membered "gamma" γ-lactams and six-membered "delta" δ-lactams can be prepared by heating appropriately substituted amino carboxylic acids to cause cyclization.

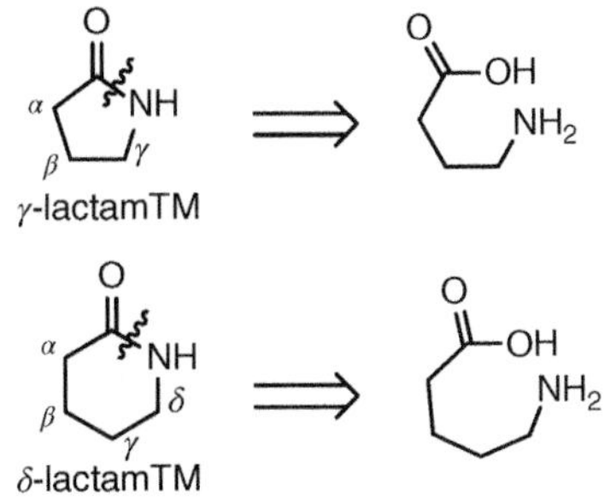

γ-lactamTM

δ-lactamTM

Retrosynthesis of Lactams

3.11.6 SYNTHESIS OF NITRILES (RC≡N)

Just as an amide can be formed by partial hydrolysis (hydration) of a nitrile, a nitrile can be formed by the reverse reaction: the dehydration of an amide. Heating an amide in the presence of a dehydrating agent such as phosphorus pentoxide ("P_2O_5" is actually the reagent P_4O_{10}) produces a nitrile. The synthesis of a nitrile can also be achieved by a carbon–carbon bond-forming reaction between a cyanide nucleophile ($^-C\equiv N$) and an alkyl halide electrophile (RX). As usual, this S_N2 substitution requires a good leaving group on an unhindered 1° or 2° carbon.

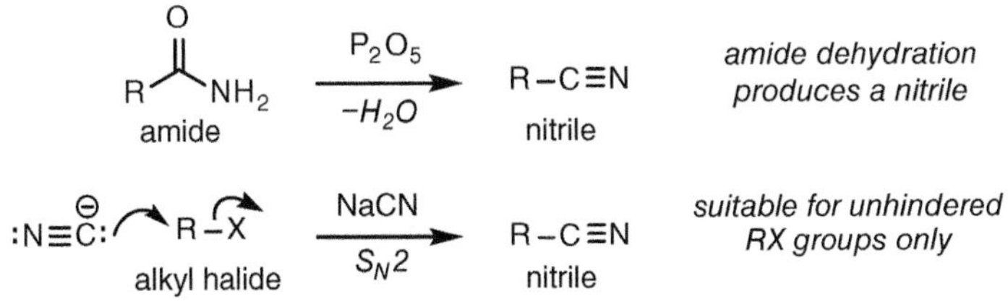

Synthesis of Nitriles

The retrosynthesis of a nitrile involves either a functional group interconversion to an amide or disconnection at the cyano carbon. The cyano carbon was the nucleophile (as the commercially available sodium cyanide, NaCN) and the other carbon was the electrophile (RX).

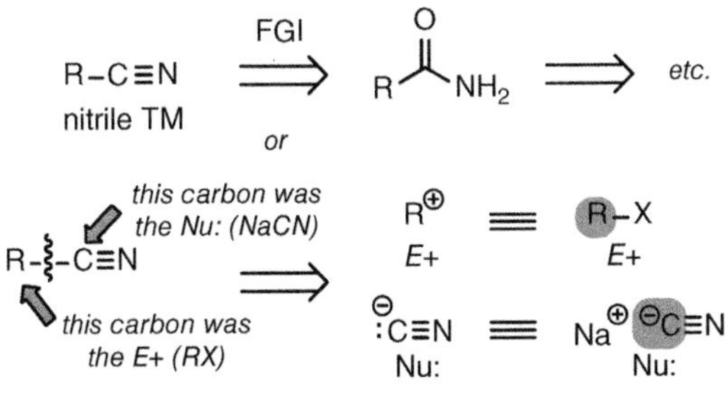

Retrosynthesis of Nitriles

Synthesis of Aromatic Nitriles (ArC≡N)

While a cyano group cannot be introduced to a benzene ring by an S_N2 substitution, it can be added by treatment of a diazonium salt (ArN_2^+) with copper(I) cyanide (CuCN). This substitution, along with the similar CuBr and CuCl reactions, is known as the Sandmeyer reaction. The retrosynthesis of an aromatic nitrile involves a disconnection at the cyano carbon and a functional group interconversion that replaces the cyano with a nitro group, since the nitro group can be converted to the diazonium salt (the synthesis of aromatic TMs are discussed further in Chapter 5).

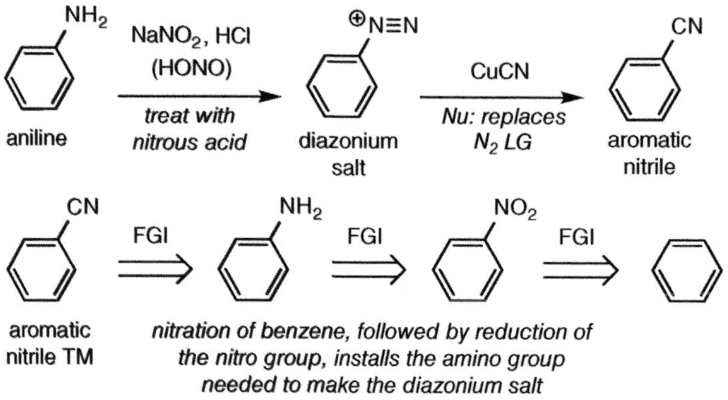

Aromatic Nitrile Synthesis and Retrosynthesis

PRACTICE PROBLEM 3.11B: CARBOXYLIC ACID DERIVATIVE SYNTHESIS

3.11B   Provide the reagents needed for each of the following transformations (more than one step may be required).

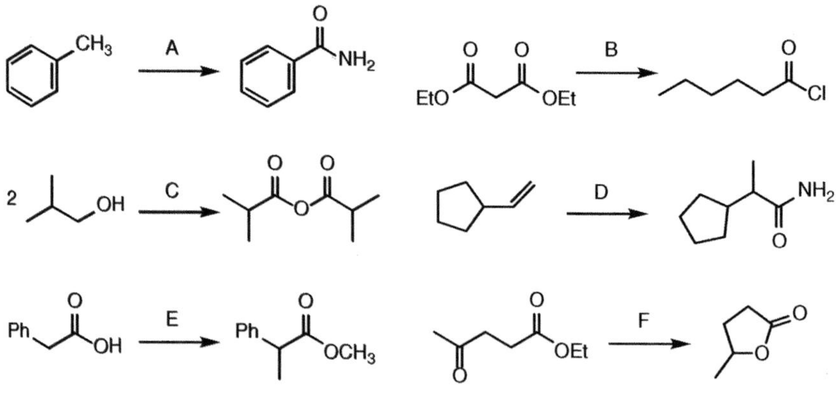

EXAMPLE: CARBOXYLIC ACID DERIVATIVE

Provide the reagents necessary to transform the given starting material into the desired product. More than one step may be required.

In this transform problem, the product is an ester with two additional carbons on the carbon chain. Recognition of the ester as a carboxylic acid derivative leads to a disconnection at the C–O bond to give a carboxylic acid and the commercially available allyl bromide. The continued retrosynthesis of this carboxylic acid is dictated by the given starting material. The new C–C bond is identified and disconnected. One of the carbons involved is an alpha carbon, so that was the required nucleophile (the enolate of acetic acid synthon is derived from diethyl malonate). The benzylic carbon must have been the electrophile; this can be achieved as an alkyl halide (benzyl bromide). An FGI of the alkyl halide concludes the retrosynthesis by working back to the given starting material (toluene).

Retrosynthesis of Ester TM

The required leaving group can be installed by free-radical halogenation of toluene to give benzyl bromide. Treatment with the anion of diethyl malonate installs the two new carbons that are required, and hydrolysis followed by decarboxylation removes the extra ester group to give a carboxylic acid product. Deprotonation, followed by S_N2 displacement of the leaving group on allyl bromide, gives the desired target molecule.

Synthesis of Ester TM

1-FG TMs

3-1. Provide the corresponding synthetic equivalent for each of the following synthons. In other words, what starting material would have the desired reactivity?

A $\overset{\ominus}{\diagup\diagdown\diagup}$ ≡

B (acetyl anion, C(=O)CH₃ with $\ominus$) ≡

C (carboxylate, C(=O)OH with $\ominus$) ≡

D $\overset{\oplus}{\diagup\diagdown\diagup}$ ≡

E (OH-bearing carbon cation) ≡

F (OH, tertiary carbon cation) ≡

G (acyl cation, C=O with $\oplus$) ≡

H (carboxylic acid acyl cation, C(=O)OH with $\oplus$) ≡

I $\overset{\ominus}{}NH_2$ ≡

3-2. Propose a possible disconnection/retrosynthesis for each of the following target molecules. There may be more than one reasonable disconnection.

A R—CH(OH)—CH(R')— (with OH and R')

B R—C(=O)—OR'

C Ph—C(=O)—CH₂—R

D $\diagup\diagdown$NH₂

E R—≡—R'

F R—CH₂—C(=O)—OH

G R—O—R'

H R'—CH=C(R)(R) (with R, R, R')

I R—CH₂—CH₂—R'

J R—CH(Br)—CH₃

K R—CH₂—CH₂—CN

L R—S—R'

3-3. Provide the reagents necessary to transform the given starting material into the desired product. If more than one step is required, show the structure of each intermediate product. Consider both regiochemistry and stereochemistry when planning the synthesis; it may help to first do a retrosynthesis of the product.

Introduction to Strategies for Organic Synthesis, Second Edition. Laurie S. Starkey.
© 2018 John Wiley & Sons, Inc. Published 2018 by John Wiley & Sons, Inc.

A 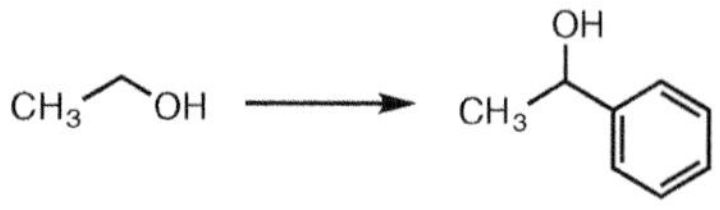

B 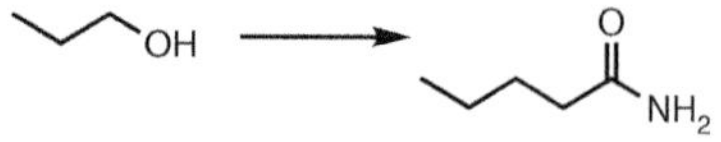

C 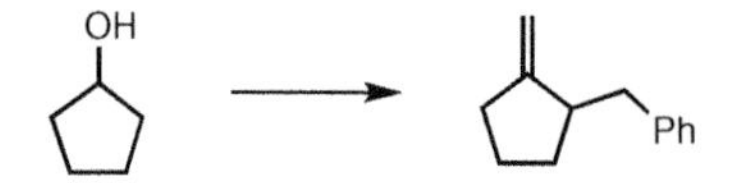

D 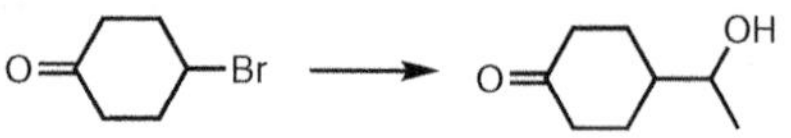

E 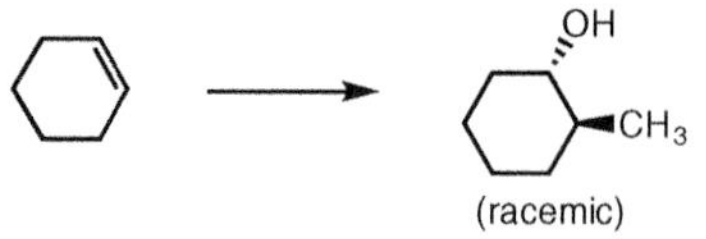

F 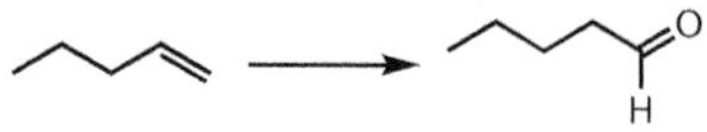

G 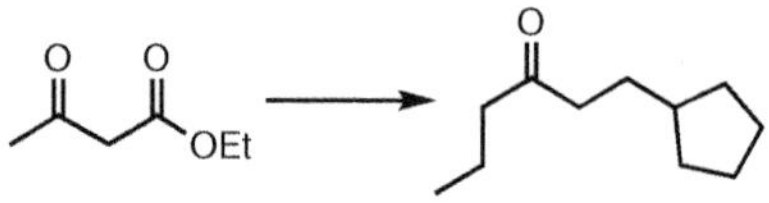

H 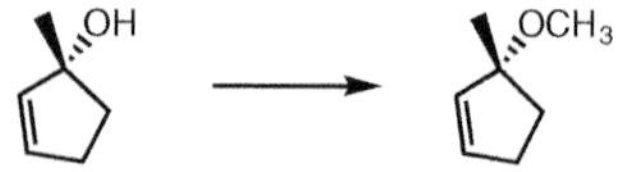

I 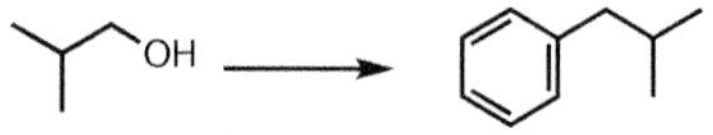

J 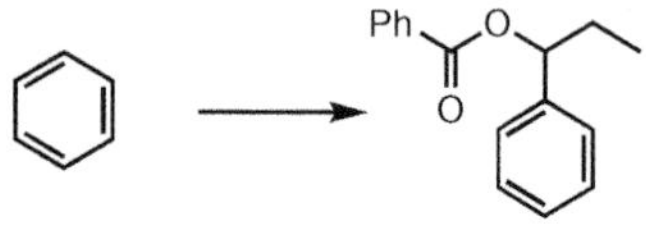

K 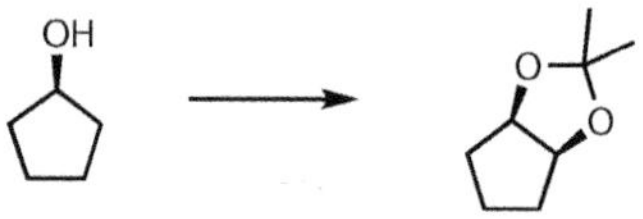

L 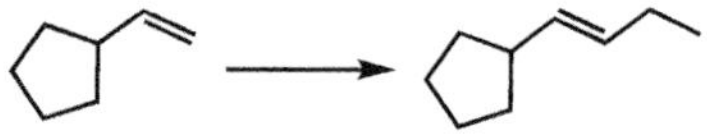

M 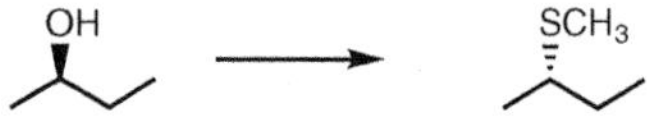

N 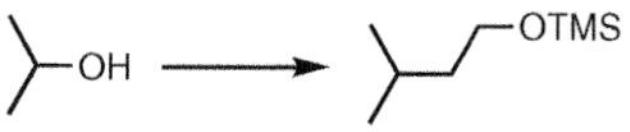

O 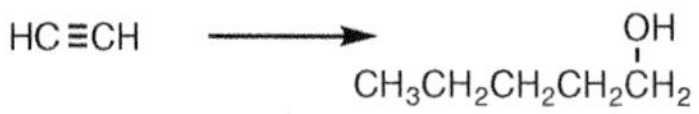

P 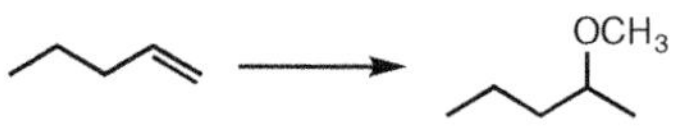

Q 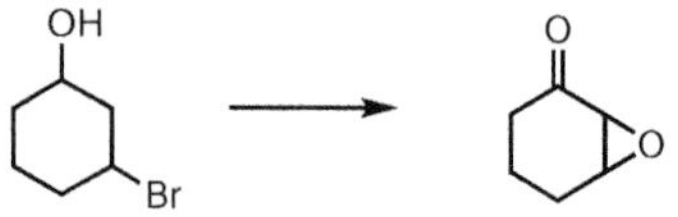

R 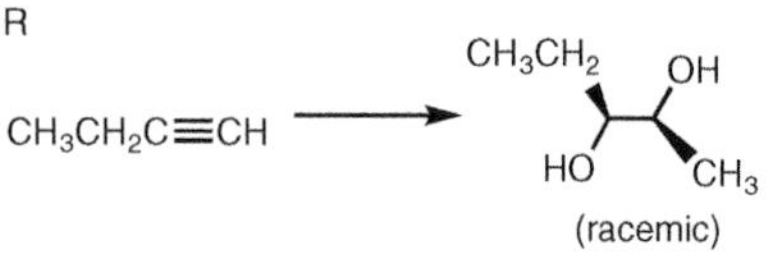

S 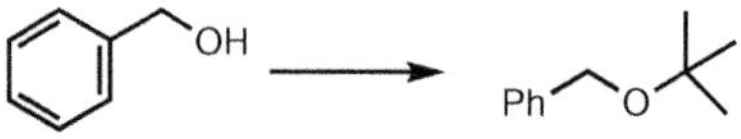

T 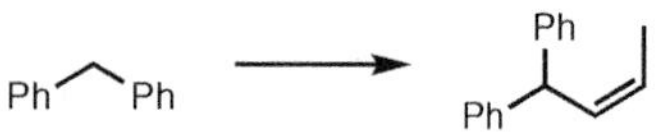

U 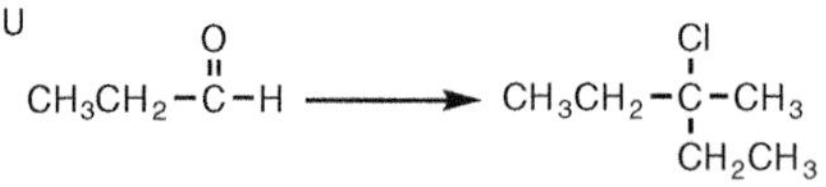

V 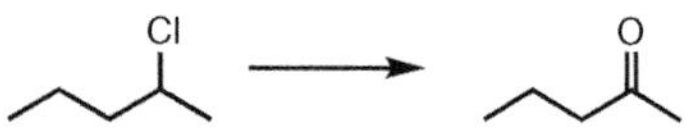

W 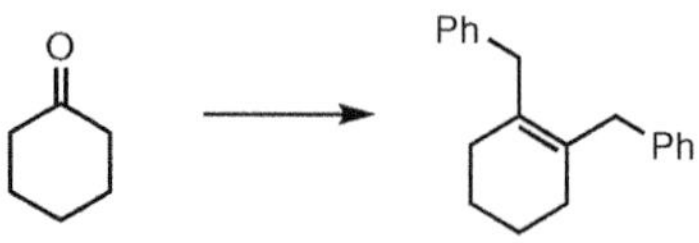

X 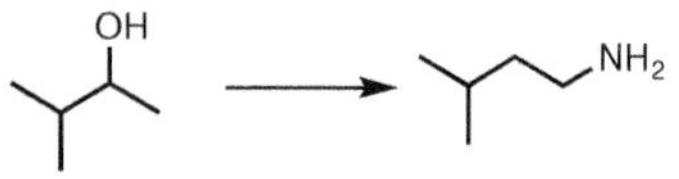

Y 

Z 

3-4. Provide a synthesis for each of the following target molecules, using NaCN and any alkyl halide(s).

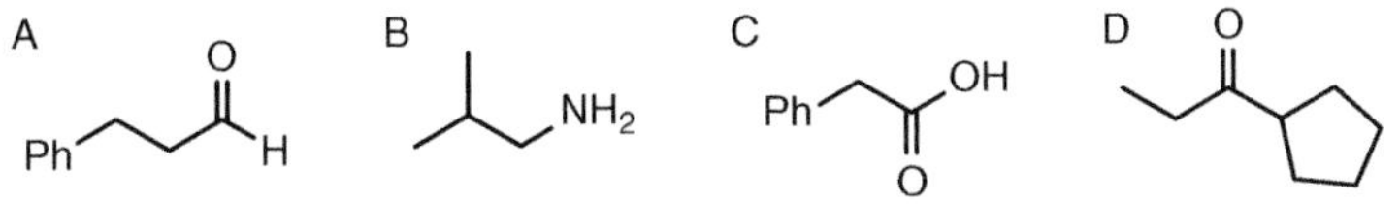

A   B   C   D

3-5. C-14 Synthesis Game.[†] Provide a synthesis for each of the following target molecules. Each synthesis must correctly incorporate the ^{14}C-labeled (*) carbon atom as shown, using the given starting materials as the only sources of carbon. Any commercially available reagents and protective groups may be used, and any previously synthesized molecule can be used as a starting point for another target molecule.

available starting materials: HC≡CH   *CH$_3$Br

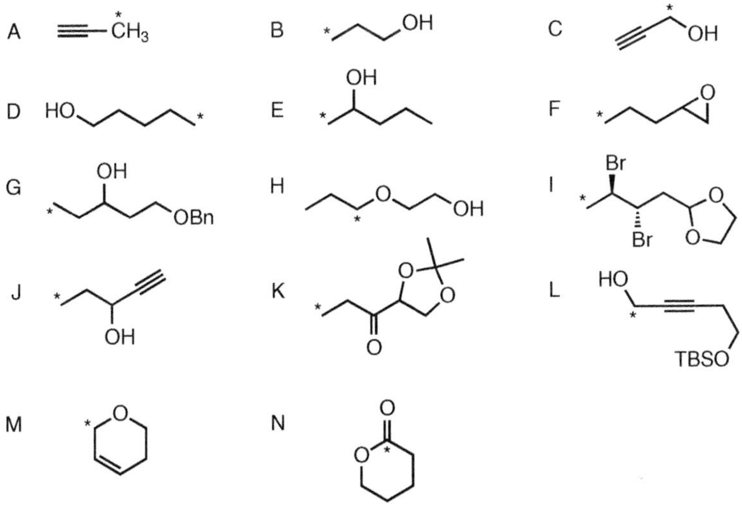

A   B   C

D   E   F

G   H   I

J   K   L

M   N

3-6. Provide the missing reagents. If more than one step is required, show the structure of each intermediate product. (*J. Org. Chem.*, **2002**, 67, 1109.)

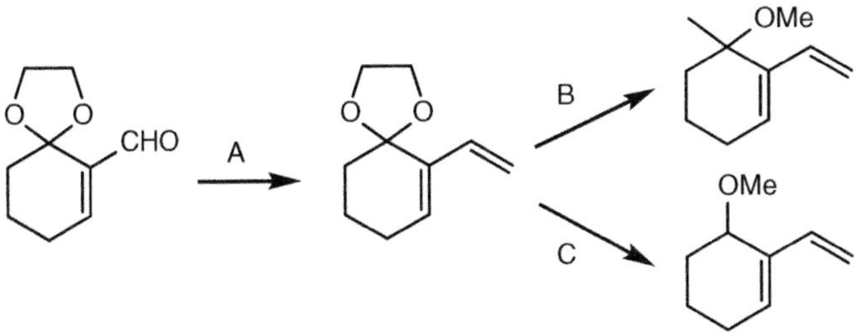

†Phil Beauchamp, an Organic Chemistry Professor at Cal Poly Pomona, developed this unique teaching and learning strategy.

3-7. The following scheme was taken from a journal article describing the synthesis of the biologically active compound kopeolin (*J. Org. Chem.*, **2014**, 79, 2268). Provide the missing reagents or products.

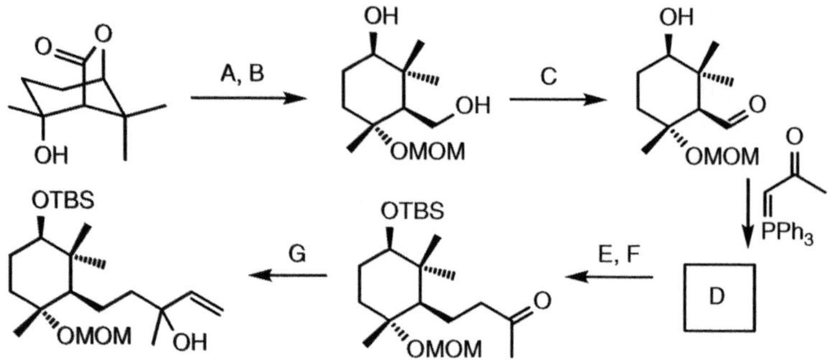

CHAPTER 4

SYNTHESIS OF TARGET MOLECULES WITH TWO FUNCTIONAL GROUPS (2-FG TMs)

When a target molecule contains two functional groups, an ideal disconnection is one that incorporates both. Certain reactions give products containing characteristic *patterns* of functional groups. These patterns and the corresponding reactions will be the focus of this chapter.

SYNTHESIS OF β-HYDROXY CARBONYLS AND α,β-UNSATURATED CARBONYLS

4.1.1 THE ALDOL REACTION

The aldol reaction of aldehydes and ketones involves the attack of an enolate (or enol) nucleophile on a carbonyl electrophile. When the alpha carbon of one compound bonds with the carbonyl carbon of another compound, a β-hydroxy carbonyl product is initially formed; this product can undergo a further reaction to give an α,β-unsaturated carbonyl product. When either of these patterns is present in a target molecule, it is an indication that the target molecule might be the product of an aldol reaction, and an aldol disconnection will be one option for retrosynthesis.

Synthesis of β-Hydroxy Carbonyls (Aldol)

The aldol reaction is a reaction between two carbonyl-containing compounds. In a "self-condensation" aldol, the same carbonyl starting material is used as both the nucleophile and the electrophile. In other cases, it is possible to have a "mixed aldol," which reacts one ketone/aldehyde with a different ketone/aldehyde. Such reactions would require some means of controlling the regiochemistry of the reaction.

Introduction to Strategies for Organic Synthesis, Second Edition. Laurie S. Starkey.
© 2018 John Wiley & Sons, Inc. Published 2018 by John Wiley & Sons, Inc.

Aldol Reaction (Ketone Self-Condensation Shown)

Mechanism of the Aldol Reaction

A base-catalyzed aldol reaction begins with deprotonation of the weakly acidic proton next to the carbonyl, on the alpha carbon. The reversible reaction with a mild base such as hydroxide ensures that only a small amount of the ketone will be deprotonated to give a small amount of an enolate intermediate.

Relatively Weakly Basic Conditions Provide Both Ketone and Enolate

The enolate is nucleophilic at the alpha carbon, and it will attack the only electrophile present: the carbonyl carbon of the remaining ketone starting material. As usual, attack on a carbonyl results in an alkoxide intermediate that can be protonated upon workup, or *in situ* if a protic solvent is present.

Base-Catalyzed Aldol Mechanism

The product of this reaction is a β-hydroxy carbonyl that contains a newly formed carbon–carbon bond between the alpha and beta carbons. This is the key bond to be identified in a β-hydroxy carbonyl target molecule; a disconnection at this bond will lead to an aldol retrosynthesis.

Retrosynthesis of β-Hydroxy Ketones/Aldehydes (Aldol)

A typical retrosynthesis of a β-hydroxy ketone or β-hydroxy aldehyde involves making a disconnection between the two functional groups, more specifically, between the alpha carbon and the carbon bearing the OH group (the beta carbon). The carbon that now has an OH on it used to be a carbonyl carbon (C=O), which was an electrophilic carbon. Therefore, the other carbon group must have been introduced as a nucleophile; this is a logical disconnection since a carbon alpha to a carbonyl can be a nucleophile as an enolate.

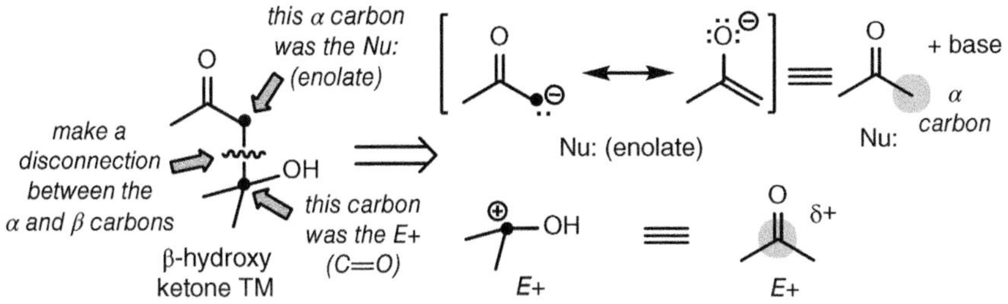

Retrosynthesis of β-Hydroxy Ketones

Synthesis of α,β-Unsaturated Ketones/Aldehydes (Aldol)

The β-hydroxy ketone or β-hydroxy aldehyde product of an aldol reaction can undergo dehydration to give an α,β-unsaturated product.

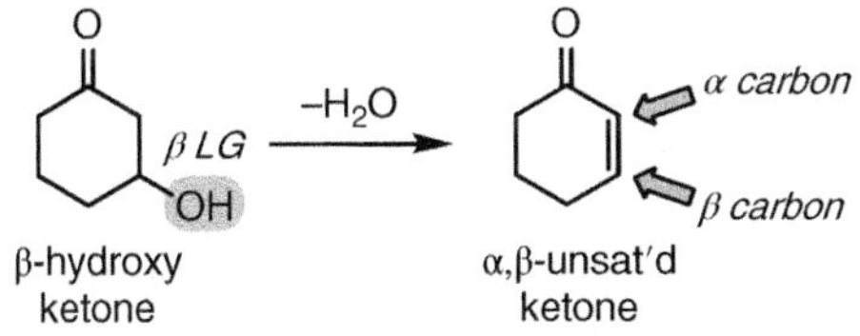

β-Hydroxy Ketones and α,β-Unsaturated Ketones are Related Compounds

The location of the OH leaving group in a position that is beta to a carbonyl, along with the formation of a stable, conjugated pi bond, makes this reaction much easier than a dehydration of an ordinary alcohol. The dehydration of an aldol product requires much milder reaction conditions than a typical alcohol and can even occur spontaneously at room temperature. For example, in the reaction of benzaldehyde with acetone, the double aldol product shown is the only one isolated; the presence of the two benzene rings makes the newly formed pi bonds highly conjugated and very stable.

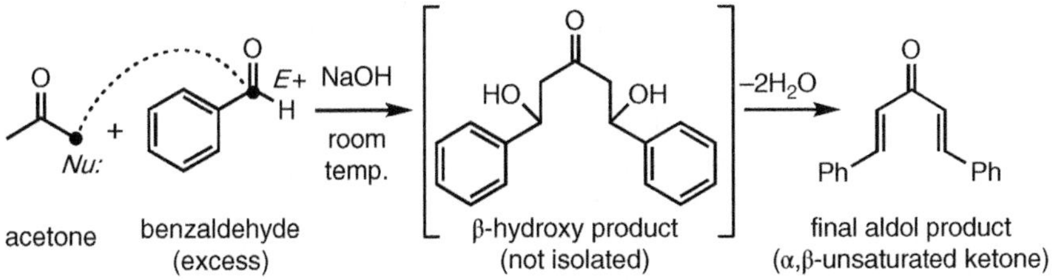

Aldol Reactions Sometimes Give Only α,β-Unsaturated Ketones/Aldehydes

Aldol reactions typically give the most stable product, since the reversible reaction undergoes equilibration and is controlled by thermodynamics. In the intramolecular aldol reaction shown below, products **A** and **B** are not favored products. Since **A** is a more reactive aldehyde and **B** is a less stable seven-membered ring, both of these products are more likely to do the reverse reaction, the retro-aldol. As this equilibrium continues, the most stable product will become the major one. In intermolecular aldol reactions, the more stable alkene product is expected.

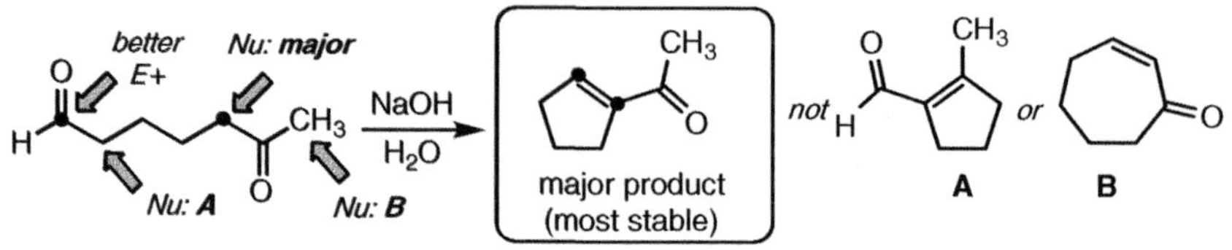

Intramolecular Aldol Reactions Give Most Stable Cyclic Product

Mechanism of the Dehydration of an Aldol Product

The two-step, base-catalyzed dehydration mechanism, called β-elimination or E1cb (stepwise elimination involving a conjugate base), begins with deprotonation of the alpha carbon, to form an enolate. This enolate intermediate then ejects the β-leaving group (hydroxide), in a mechanism similar to the collapse of a CTI. Although hydroxide is a poor leaving group in an ordinary elimination mechanism (such as E1 or E2), it is an acceptable leaving group in the β-elimination mechanism since formation of the stable carbonyl group is a driving force.

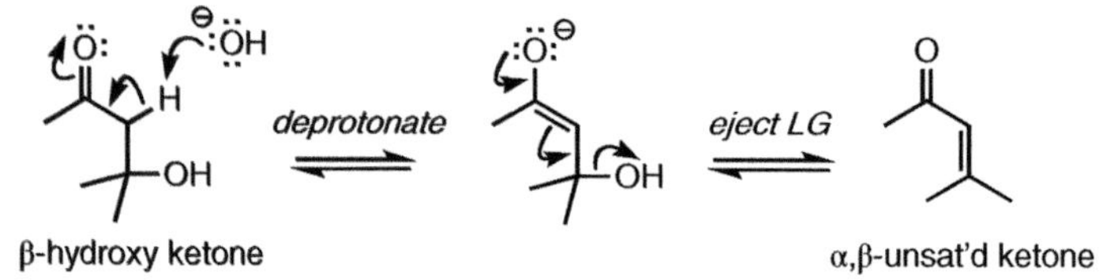

β-hydroxy ketone   α,β-unsat'd ketone

Mechanism for Dehydration of an Aldol Product

Retrosynthesis of α,β-Unsaturated Ketones/Aldehydes (Aldol)

Dehydration of an aldol product is no more than a functional group inter-conversion, so the retrosynthesis of a α,β-unsaturated ketone or aldehyde involves making the same disconnection as used for a β-hydroxy ketone or aldehyde: between the alpha and beta carbons. It may help to start the retrosynthesis by adding water back in (undoing the dehydration step) to give the more recognizable β-hydroxy carbonyl compound before making the disconnection.

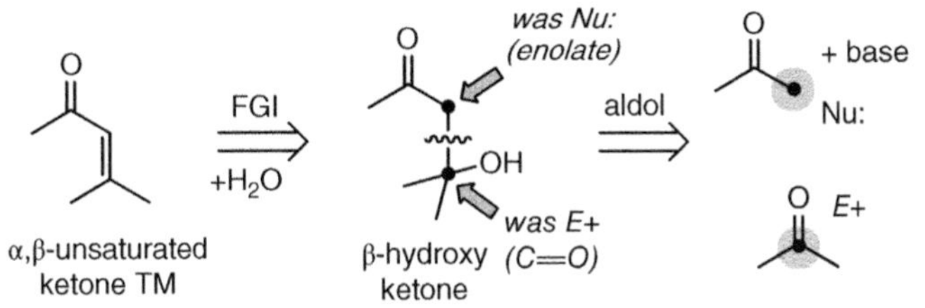

Retrosynthesis of α,β-Unsaturated Ketones

PRACTICE PROBLEM 4.1A: ALDOL REACTION

4.1A For each of the following aldol reactions, provide the missing starting material(s) or products.

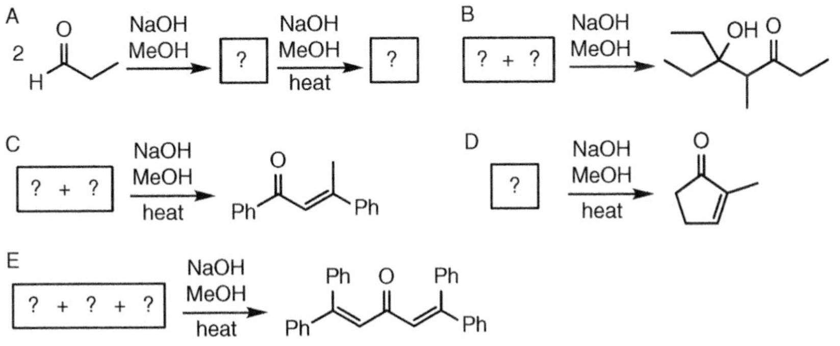

4.1.2 MIXED ALDOL AND MANNICH REACTIONS

The Mixed Aldol Reaction and Regiocontrol Involving Enolates

The aldol examples shown above are described as "self" condensations because they involve the reaction of a single carbonyl compound acting as both the nucleophile and the electrophile. The reaction of two different carbonyls presents a problem of *chemoselectivity*: which carbonyl will act as the nucleophile, and which will act as the electrophile? Mixing two ketones in the presence of a mild base, for example, could give a mixture of four aldol products: two cross-condensations and two self-condensations. Since acetone and 3-pentanone have similar reactivities, we would expect all possible combinations to occur.

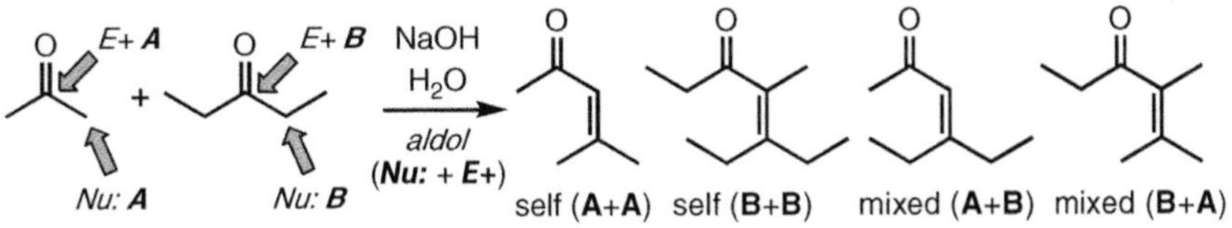

Mixed Aldol Becomes "Messy" Aldol if Carbonyls Have Similar Reactivity

To exert control over a mixed aldol, a stepwise approach is required. A strong base such as LDA (−78°C) can be used to irreversibly form an enolate (Nu:), to which an electrophile can be added. Since a single nucleophile is reacting with a single electrophile, a single aldol product can be expected.

Stepwise Approach Controls a Mixed Aldol

Complimentary methods are available if the "thermodynamic" enolate (the more stable enolate) is required, such as forming the silyl enol ether. The silyl enol ether is an enol/enolate equivalent that will react with carbonyl electrophiles in the presence of a Lewis acid catalyst, such as $TiCl_4$.

Choice of Base/Reaction Conditions Determines Regiochemistry

Certain mixed aldols are expected to give a single product, without requiring special measures to control regiochemistry. If the two carbonyl compounds have different reactivities, then it might be possible for the aldol reaction to take place by simply mixing them together in the presence of a mild base. For example, if only one of the carbonyl starting materials can form an enolate (because the other has no alpha protons) then only one major product can be expected.

Mixed Aldols are Reasonable if Starting Material Reactivities Differ

When determining which carbonyl will be attacked by the enolate, one should look for the best electrophile (the one that is most electron-deficient). Determination of the reactivity of a carbonyl compound is based on the electron-donation or electron-withdrawal of the groups attached and, to a lesser extent, the sterics involved. Ketones are poorer electrophiles than aldehydes because alkyl groups are not only electron-donating but also bulkier than hydrogens.

acid chloride   aldehyde   aromatic ketone   ketone   ester   amide

most electron-deficient
best electrophile

most electron-rich
worst electrophile

The Best Electrophiles are the Most Electron-Deficient Carbonyls

Synthesis of β-Dialkylamino Ketones (Mannich Reaction)

In a mechanism analogous to the aldol, the Mannich reaction involves the combination of an enol nucleophile with an iminium ion electrophile to give a β-amino ketone product. The reaction is performed by mixing a ketone with a 2° amine (R_2NH) and an aldehyde (commonly formaldehyde). The iminium ion is formed *in situ* by combining the more electrophilic aldehyde with the nucleophilic amine. As it is generated, the highly electrophilic iminium carbon (e.g., $H_2C=NMe_2^+$) is attacked by the nucleophilic enol of the ketone that is also present in the reaction mixture.

Mannich Reaction

Retrosynthesis of β-Dialkylamino Ketones

Ultimately, a complete retrosynthesis of a β-amino ketone leads to the three components needed in a Mannich reaction: a ketone, an aldehyde, and an amine. The retrosynthesis of a β-amino ketone begins with making a disconnection between the alpha carbon and the carbon bearing the amino group (the beta carbon). This aldol-like disconnection leads to a nucleophilic alpha carbon (enol) and an electrophilic imine carbon (C=N). Further disconnection of the imine affords the two necessary building blocks for its formation: an aldehyde and an amine.

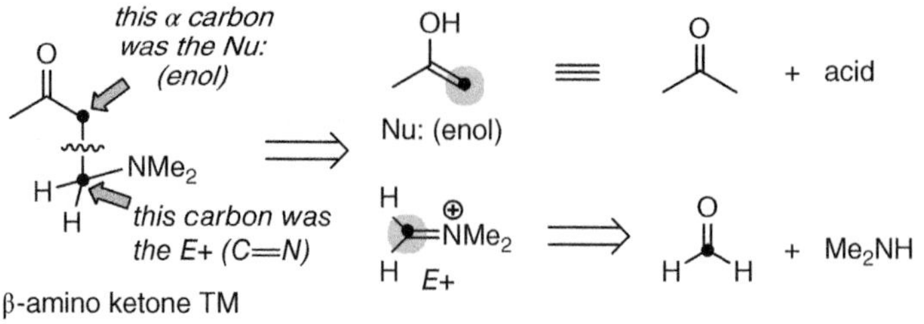

β-Dialkylamino Ketone Retrosynthesis

Synthesis of α,β-Unsaturated Ketones (via Mannich Bases)

The β-dialkylamino ketone product, known as a Mannich base, can undergo an elimination reaction upon heating to afford an α,β-unsaturated ketone. Alternatively, treatment of a β-amino ketone with an excess of methyl iodide gives a quaternary ammonium salt that can be eliminated with base (Hofmann elimination).

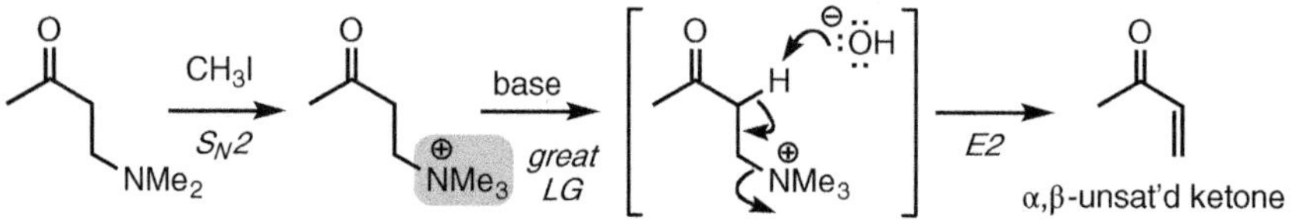

Hofmann Elimination of β-Amino Group Gives an α,β-Unsaturated Ketone

This strategy is especially useful when preparing a target molecule containing a vinyl ketone or an unstable exocyclic methylene group. An ordinary aldol disconnection would lead to a synthesis involving highly reactive formaldehyde, but the Mannich strategy is a higher yielding one. The iminium ion intermediate ($H_2C=NMe_2^+$) can be described as a synthetic equivalent for formaldehyde and is likely to be considered when disconnecting an α,β-unsaturated ketone containing a methylene unit ($=CH_2$).

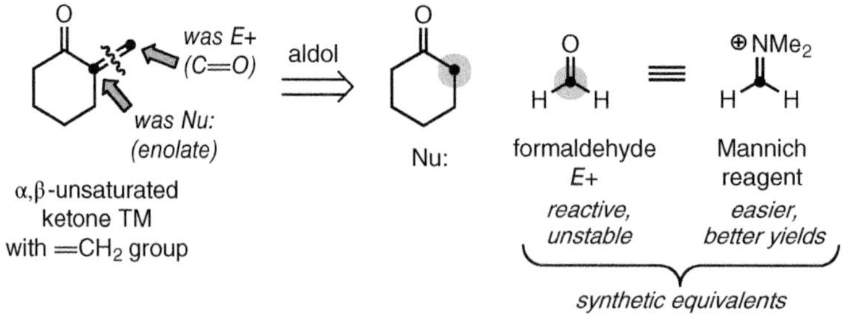

Retrosynthesis of α,β-Unsaturated Ketone via Mannich

4.1.3 SYNTHESIS OF α,β-UNSATURATED ESTERS USING THE WITTIG REACTION

Synthesis of α,β-Unsaturated Esters (Wittig)

The Wittig reaction, which combines a ketone or an aldehyde with a Wittig reagent, is used to create alkenes. If the Wittig reagent itself contains an ester group, then the resulting product will be an α,β-unsaturated ester. Such Wittig reagents are known as "stabilized ylides" since the ester electron-withdrawing group (EWG) offers resonance stabilization to the negative charge. The related Horner–Wadsworth–Emmons (HWE) reagent, a phosphonate-stabilized ylide that is prepared from P(OEt)$_3$ rather than PPh$_3$, has the advantage of being more reactive (reacts with both aldehydes and ketones) and offering better stereoselectivity.

stabilized Wittig reagent   HWE reagent

Wittig Reagent Containing an Ester EWG, a "Stabilized" Ylide

While the Wittig reaction typically gives the (Z)-alkene as the major product, *stabilized* ylides usually give the (E) stereoisomer.

stabilized Wittig
reagent   aldehyde   (E)-α,β-unsat'd ester

HWE reagent   ketone *or*
                               aldehyde   α,β-unsat'd ester

Stabilized Ylides Produce α,β-Unsaturated Esters

Retrosynthesis of α,β-Unsaturated Esters (Wittig)

When considering an α,β-unsaturated ester target molecule as the product of a Wittig reaction, the disconnection will focus on the alkene. It must be determined which of the alkene carbons started out as the Wittig reagent and which one was originally the carbonyl. The better disconnection places the ester group of the target molecule on the readily prepared, stabilized Wittig reagent as shown in "a." The alternate retrosynthesis "b" would result in a more complicated starting material and would produce the (Z)-alkene as the major product.

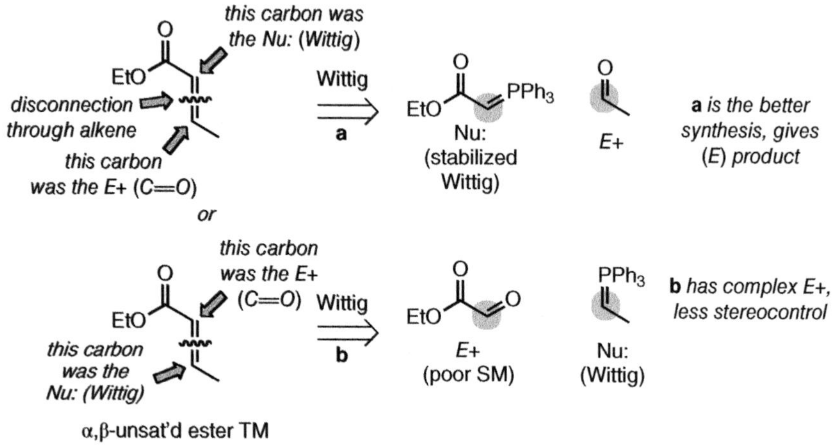

Retrosynthesis of α,β-Unsaturated Esters via Wittig

PRACTICE PROBLEM 4.1B: ALTERNATIVES TO ALDOL

4.1B   Provide the reagents needed for each of the following transformations (more than one step may be required).

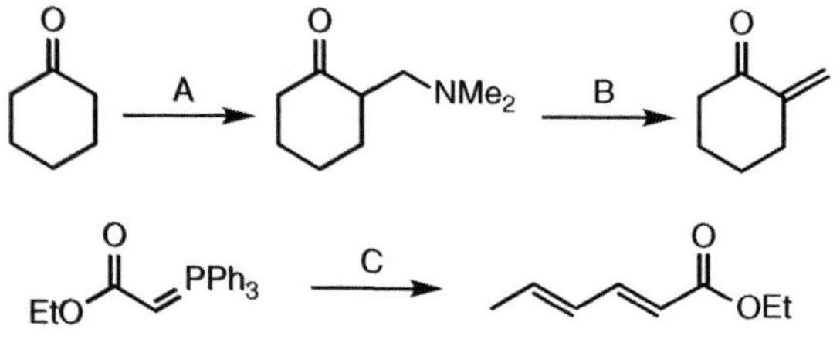

EXAMPLE: α,β-UNSATURATED CARBONYL

Synthesize the following target molecule using readily available starting materials and reagents. The synthesis must involve the formation of a new C—C bond.

TM

Since the target molecule is an α,β-unsaturated carbonyl, one possible approach would be an aldol reaction. A disconnection can be made directly through the C=C double bond, or a stepwise approach can be taken by first imagining the target molecule as a β-hydroxy structure that can then be disconnected between the alpha and the beta carbons. Either approach leads to the same aldol starting materials: a nucleophilic alpha carbon (enolate) and an electrophilic carbonyl carbon (benzaldehyde).

Retrosynthesis of TM via Aldol

The resulting mixed aldol should be a reasonable one, since benzaldehyde has no alpha protons and the only enolate (Nu:) present would be from methyl acetate. Sodium methoxide is a good choice of base, since it would not give a new product if an addition/elimination (acyl substitution) mechanism were to take place at the ester carbonyl (i.e., no hydrolysis or transesterification will take place if the alkoxide base matches the alkoxy group of the ester). "Heat" is added to the reaction conditions to imply that the α,β-unsaturated product is desired, rather than the β-hydroxy product. However, since the resulting alkene is in conjugation with both the carbonyl and the phenyl groups, this dehydration is likely to occur spontaneously even at room temperature.

Synthesis of TM via Aldol

Another possible approach is to view the target molecule as the product of a Wittig reaction with a stabilized Wittig reagent (or HWE reagent).

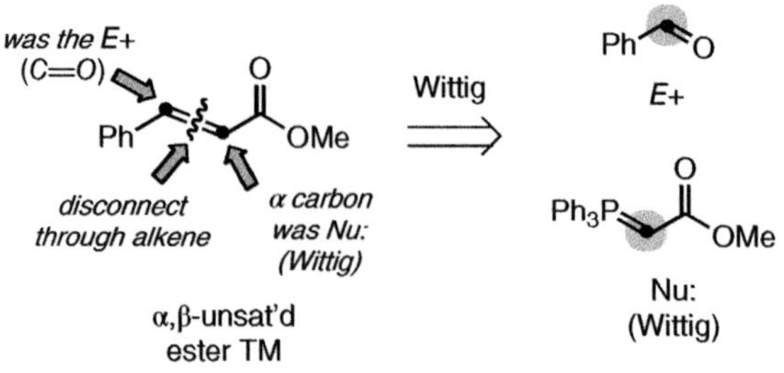

Retrosynthesis of TM via Wittig

This synthetic approach begins with the preparation of the Wittig reagent from methyl bromoacetate. Treatment of this reagent with benzaldehyde produces the target molecule.

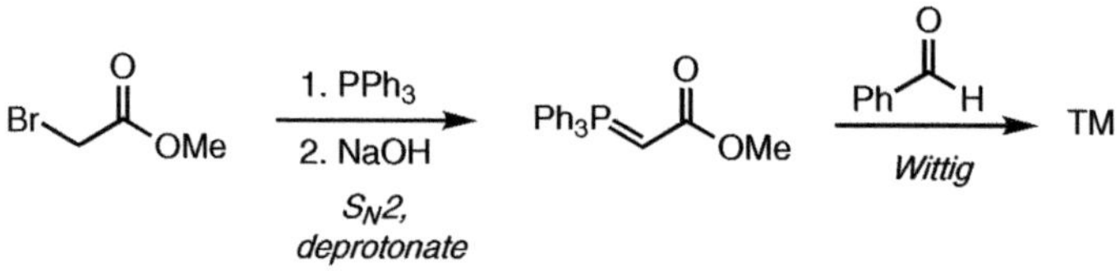

Synthesis of TM via Wittig

Since the Wittig synthesis is a longer one that begins with a more complex starting material, the simpler aldol synthesis is likely to be the preferred route in this case.

MORE ENOLATE REACTIONS: SYNTHESIS OF 1,3-DICARBONYLS, 1,5-DICARBONYLS, AND CYCLOHEXENONES

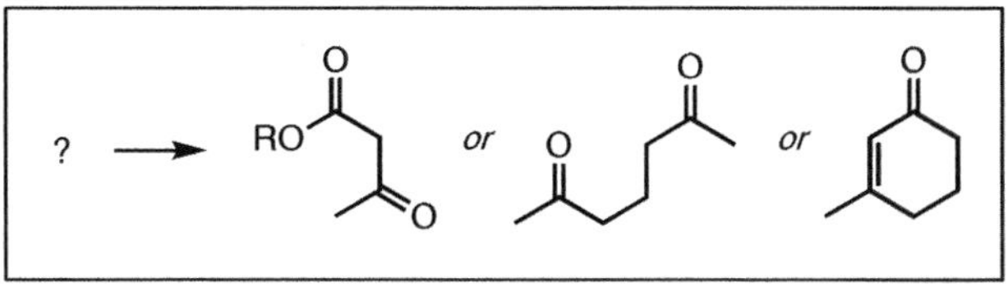

4.2.1 THE CLAISEN CONDENSATION

Synthesis of β-Keto Esters

The self-aldol reaction of aldehydes and ketones gives either β-hydroxy carbonyl or α,β-unsaturated carbonyl products. When the same reaction mechanism is applied to esters, the reaction is called the Claisen condensation. Like the aldol reaction, the Claisen condensation involves the attack of an enolate nucleophile on a carbonyl electrophile. However, subsequent elimination of the leaving group creates a β-keto ester product. If this 1,3-dicarbonyl pattern is present in a target molecule, it is an indication that the TM might be the product of a Claisen condensation, and a Claisen disconnection will be one option for retrosynthesis.

Introduction to Strategies for Organic Synthesis, Second Edition. Laurie S. Starkey.
© 2018 John Wiley & Sons, Inc. Published 2018 by John Wiley & Sons, Inc.

Claisen Condensation

Mechanism of the Claisen Condensation

The base-promoted Claisen condensation mechanism has the same first two steps as the aldol reaction: deprotonation of the alpha carbon and then attack of the resulting enolate on a carbonyl carbon. Again, reaction with an alkoxide base ensures that only a small amount of the ester will be deprotonated (Nu:) so there will be plenty of remaining ester starting material ($E+$). The intermediate formed upon attack of the ester carbonyl is not an ordinary intermediate; it is a charged tetrahedral intermediate (CTI) that "collapses" back to a carbonyl by ejecting the alkoxide leaving group. Overall, an "addition–elimination" reaction takes place at the ester carbonyl, to substitute the leaving group (OR) with a nucleophile (alpha carbon).

Collapse of the CTI forms the neutral β-keto ester product that will eventually be isolated, but the reaction mechanism does not stop here. Since the 1,3-dicarbonyl product has a highly acidic alpha proton (α to two EWGs), it becomes deprotonated in the basic reaction conditions to give a stabilized enolate. This deprotonation step is a necessary one, as it drives the equilibrium in the forward direction; the Claisen condensation will not occur if there are not at least two alpha protons present in the ester starting material, and the acid-catalyzed Claisen condensation does not exist. Upon treatment with a mild aqueous workup, the enolate is protonated and the neutral β-keto ester product can be isolated.

Mechanism of Claisen Condensation

The product of this reaction is a β-keto ester that contains a newly formed carbon–carbon bond between the alpha carbon and one of the carbonyl carbons. This is the key bond to be identified in a β-keto ester target molecule; a disconnection at this bond will lead to a Claisen retrosynthesis.

Retrosynthesis of β-Keto Esters

A typical retrosynthesis of a β-keto ester involves making a disconnection from one of the carbonyls to the alpha carbon between the two carbonyls. The alpha carbon will be introduced as a nucleophilic enolate. The β-keto carbon used to be a carbonyl ($E+$), but how can it *still* be a carbonyl after being attacked by a nucleophile? The β-keto carbonyl must have had a leaving group attached to it, in order for the addition–elimination (acyl substitution) mechanism to take place. The alkoxy group is the preferred leaving group to select, since esters are stable, easy to work with, and commercially available.

Retrosynthesis of β-Keto Esters (1,3-Dicarbonyl TMs)

Because the disconnection in the example above leads to two identical pieces, it results in a simple synthesis and an ideal retrosynthesis (the more nearly equal the resulting pieces, the better the disconnection). Otherwise, the regioselectivity of a mixed Claisen reaction can be controlled in a stepwise synthesis that employs LDA, as done in a mixed aldol reaction.

Stepwise Approach Controls a Mixed Claisen

Example: β-Keto Ester (1,3-Dicarbonyl) TM

Synthesize the following target molecule using readily available starting materials and reagents. The synthesis must involve the formation of a new C—C bond.

TM

Since the TM is a 1,3-dicarbonyl, a disconnection should be made at the alpha carbon between the two carbonyls. Disconnection on either side of the alpha carbon will result in different retrosyntheses, so there is more than one possible solution.

Retrosynthesis of TM: More Than One Reasonable Disconnection is Possible

Retrosynthesis "a" is not a traditional Claisen condensation, since it involves a ketone enolate attacking a non-ester electrophile (diethyl carbonate). This illustrates how the Claisen retrosynthetic strategy can be applied to any 1,3-dicarbonyl, not just β-keto esters. The second retrosynthesis, "b," is an example of an intramolecular Claisen condensation (called a Dieckmann condensation). It should be noted that this disconnection does not greatly simplify the TM, as the starting material has a carbon chain that is still the same length. Still, the Dieckmann condensation is a good method for synthesizing five- and six-membered cyclic β-keto esters.

Both of the syntheses shown are reasonable. The reaction conditions are identical and both have commercially available starting materials. Since no regiocontrol is needed for the deprotonation, the starting materials can simply be mixed together with sodium ethoxide in ethanol (note that the alkoxide base is chosen to match the ester or carbonate leaving group). A mild aqueous workup is needed to neutralize the target molecule, which will be formed as the stabilized enolate in these basic reaction conditions.

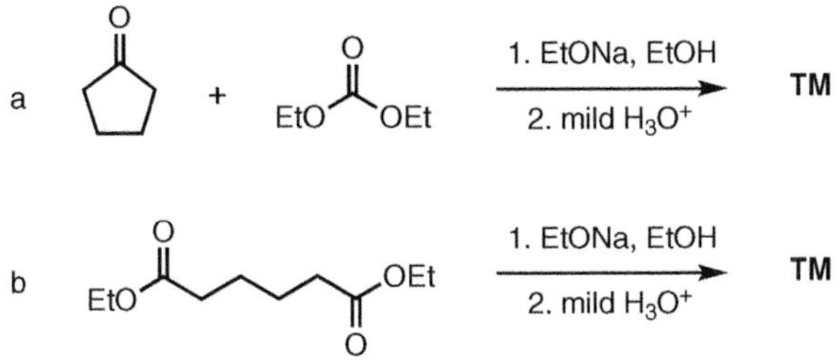

Each Retrosynthesis Leads to a Simple Synthesis

4.2.2   THE MICHAEL REACTION

Synthesis of 1,5-Dicarbonyls

The Michael reaction involves the attack of a stabilized enolate (or enol) nucleophile onto the beta carbon of an α,β-unsaturated carbonyl electrophile. The resulting enolate intermediate is protonated to give a 1,5-dicarbonyl product. If this pattern is present in a target molecule, it is an indication that the TM might be the product of a Michael reaction, and a Michael disconnection will be one option for retrosynthesis.

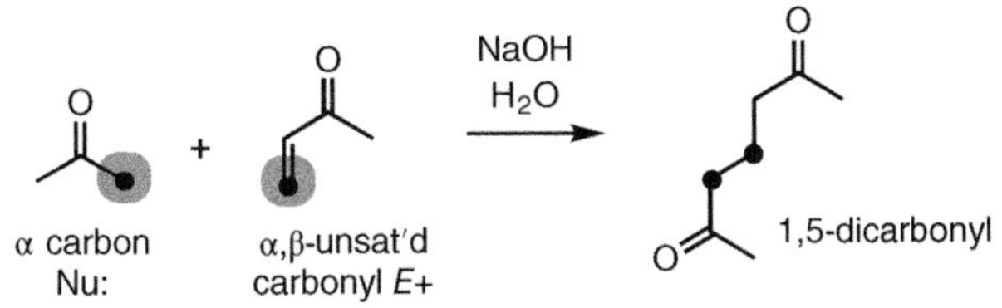

Michael Reaction

Mechanism of the Michael Reaction

Like the aldol and the Claisen, the Michael reaction also involves an enolate, so the mechanism begins with the deprotonation of carbon that is alpha to an EWG. In the Michael reaction, however, the electrophile is not an ordinary carbonyl but an α,β-unsaturated carbonyl. Attack of the enolate nucleophile occurs at the beta carbon of the α,β-unsaturated carbonyl (called a 1,4-addition or conjugate addition or a Michael addition) to give an enolate intermediate. Protonation at the alpha carbon of the enolate gives the final product, a 1,5-dicarbonyl compound.

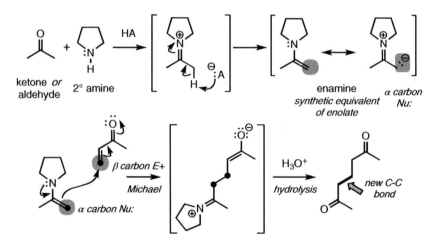

Base-Catalyzed Michael Reaction Mechanism

Ordinary enolates are very reactive (strong nucleophiles and strong bases) and can give mixtures of 1,2- and 1,4-addition reactions with α,β-unsaturated carbonyl compounds. The Michael reaction requires a "softer" enolate, such as a stabilized enolate with two electron-withdrawing groups. Otherwise, an enamine can be used as the synthetic equivalent of an enolate (followed by hydrolysis to regenerate the carbonyl).

Preparation and Use of Enamines

The product of this reaction is a 1,5-dicarbonyl that contains a newly formed carbon–carbon bond between the alpha carbon of one of the carbonyls and the beta carbon of the other. This is the key bond to be identified in a 1,5-dicarbonyl target molecule; a disconnection at this bond will lead to a Michael retrosynthesis.

Retrosynthesis of 1,5-Dicarbonyl Compounds

A typical retrosynthesis of a 1,5-dicarbonyl compound involves making a disconnection from one of the alpha carbons between the two carbonyls. The alpha carbon will be introduced as a nucleophilic enolate; the other carbon needs to be electrophilic. Since the electrophilic carbon is beta to a carbonyl, the starting material is derived by installing a double bond between the alpha and beta carbons to make an α,β-unsaturated carbonyl.

Retrosynthesis of 1,5-Dicarbonyl TMs

While it is not unreasonable to consider placing a *leaving group* at the electrophilic carbon, this would generate an unstable combination of functional groups. Carbonyl compounds with β-leaving groups easily undergo β-elimination, so they do not make good starting materials.

Leaving Groups in the Beta Position are Easily Eliminated

Additional Applications of the Michael Reaction

When a stabilized enolate undergoes a 1,4-conjugate addition with an enone, it is called a Michael reaction. However, many related reactions with analogous mechanisms are also described as Michael additions. Any C=C double bond containing one or more electron-withdrawing groups is referred to as a "Michael acceptor" and soft nucleophiles that prefer 1,4-addition, including stabilized enolates and organocuprates, are described as "Michael donors."

Michael acceptors (*E*+)   *general structure*

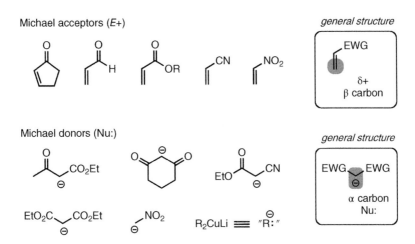

1,4-Addition Reactions: Michael Acceptors and Michael Donors

Example: 1,5-Dicarbonyl TM

Synthesize the following target molecule using readily available starting materials and reagents. The synthesis must involve the formation of a new C—C bond.

TM

Since the TM is a 1,5-dicarbonyl, a disconnection can be made at either alpha carbon between the two carbonyls. In this case, identification of the more stable enolate starting material will lead to the better synthesis.

The Better Synthesis Involves the More Stable Enolate

Retrosynthesis "b" is preferred since it has a more stable enolate nucleophile containing two electron-withdrawing groups (EWGs). Such a stabilized enolate is required for the Michael reaction. In addition, retrosynthesis "b" results in simpler starting materials that are more nearly equal in size.

The synthesis of the target molecule involves simply mixing the starting materials together with sodium ethoxide in ethanol (note that the alkoxide base is chosen to match the ester leaving group). Recall that an α,β-unsaturated compound such as phenyl vinyl ketone can be prepared via the Mannich reaction.

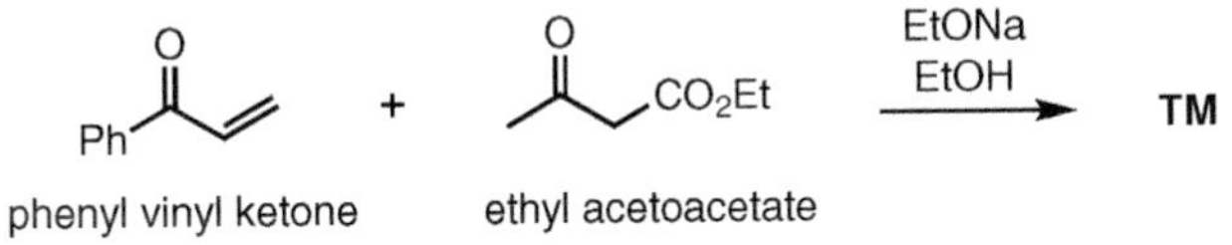

phenyl vinyl ketone   ethyl acetoacetate

Simple Synthesis of TM via Michael Reaction

PRACTICE PROBLEM 4.2: CLAISEN AND MICHAEL REACTIONS

4.2 For each of the following reactions, provide the missing starting material(s) or products.

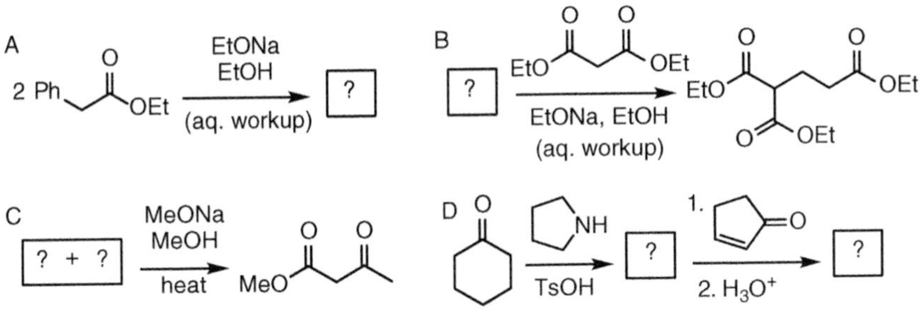

4.2.3 SUMMARY OF ENOLATE SYNTHESES

When the TM has a carbonyl, always look to the alpha carbon for a potential disconnection site. The enolate nucleophile (alpha carbon) can react with a variety of electrophiles (ketone, aldehyde, ester, or α,β-unsaturated C=O) to give a variety of functional group patterns.

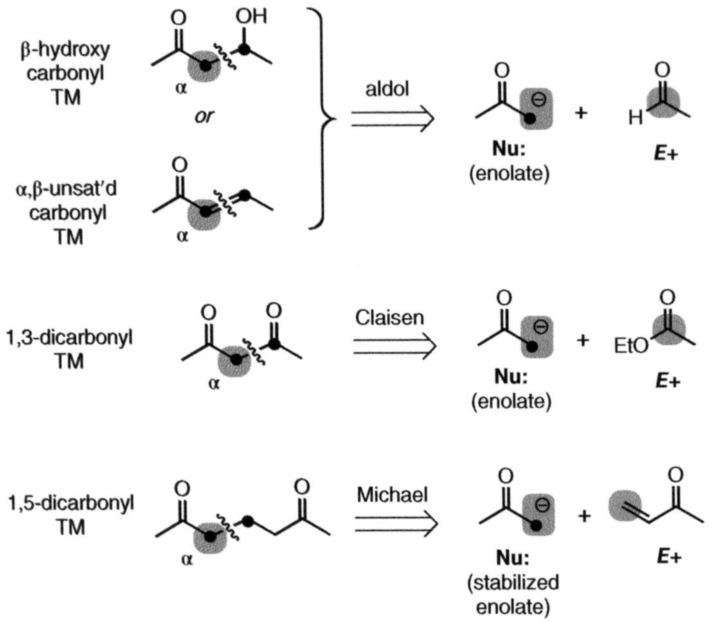

Disconnect at Alpha Carbon of Carbonyl-Containing TMs

4.2.4 ROBINSON ANNULATION

Synthesis of Cyclohexenone Derivatives

The Robinson annulation combines two of the reactions above to create a cyclic product. It begins with the Michael addition of an enolate nucleophile (often a cyclic ketone) onto methyl vinyl ketone (MVK), or a derivative of MVK. The resulting 1,5-dicarbonyl product can undergo an intramolecular aldol reaction with dehydration to give a cyclohexenone structure. If this pattern is present in a target molecule, it is an indication that the TM could be the result of a Robinson annulation.

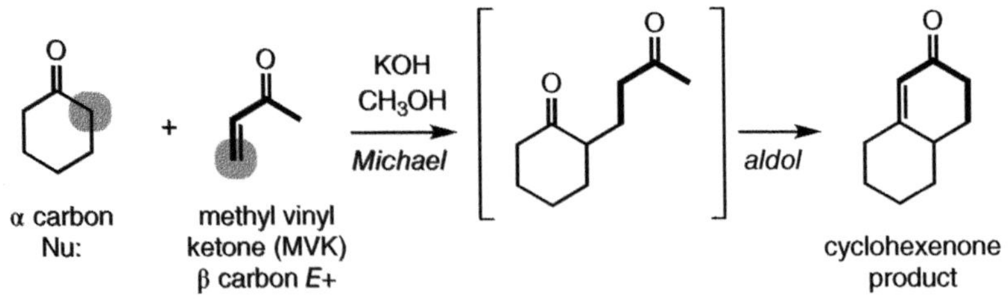

Robinson Annulation

Mechanism of the Robinson Annulation

The Robinson annulation involves two reactions occurring in tandem: a Michael reaction followed by an aldol condensation (loss of water is normally expected in this reaction so the aldol product is typically dehydrated to give an α,β-unsaturated cyclohexenone product). Reaction of an enolate as a nucleophile attacking the beta carbon of methyl vinyl ketone as the electrophile (a Michael reaction) forms the first carbon–carbon bond in the Robinson annulation and results in a 1,5-dicarbonyl product. The methyl group from MVK serves as the nucleophile for the second part of the reaction when it finds a carbonyl electrophile six atoms away to undergo an intramolecular aldol reaction. After dehydration, an α,β-unsaturated cyclohexenone product is formed. Ultimately, two new carbon–carbon bonds are formed within the cyclohexenone moiety.

Robinson Annulation Involves Tandem Reactions

Retrosynthesis of Cyclohexenones

Identification of the four-carbon MVK unit within the cyclohexenone target molecule leads to the appropriate disconnections. The first retrosynthetic step is a retro-aldol, so a disconnection is made between the alpha and beta carbons. The alpha carbon will be introduced as an enolate nucleophile; the other carbon was a carbonyl electrophile.

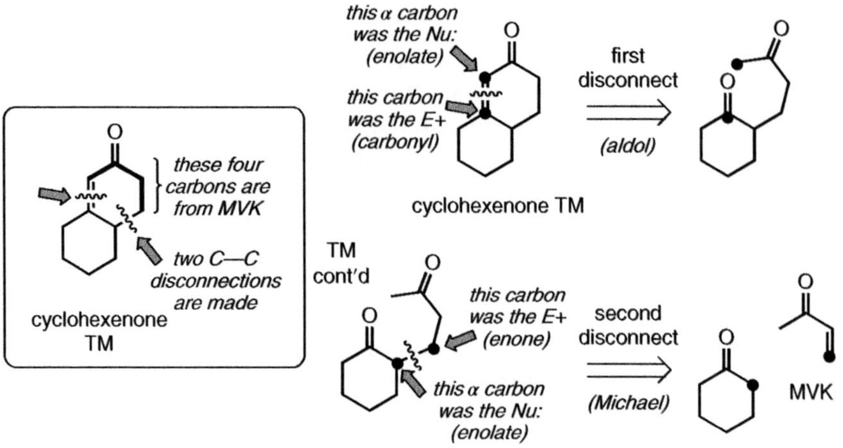

this α carbon
was the Nu:
(enolate)

this carbon
was the E+
(carbonyl)

first
disconnect

(aldol)

cyclohexenone TM

these four
carbons are
from MVK

two C—C
disconnections
are made

cyclohexenone
TM

TM
cont'd

this carbon
was the E+
(enone)

second
disconnect

this α carbon
was the Nu:
(enolate)

(Michael)

MVK

Retrosynthesis of Cyclohexenones

Example: Cyclohexenone TM

Synthesize the following target molecule using readily available starting materials and reagents. The synthesis must involve the formation of a new C—C bond.

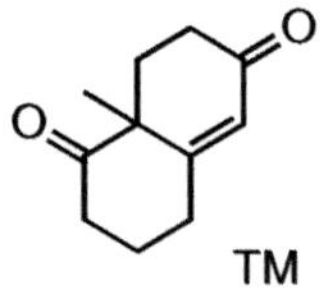

TM

Since the TM contains a cyclohexenone moiety, two disconnections can be made around the four-carbon methyl vinyl ketone unit within the six-membered cyclohexenone ring. The first disconnection is through the double bond of the cyclohexenone (aldol disconnection). The second disconnection is from an alpha carbon (Nu:) to the beta carbon of the MVK unit ($E+$) (Michael disconnection).

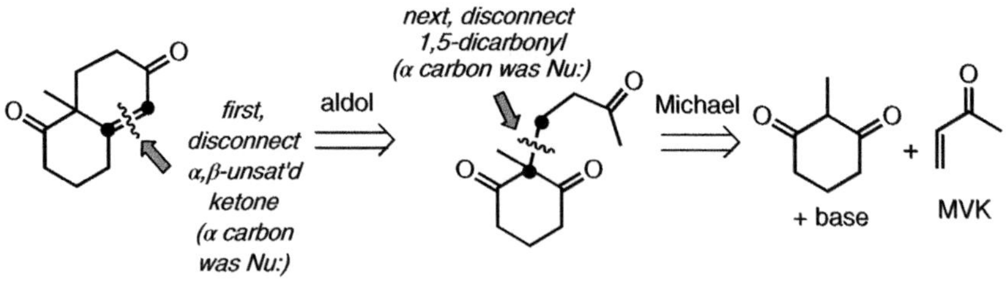

first,
disconnect
α,β-unsat'd
ketone
(α carbon
was Nu:)

aldol

next, disconnect
1,5-dicarbonyl
(α carbon was Nu:)

Michael

+ base

+

MVK

Retrosynthesis of Cyclohexenone TM

Synthesis of the target molecule involves combining the diketone shown with MVK in the presence of base; addition of heat facilitates the final dehydration.

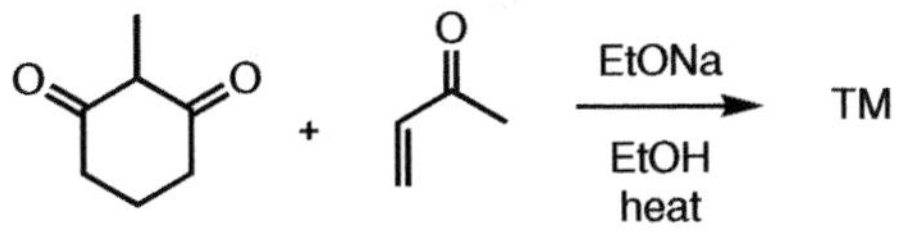

Synthesis of Cyclohexenone TM

"ILLOGICAL" 2-GROUP DISCONNECTIONS: UMPOLUNG (POLARITY REVERSAL)

4.3.1 SYNTHESIS OF TMs WITH A 1,2-DIOXYGENATED PATTERN

Up to this point, the disconnections for target molecules containing two functional groups have been logical ones leading back to recognizable nucleophiles and electrophiles. However, when two adjacent carbon atoms in a target molecule are attached to oxygen atoms, the resulting disconnection does not result in a logical electrophile and nucleophile.

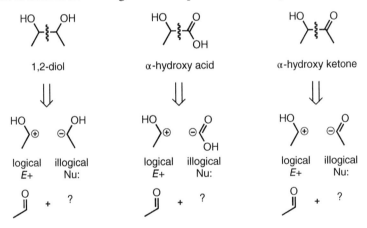

1,2-Dioxygenated TMs: Illogical Disconnections

An example of an "illogical" synthon is a negatively charged carbonyl nucleophile. Since carbonyls are normally electrophilic, a reversal of polarity (called "umpolung") must occur in order to accomplish the synthesis. In the following sections, we will explore these 1,2-dioxygenated patterns and learn appropriate synthetic strategies to achieve such TMs.

Introduction to Strategies for Organic Synthesis, Second Edition. Laurie S. Starkey.
© 2018 John Wiley & Sons, Inc. Published 2018 by John Wiley & Sons, Inc.

1,2-Diol TMs

Rather than attempting an illogical disconnection of a 1,2-diol TM, a retro-synthesis involving a functional group interconversion should be considered instead. Dihydroxylation of an alkene affords 1,2-diols, so this would be an effective approach to such target molecules. Depending on the reagents chosen, dihydroxylation of an alkene can occur with either syn (OsO_4) or anti (via epoxide) stereochemistry.

syn dihydroxylation   *anti dihydroxylation*
                                                          (via epoxide)

Synthesis of 1,2-Diols from Alkenes

The retrosynthesis of a 1,2-diol leads to an alkene, and the resulting alkene can be further analyzed retrosynthetically (e.g., the alkene can be prepared by a Wittig reaction or by dehydration of an alcohol, if it is not commercially available).

1,2-diol TM   alkene TM

Retrosynthesis of 1,2-Diols by FGI

α-Hydroxy Carboxylic Acid TMs: Umpolung

Disconnection of an α-hydroxy carboxylic acid results in an illogical syn-thon: a carboxylic acid moiety that is *nucleophilic* at the carbonyl carbon.

α-hydroxy acid   *need to make*
                                         carbonyl carbon
                                         nucleophilic!
                                         (umpolung)

Disconnection of α-Hydroxy Acids: Umpolung (Polarity Inversion)

The required umpolung can be achieved by using cyanide as a synthetic equivalent of the carboxyl anion, since the cyano group can be converted to a carboxylic acid by hydrolysis.

                              synthetic
                              equivalent

$^-$CN Is Synthetic Equivalent of $^-CO_2H$

Therefore, the synthesis of an α-hydroxycarboxylic acid can be achieved by addition of cyanide to a ketone or an aldehyde, followed by hydrolysis of the resulting cyanohydrin.

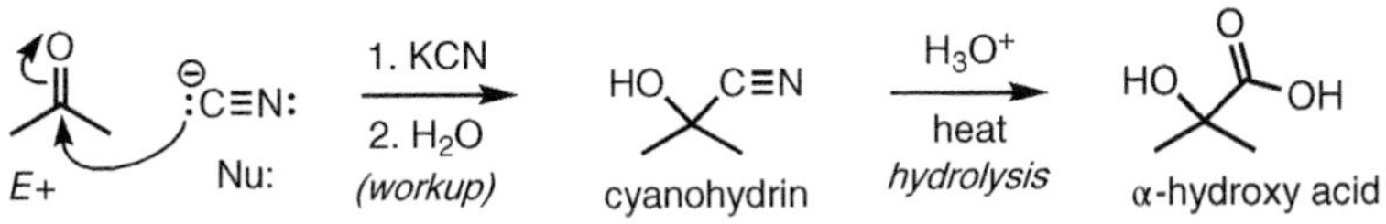

Synthesis of α-Hydroxy Carboxylic Acids

The retrosynthesis of a target molecule containing an α-hydroxycarboxylic acid begins with a functional group interconversion from a carboxylic acid to a nitrile. Disconnection of the cyanohydrin at the cyano group is now logical, affording recognizable synthons and reagents.

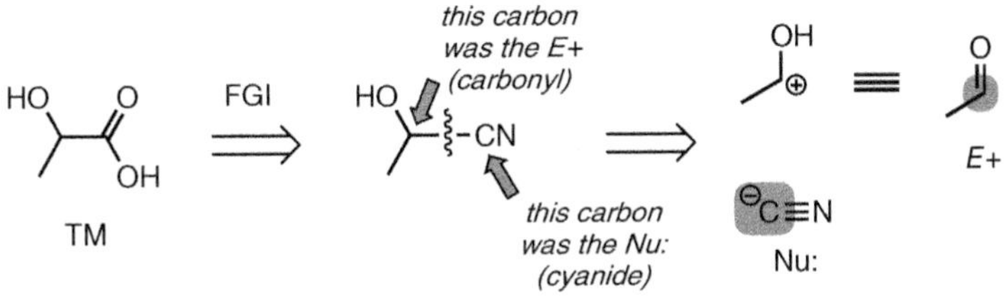

Retrosynthesis of α-Hydroxy Carboxylic Acids

α-Amino Acid TMs: The Strecker Synthesis

The methodology used to prepare α-hydroxy carboxylic acids is applied in the Strecker synthesis to make α-amino acids. In this synthesis, cyanide is once again used as a synthetic equivalent for the carboxyl group, but it attacks an imine rather than a carbonyl. The imine is formed *in situ* by combining ammonia with a ketone or an aldehyde, and hydrolysis of the nitrile affords the carboxylic acid.

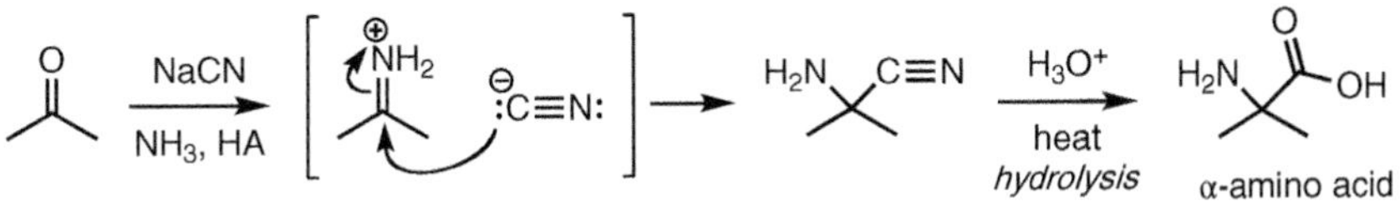

Strecker Synthesis of α-Amino Acids

α-Hydroxy Ketone TMs: The Dithiane Anion

Disconnection of an α-hydroxy ketone or aldehyde also results in an illogical synthon that requires a *nucleophilic* carbonyl carbon.

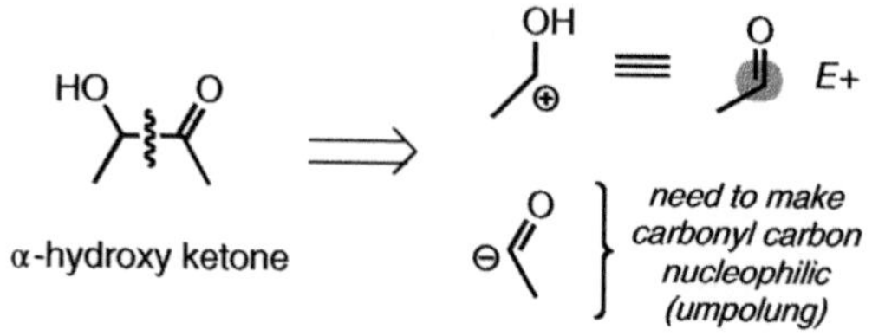

Illogical Disconnection of an α-Hydroxy Ketone/Aldehyde

One method that achieves the required umpolung is using a 1,3-dithiane anion as a synthetic equivalent of the acyl anion (the Henry–Nef strategy using a nitroalkane anion offers an analogous approach). While the acyl anion is unstable and cannot be prepared, the 1,3-dithiane anion is stable due to the inductive withdrawal of electron density by the two sulfur atoms. The 1,3-dithiane anion can be used as a nucleophile and then converted to a carbonyl by hydrolysis.

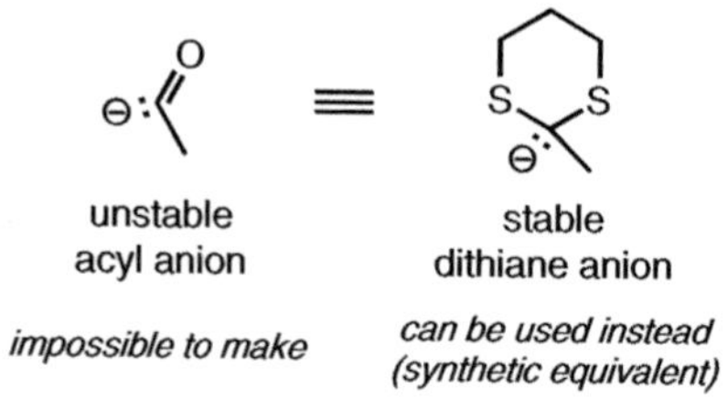

1,3-Dithiane Anion is the Synthetic Equivalent of an Acyl Anion

To make a 1,3-dithiane anion, an aldehyde is first converted to a thioacetal by reaction with 1,3-propanedithiol and a Lewis acid such as BF_3 (the formaldehyde derivative, called 1,3-dithiane, is commercially available). The resulting thioacetal is then deprotonated with a strong base such as *n*-butyllithium.

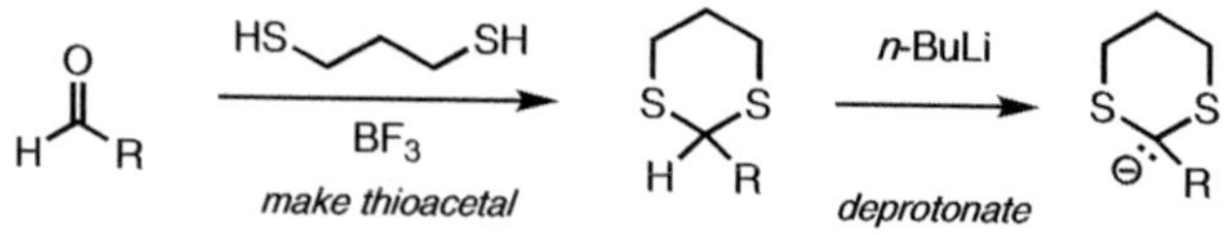

Preparation of a 1,3-Dithiane Anion

Reaction of a 1,3-dithiane anion nucleophile with a ketone or aldehyde electrophile, followed by a mercury-assisted hydrolysis, affords an α-hydroxy ketone.

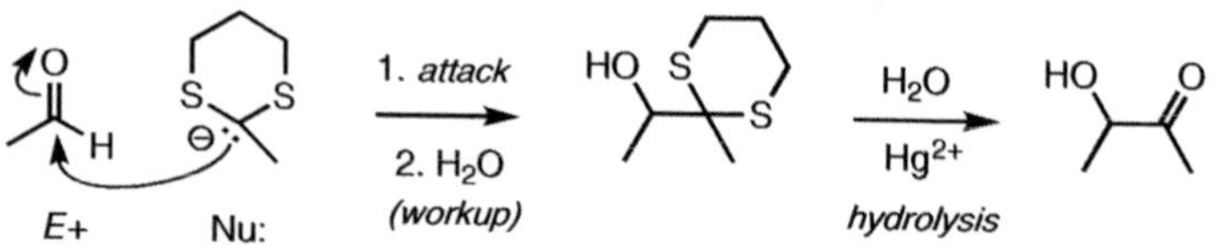

Synthesis of an α-Hydroxy Ketone

The retrosynthesis of a target molecule containing an α-hydroxy ketone begins with a functional group interconversion from a ketone to a 1,3-dithiane. Disconnection from the alcohol carbon is now logical, affording recognizable synthons and reagents.

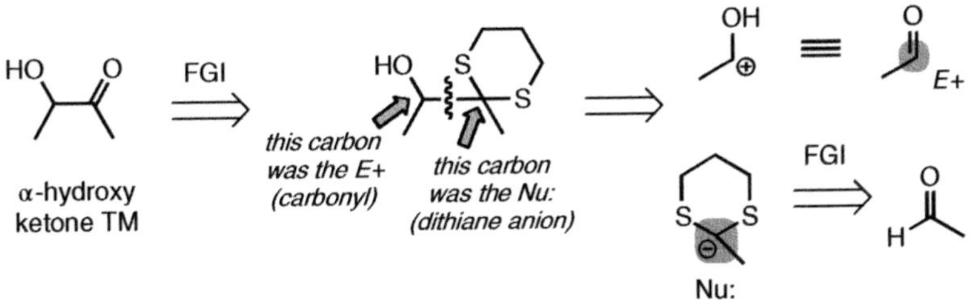

HO
O
FGI
α-hydroxy ketone TM
HO
S
S
this carbon was the E+ (carbonyl)
this carbon was the Nu: (dithiane anion)
OH
O
E+
FGI
O
S
S
H
Nu:

Retrosynthesis of α-Hydroxy Ketones

Example: α-Hydroxy Ketone TM

Provide the reagents necessary to transform the given starting material into the desired product. More than one step may be required.

OH
?
O
OH

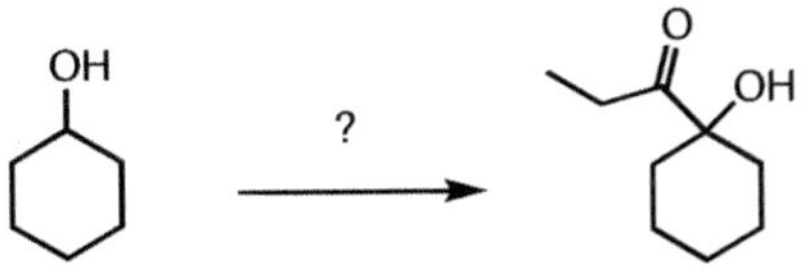

Disconnection of the target molecule at the newly formed C—C bond gives an illogical synthon (an acyl anion, for which a 1,3-dithiane anion is the synthetic equivalent). An alternate, stepwise approach to the retrosynthesis involves first doing a functional group interconversion, followed by a logical disconnection. Either method leads to the same combination of a 1,3-dithiane anion nucleophile and a cyclohexanone electrophile.

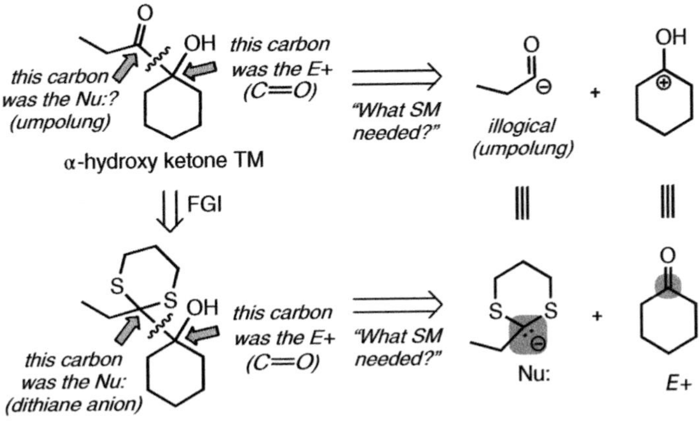

O
OH
this carbon was the E+ (C=O)
this carbon was the Nu:? (umpolung)
"What SM needed?"
O
illogical (umpolung)
+
OH
α-hydroxy ketone TM
FGI
S
S
OH
this carbon was the E+ (C=O)
"What SM needed?"
this carbon was the Nu: (dithiane anion)
S
S
Nu:
+
O
E+

Retrosynthesis of α-Hydroxy Ketone TM

The first step in the synthesis is an oxidation of the alcohol starting material (via DMP or Swern conditions) to give the needed carbonyl. Reaction of this ketone with the 1,3-dithiane anion shown, followed by hydrolysis, affords the desired TM.

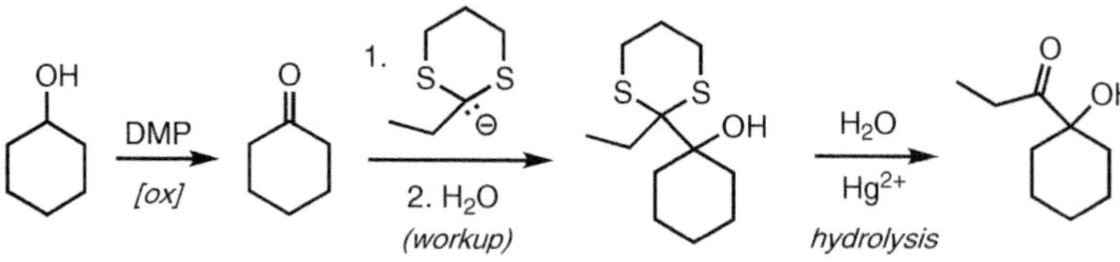

Synthesis of α-Hydroxy Ketone TM

The 1,3-dithiane anion can be prepared in two steps from the corresponding aldehyde (formation of thioacetal, followed by deprotonation). Another option is to start with formaldehyde. Alkylation of its 1,3-dithiane anion will introduce the required ethyl group. This method can be used to prepare a variety of 1,3-dithiane compounds and can even be used to synthesize aldehydes and ketones (one or two alkylation steps, followed by hydrolysis).

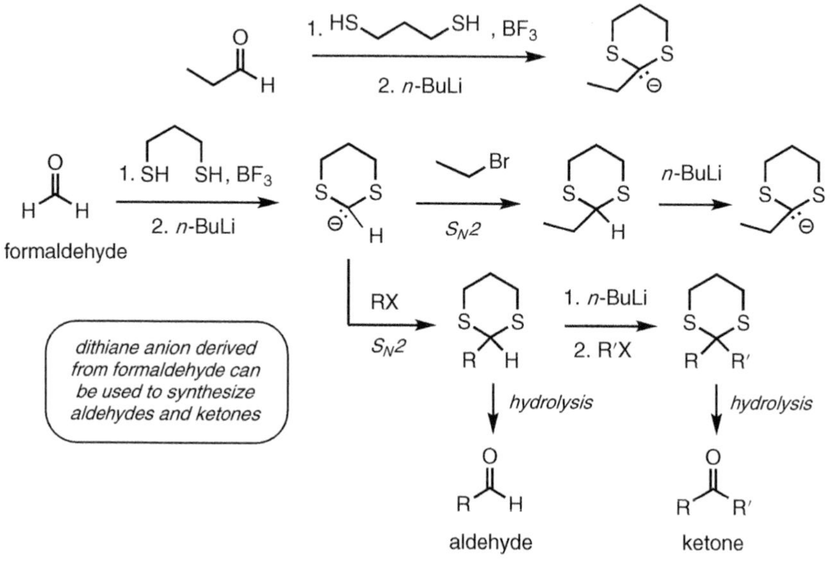

Synthesis and Uses of 1,3-Dithiane Anion

4.3.2 SYNTHESIS OF TMs WITH A 1,4-DIOXYGENATED PATTERN

A 1,4-dioxygenated target molecule, such as a γ-"gamma"-hydroxy ketone or a 1,4-diketone, is another pattern with an apparently illogical disconnection.

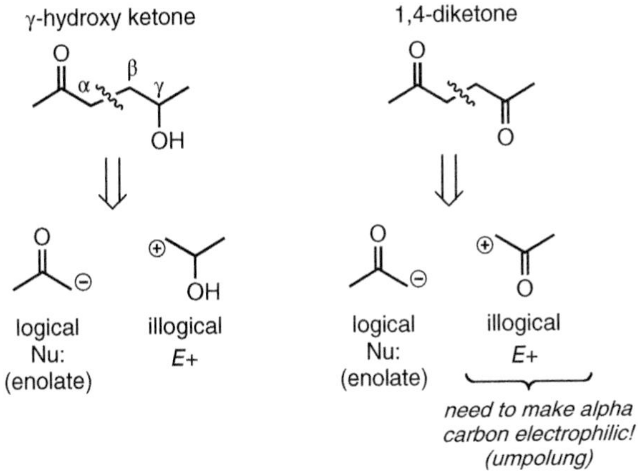

γ-hydroxy ketone

1,4-diketone

1,4-Dioxygenated TMs: Illogical Disconnections

In the upcoming sections, we will explore these 1,4-dioxygenated patterns and learn appropriate synthetic strategies to achieve such TMs.

γ-Hydroxy Carbonyl TMs

The disconnection of a γ-hydroxy carbonyl at the alpha carbon seems illogical because it is different from the aldol disconnection of a β-hydroxy carbonyl. The alpha carbon is the nucleophile as expected, but the electrophilic synthon has a positive charge not on the same carbon as the hydroxyl but on the adjacent carbon. This reactivity can be achieved by using an epoxide ring, because when a nucleophile attacks an epoxide, the hydroxyl group of the product is on the carbon next to the carbon that was attacked.

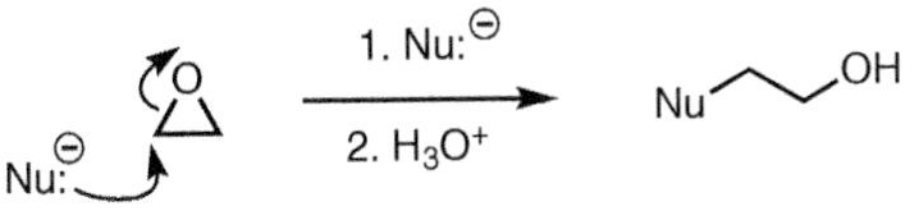

Synthesis Using an Epoxide

This is not the typical disconnection for alcohols, but it is a reasonable one.

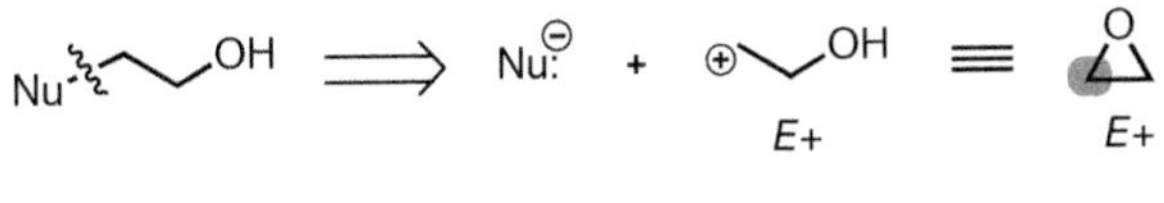

A Possible Retrosynthesis of Alcohols

This same disconnection can be used in combination with a disconnection at an alpha carbon (giving an enolate nucleophile) to make γ-hydroxy carbonyl TMs.

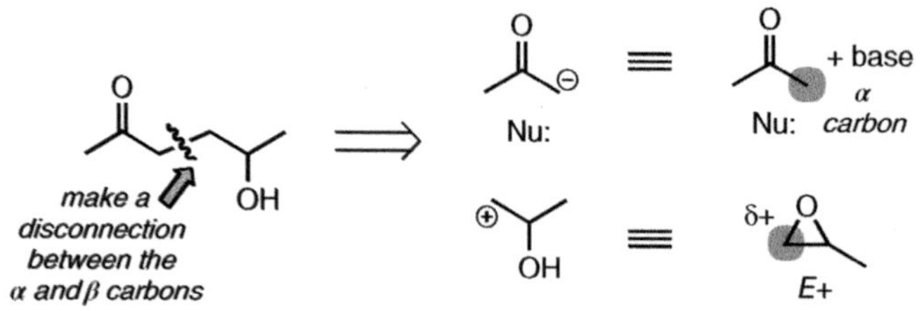

Retrosynthesis of γ-Hydroxy Carbonyl TMs

Synthesis of a γ-hydroxy carbonyl requires a stepwise approach: formation of an enolate (via LDA, enamine, acetoacetate ester, etc.), followed by addition of the epoxide and a final workup step.

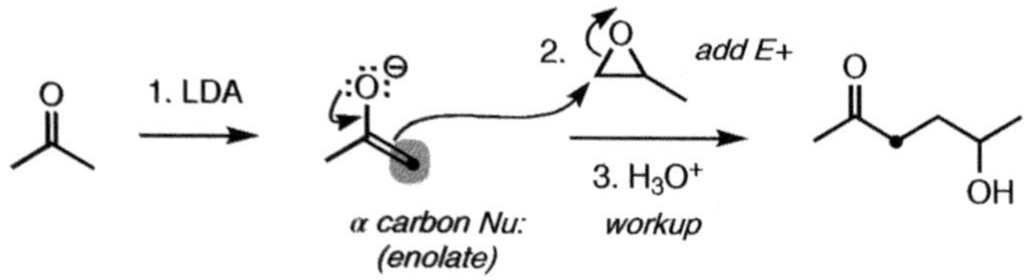

Synthesis of a γ-Hydroxy Carbonyl TM

1,4-Dicarbonyl TMs

The 1,4-dicarbonyl pattern is illogical because such a target molecule has two alpha carbons connected to one another. Disconnection between the two alpha carbons leads to one alpha carbon as the nucleophile (logical) and requires the other alpha carbon to be an electrophile (illogical). This reversal of polarity (umpolung) is achieved by placing a leaving group, such as a halogen, on the alpha carbon.

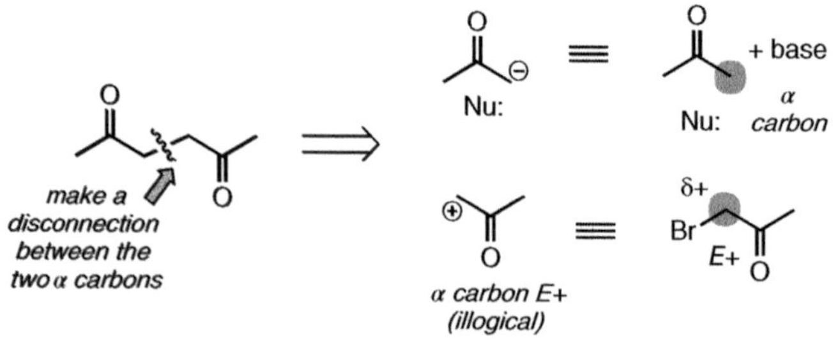

Retrosynthesis 1,4-Dicarbonyls

α-Brominated ketones can be prepared by reaction of the ketone with bromine in acidic reactions conditions (such as Br_2 with HBr or HOAc). When a nucleophile attacks α-halogenated ketones, an S_N2 substitution mechanism is preferred over addition to the carbonyl.

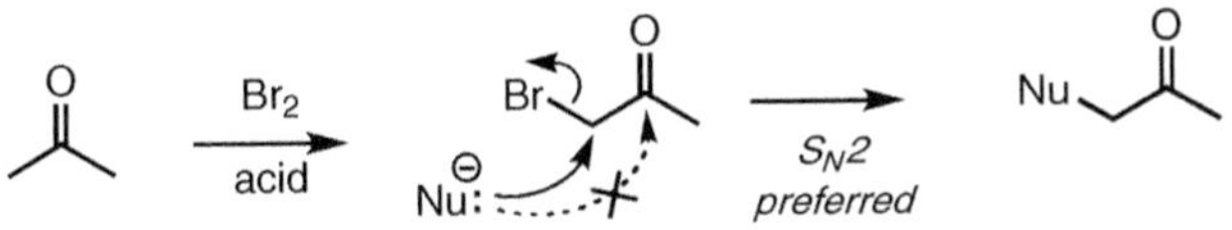

Preparation and Reactivity of α-Bromoketones

Synthesis of a 1,4-dicarbonyl also requires a stepwise approach: formation of an enolate, followed by addition of the α-bromoketone. However, since the bromine increases the acidity of the alpha proton, an ordinary enolate would more likely act as a base rather than a nucleophile so the synthesis would not work. Instead, a stabilized enolate (or an enamine equivalent) must be used for the S_N2 displacement.

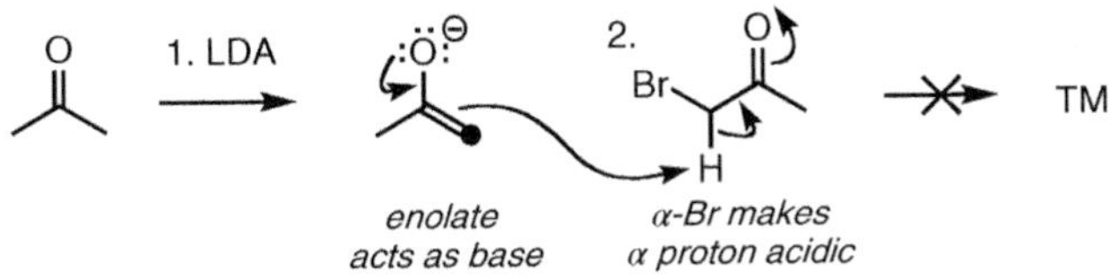

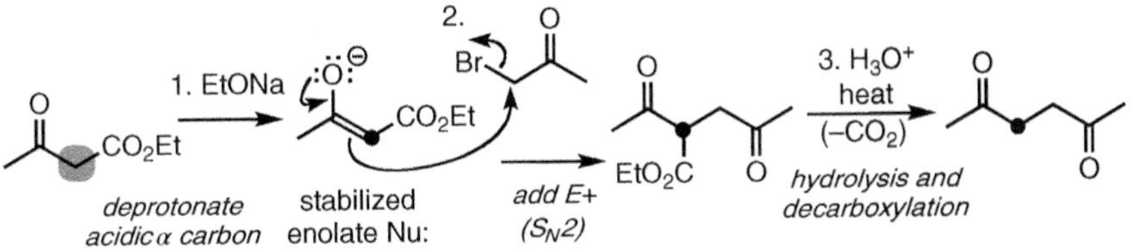

Synthesis of a 1,4-Dicarbonyl TM Requires Stabilized Enolate

Example: 1,4-Dioxygen TM

Provide the reagents necessary to transform the given starting material into the desired product. More than one step may be required.

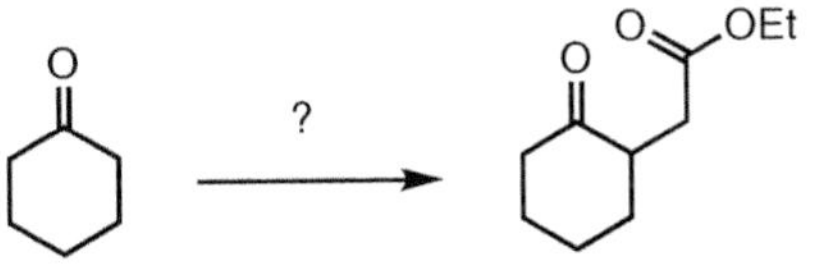

The required disconnection at the newly formed C—C bond is the typical one for a target molecule containing a 1,4-dicarbonyl pattern: between the

two alpha carbons. One of these carbons used to be the nucleophile (eno-late) and the other used to be an electrophile (α-bromo compound). Either combination is reasonable, and both are shown here.

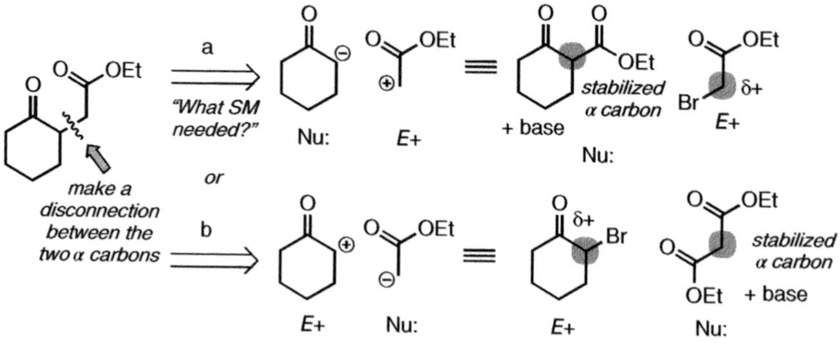

Retrosynthesis of 1,4-Dicarbonyl TM

The first step in transformation "a" is a mixed Claisen condensation of cyclohexanone with diethyl carbonate to add the needed EWG to the alpha carbon. Deprotonation, followed by alkylation with ethyl bromoacetate provides the required carbon framework. Hydrolysis of both esters, followed by decarboxylation of the β-keto acid moiety, removes the activating group. The resulting carboxylic acid must be re-esterified to give the desired target molecule.

Transformation "b" converts the cyclohexanone alpha carbon to an electrophile by brominating that position. An S_N2 with the stabilized diethyl malonate anion add the required two-carbon chain. As in "a," the synthesis of the target molecule concludes with hydrolysis of the diester, decarboxylation, and re-esterification.

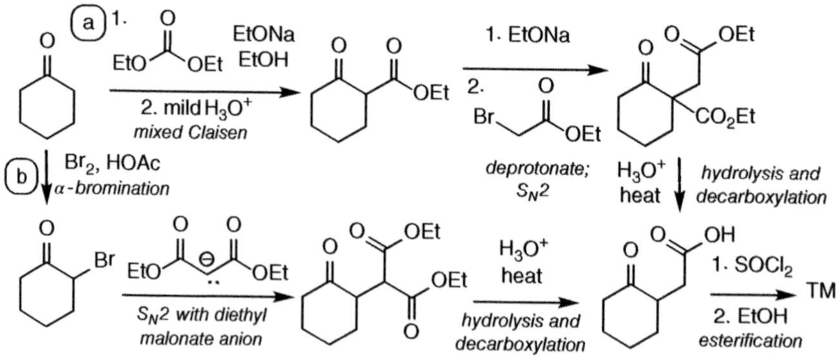

Synthesis of 1,4-Dicarbonyl TM

PRACTICE PROBLEM 4.3: "ILLOGICAL" 2-GROUP PATTERNS

4.3 Provide the reagents needed for each of the following transformations (more than one step may be required).

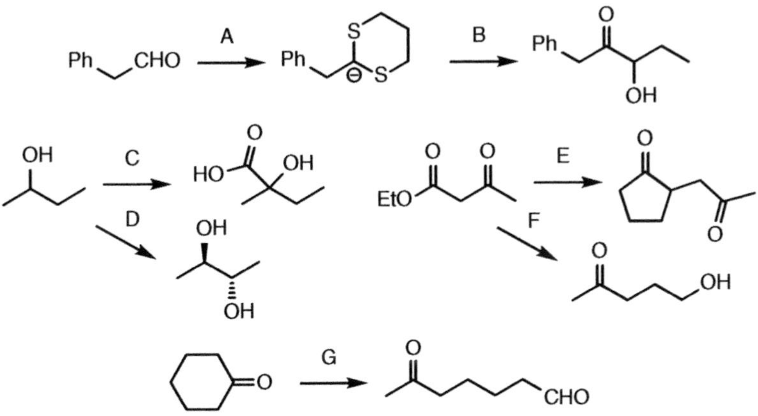

4.3.3 SYNTHESIS OF TMs WITH A 1,6-DICARBONYL PATTERN

Since attempting to disconnect a 1,6-dicarbonyl leads to illogical synthons, a retrosynthesis involving a functional group interconversion should be considered instead.

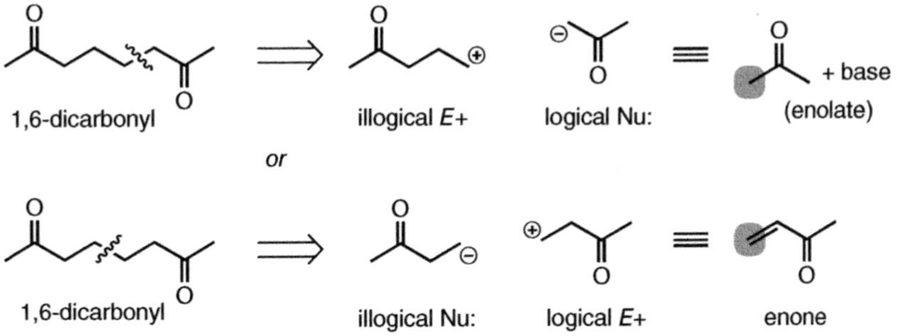

1,6-Dicarbonyl TM: Illogical Disconnections

Since ozonolysis of a cyclohexene affords 1,6-dicarbonyls, this would be an effective approach to such target molecules. Use of reductive or oxidative workups can give rise to various combinations of aldehydes, ketones, and carboxylic acids.

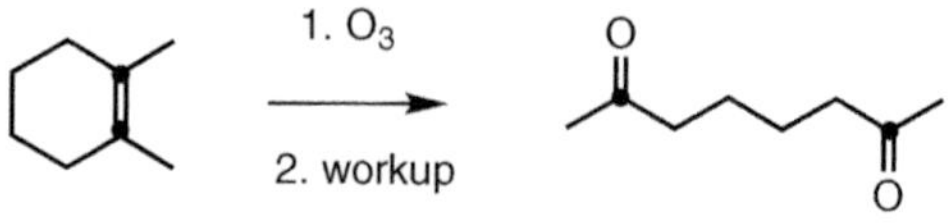

Synthesis of 1,6-Dicarbonyls from Cyclohexenes

One possible retrosynthesis of a 1,6-dicarbonyl involves a functional group interconversion to a cyclohexene. The cyclohexene derivative can be prepared in a variety of ways (e.g., by a Diels–Alder reaction), so the retrosynthesis can continue from there. The synthesis of target molecules containing six-membered rings will be discussed in a later section.

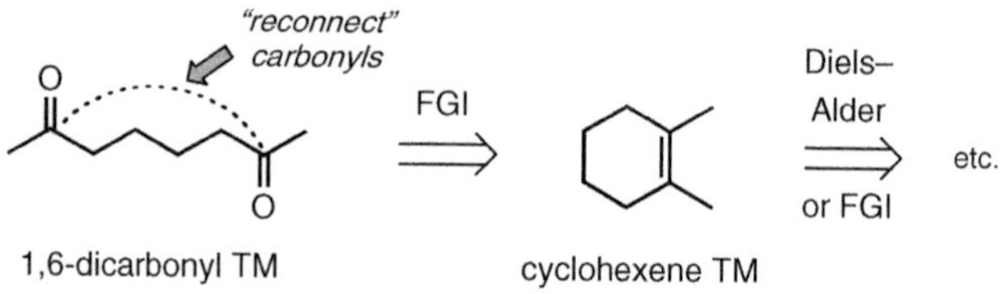

Retrosynthesis of 1,6-Dicarbonyls by FGI

2-FG TMs

4-1. Provide the corresponding synthetic equivalent for each of the following synthons. In other words, what starting material would have the desired reactivity?

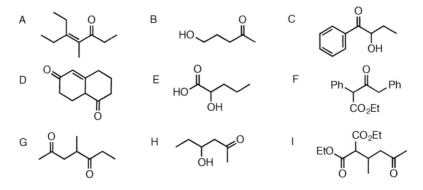

A

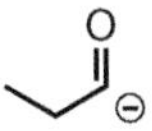

B 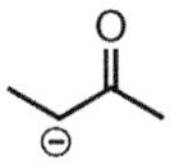
(2 synthons possibie)

C 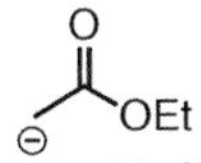
reacts with C=O
(2 synthons possibie)

D

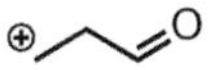

E

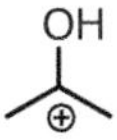

F

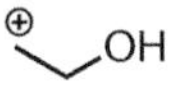

G

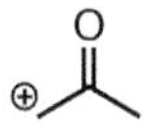

H

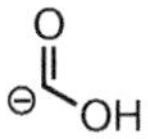

I

J

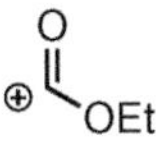

K 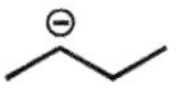
adds 1,4–

L

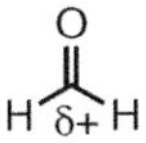

4-2. Propose a possible disconnection/retrosynthesis for each of the following target molecules. Consider the pattern of functional groups when determining the best site for a disconnection.

A

B

C

D

E

F

G

H

I

Introduction to Strategies for Organic Synthesis, Second Edition. Laurie S. Starkey.
© 2018 John Wiley & Sons, Inc. Published 2018 by John Wiley & Sons, Inc.

4-3. Starting with cyclohexanone, provide a synthesis for each of the following target molecules. It may help to first do a retrosynthesis of the TM.

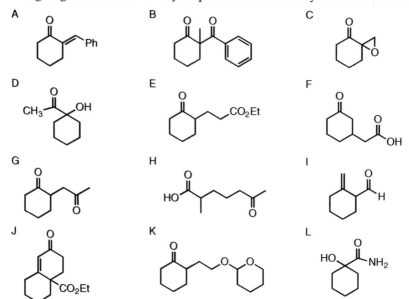

4-4. Provide the reagents necessary to transform the given starting material into the desired product. If more than one step is required, show the structure of each intermediate product. It may help to first do a retrosynthesis of the product.

I

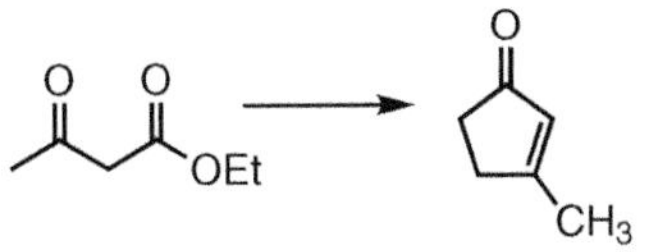

J

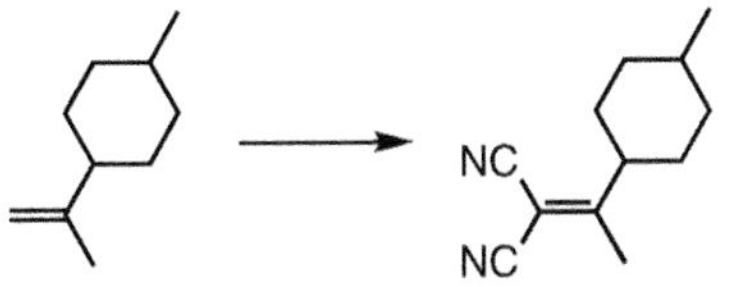

4-5. Provide a synthesis for each of the following target molecules. Each synthesis must form at least one new carbon–carbon bond.

A

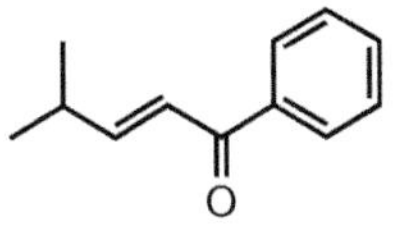

B

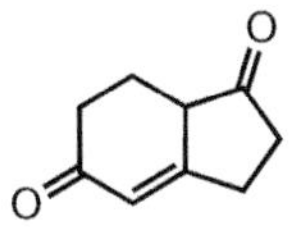

from acyclic
starting materials

C

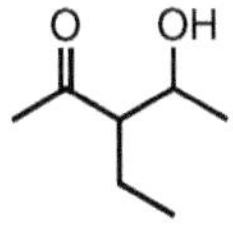

D

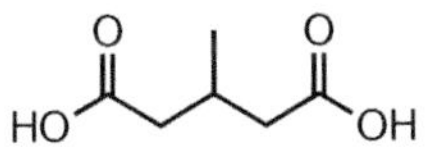

E

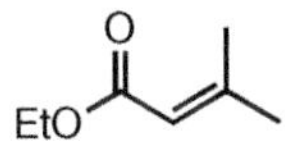

F

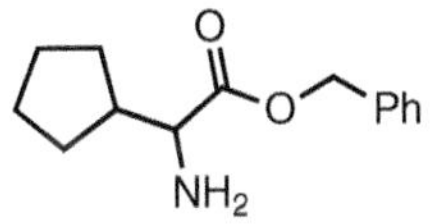

G

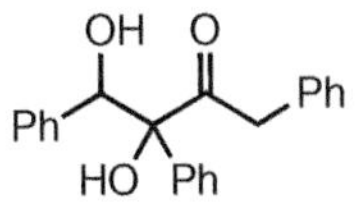

H

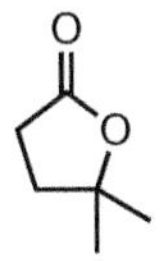

I

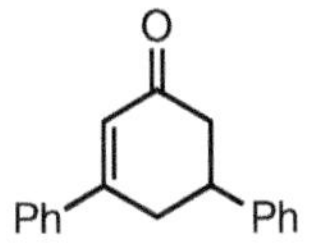

from starting materials with
no more than eight carbons

4-6. The following three components (acetophenone, 2-butanone, and LDA) are needed to synthesize the given TM. In which order should these components be added to the reaction flask to generate the desired product?

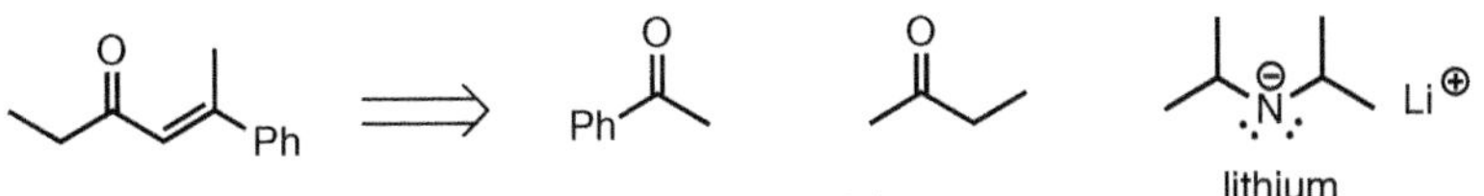

TM

acetophenone   2-butanone

lithium
diisopropylamide
(LDA)

4-7. Provide a complete mechanism for the following transformation.

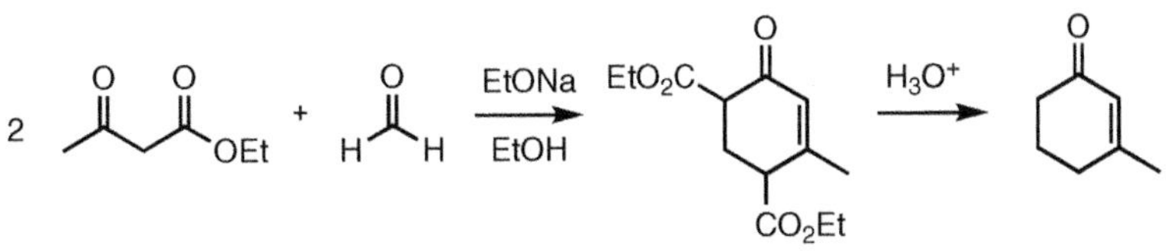

4-8. Provide a complete mechanism for the following transformation.

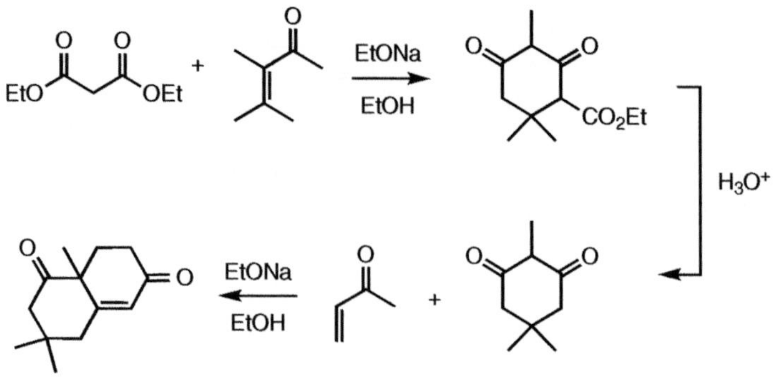

CHAPTER 5

SYNTHESIS OF AROMATIC TARGET MOLECULES

The synthesis of an aromatic target molecule typically begins with benzene, toluene, or another simple, commercially available benzene derivative. Such aromatic compounds can undergo various reactions to give more highly substituted targets.

Introduction to Strategies for Organic Synthesis, Second Edition. Laurie S. Starkey.
© 2018 John Wiley & Sons, Inc. Published 2018 by John Wiley & Sons, Inc.

ELECTROPHILIC AROMATIC SUBSTITUTION (ArH + E+ → ArE)

5.1.1 MECHANISM OF THE ELECTROPHILIC AROMATIC SUBSTITUTION REACTION

The electrophilic aromatic substitution reaction involves the reaction of a highly reactive electrophile with benzene (or any other aromatic ring system, including benzene derivatives, naphthalene, heteroaromatic compounds, etc.). The benzene ring nucleophile adds the electrophile to one of its pi bonds, producing a resonance-stabilized carbocation intermediate (called a sigma complex). This rate-determining step is quite unfavorable since it causes the loss of aromaticity, so it should be noted that this reaction only occurs with special, reactive electrophiles. Rapid deprotonation of the carbocation regenerates the aromatic ring to give a substituted benzene product. The overall reaction in an electrophilic aromatic substitution reaction is that an electrophile replaces one of the starting aromatic ring's protons.

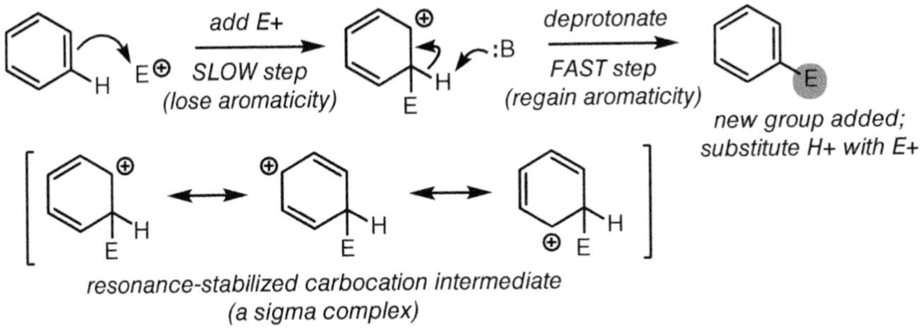

Mechanism of Electrophilic Aromatic Substitution

Introduction to Strategies for Organic Synthesis, Second Edition. Laurie S. Starkey.
© 2018 John Wiley & Sons, Inc. Published 2018 by John Wiley & Sons, Inc.

Electrophiles for Electrophilic Aromatic Substitution

The electrophilic aromatic substitution reaction can be used to add bromo (—Br), chloro (—Cl), nitro (—NO$_2$), and sulfonic acid (—SO$_3$H) groups to aromatic ring systems. It can also add alkyl (—R) and acyl (—COR) groups, as in the Friedel–Crafts alkylation and acylation reactions (these two reactions are presented in Sections 3.8 and 3.9, respectively). Each electrophilic aromatic substitution reaction requires reagents that will generate a highly electrophilic species that is typically positively charged. The combination of Br$_2$/FeBr$_3$ generates a Br$^+$ electrophile (or the equivalent Br—Br$^+$—FeBr$_3^-$). Likewise, Cl$_2$/FeCl$_3$ produces Cl$^+$; electrophilic aromatic substitution to introduce fluorine or iodine is possible but less common. Nitration can be accomplished with a mixture of nitric and sulfuric acids (HNO$_3$, H$_2$SO$_4$), and sulfonation involves reaction with "fuming" sulfuric acid (SO$_3$, H$_2$SO$_4$). A variety of methods are available to generate the carbocation needed in a Friedel–Crafts alkylation, and treatment of an acid chloride with a Lewis acid (RCOCl/AlCl$_3$) affords the acylium ion (RC≡O$^+$) needed for the corresponding acylation reaction. Formylation (addition of —CHO) can be accomplished with either the Vilsmeier–Haack or Gattermann–Koch reactions (see Section 3.9).

Reaction	Conditions	Electrophile (E+)	Mech. to make E+ (1st steps in Elect. Ar. Sub.)
halogenation –X	Br$_2$, FeBr$_3$ Cl$_2$, FeCl$_3$	Br$^{\oplus}$ Cl$^{\oplus}$	*great LG*
nitration –NO$_2$	HNO$_3$, H$_2$SO$_4$	O=N$^{\oplus}$=O	*great LG*
sulfonation –SO$_3$H	SO$_3$, H$_2$SO$_4$		*great LG*
Friedel–Crafts alkylation –R	RCl, AlCl$_3$	R$^{\oplus}$ carbocation	*great LG*
Friedel–Crafts acylation		R–C≡O$^{\oplus}$ acylium ion	*great LG*
formylation	CO, HCl, AlCl$_3$ *or* Me$_2$N +POCl$_3$	H–C≡O$^{\oplus}$ *or* $^{\oplus}$NMe$_2$	Gattermann–Koch Vilsmeier–Haack (for activated rings)

Generation of Electrophiles for Electrophilic Aromatic Substitution

5.1.2 ELECTROPHILIC AROMATIC SUBSTITUTION ON SUBSTITUTED BENZENES

When a substituent is already present on an aromatic ring undergoing electrophilic aromatic substitution, three different products are possible: ortho, meta, or para. The observed major product(s) depends on whether the substituent is an electron-donating group (EDG), an electron-withdrawing group (EWG), or a halogen.

	EDG	EWG	X
type of group	electron-donating group	electron-withdrawing group	halogen
reactivity (vs. PhH)	activated	deactivated (strongly)	deactivated (weakly)
regio- selectivity	ortho, para directing	meta directing	ortho, para directing
examples	–OH –NH$_2$ –OR –R	–NO$_2$ –CN –$\overset{\oplus}{N}R_3$ –SO$_3$H –CF$_3$	–Cl –Br –F –I

Summary of Substituent Effects on Electrophilic Aromatic Substitution

Effects of Electron-Donating Groups (EDG)

Electron-donating groups (EDG) include alkyl groups (−R) and any oxygen or nitrogen group, such as amino (−NH$_2$), hydroxyl (−OH), alkoxy (−OR), esters attached at the oxygen (−OCOR), and amides attached at the nitrogen (−NHCOR). Since all of these groups add electron density to the benzene ring (either inductively or by resonance with a lone pair of electrons), an EDG-substituted ring is more electron-rich and, therefore, is a more reactive nucleophile. Such rings, described as being "activated" toward electrophilic aromatic substitution, react faster than ordinary benzene and require milder reaction conditions. Because electron-donating groups can stabilize an adjacent carbocation, the intermediate sigma complex is better stabilized when the electrophile adds to the ortho or para positions. Electron-donating groups are known as "ortho/para directors," and para is often the major product because the ortho intermediate is destabilized by steric strain (and therefore produced more slowly). Although both ortho and para products are usually formed, when working on multistep synthesis problems it may be assumed that these isomers can be separated.

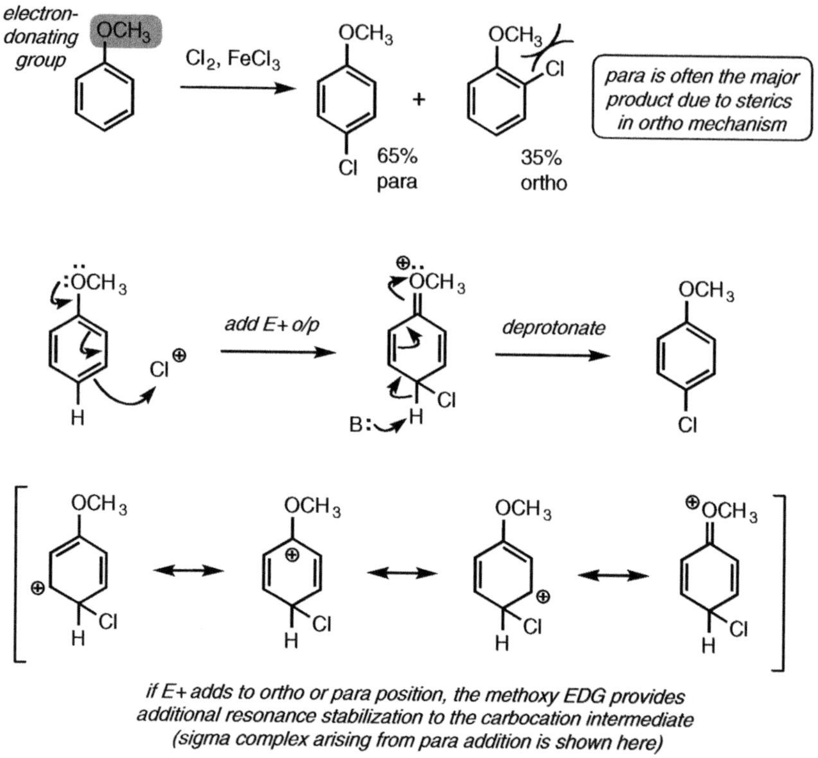

if E+ adds to ortho or para position, the methoxy EDG provides
additional resonance stabilization to the carbocation intermediate
(sigma complex arising from para addition is shown here)

EDGs are Ortho/Para-Directing for Electrophilic Aromatic Substitution

Effects of Electron-Withdrawing Groups (EWG)

Electron-withdrawing groups (EWG) include nitro ($-NO_2$), carbonyls (as in ketones $-COR$, esters $-CO_2R$, etc.), cyano ($-CN$), sulfonyl ($-SO_3H$), trifluoromethyl ($-CF_3$), and ammonium ($-NR_3^+$). Electron-withdrawing groups remove electron density from the benzene ring, either inductively or by resonance with a pi bond to oxygen or nitrogen. An EWG-substituted ring is electron-deficient, making it a poor nucleophile. Such rings, described as being "deactivated" toward electrophilic aromatic substitution, react more slowly than benzene itself and often require stronger reaction conditions. Some electrophilic aromatic substitution reactions, such as the Friedel–Crafts reactions, are so slow on rings bearing EWGs that they are described as "no reaction."

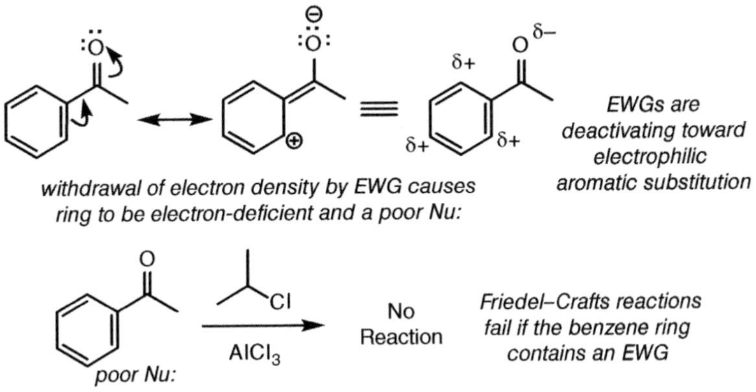

withdrawal of electron density by EWG causes
ring to be electron-deficient and a poor Nu:

EWGs are
deactivating toward
electrophilic
aromatic substitution

poor Nu:

No
Reaction

AlCl₃

Friedel–Crafts reactions
fail if the benzene ring
contains an EWG

Electron-Withdrawing Groups (EWGs) are Strongly Deactivating

Because electron-withdrawing groups would *destabilize* an adjacent carbocation, addition of the electrophile to the meta position gives a more stable sigma complex intermediate. Since meta is the major product obtained, electron-withdrawing groups are known as "meta directors."

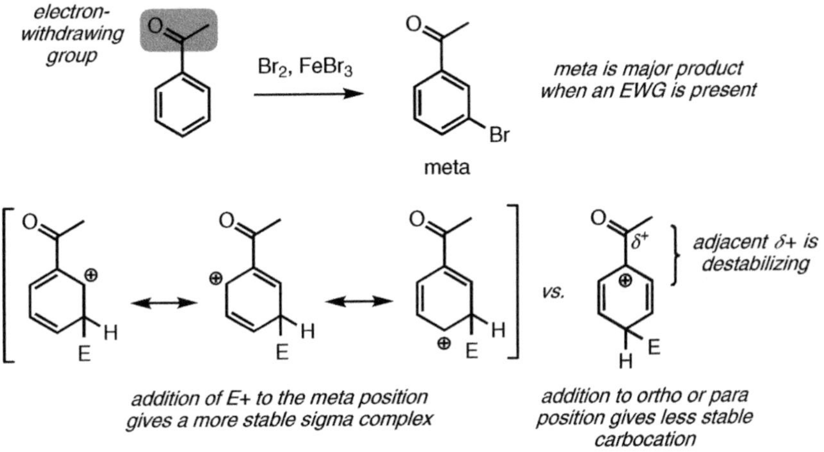

electron-withdrawing group

Br₂, FeBr₃

meta is major product
when an EWG is present

meta

addition of E+ to the meta position
gives a more stable sigma complex

addition to ortho or para
position gives less stable
carbocation

adjacent δ+ is
destabilizing

vs.

EWGs are Meta-Directing for Electrophilic Aromatic Substitution

Effects of Halogens (F, Cl, Br, I)

Halogens cannot be simply described as electron-donating or electron-withdrawing groups, since they have characteristics of both. Bromine is an ortho/para director because it can offer resonance stabilization to an adjacent carbocation with its lone pairs of electrons (like EDGs). However, this resonance donation is weak due to the poor overlap of the smaller p orbital on carbon with the larger p orbital on the halogen. The electronegative halogens also withdraw electrons inductively, so halo-substituted rings are deactivating overall toward electrophilic aromatic substitution (like EWGs).

Effects of Halogens on Electrophilic Aromatic Substitution

Directing Power of Substituents

If a benzene ring contains more than one substituent, these groups may compete as they direct the regiochemistry of the incoming electrophile. In some cases, the multiple groups direct to the same position so there is no question about which product to predict. If the groups direct to different positions, however, then the directing power of the substituents must be considered. In such a competition, any strongly activating group (−NH₂, −OH) wins over all other groups. Any activating group with lone pairs (−OR, −NHCOR, −OCOR) has more directing power over those without (−R). The weakly activating alkyl groups (−R) have very similar directing powers as the weakly deactivating halides (−X). Strongly deactivating groups (EWGs) will always lose to any other substituent. In the example shown, the methoxy group is a stronger director than the methyl group, so the incoming sulfonyl group will add ortho or para to the methoxy. Only one ortho position is available (the other is blocked by the methyl group), and that is expected to be the minor product, with para as the major product.

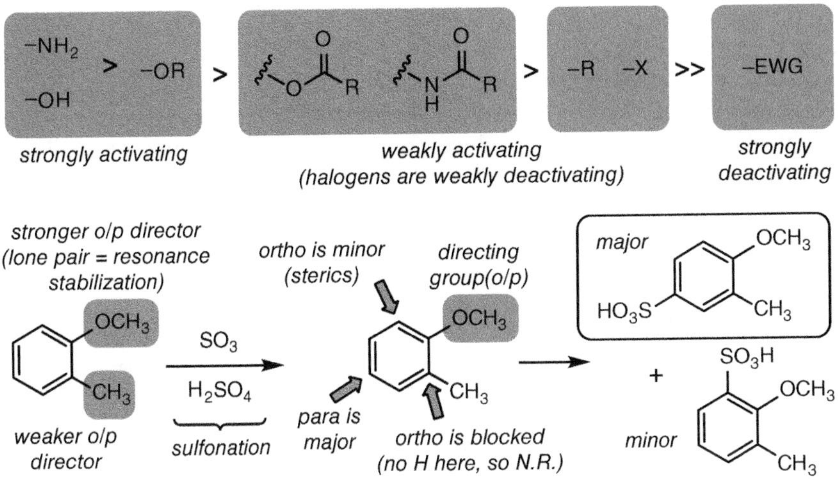

Directing Power of Substituents

Reaction with Aniline (PhNH$_2$): Use of Protective Groups

The electrophilic aromatic substitution reaction requires strong electrophiles and typically involves strongly acidic reaction conditions. Since amines are basic, they will be protonated in such conditions. When treated with acid, the amino group of aniline (an ortho/para directing EDG) would be converted to a positively charged ammonium group (a meta-directing EWG), thus affecting the outcome of the substitution reaction.

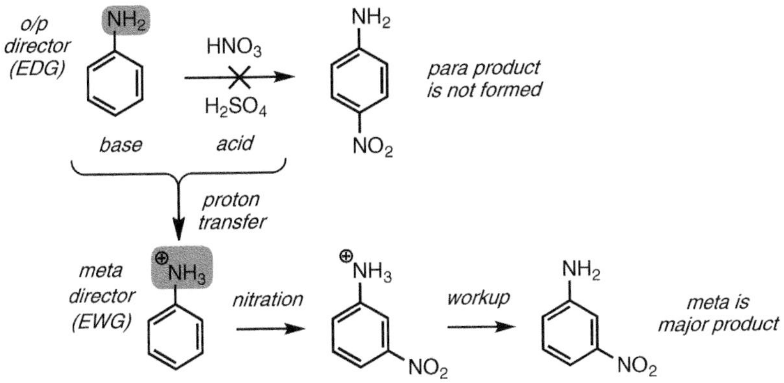

Nitration of Aniline Gives Meta Product

To avoid this acid–base reaction, the amino group must be protected: typically it is converted to an amide by reaction with acetyl chloride (AcCl) and base. The amide nitrogen is much less basic, since its lone pair is delocalized by resonance with the carbonyl. Therefore, the amide will not be significantly protonated by acidic reaction conditions, and it remains an ortho/para director for the substitution reaction. The amide protective group can eventually be removed by hydrolysis $(NaOH, H_2O)$. Such a protective group might also be employed to reduce the reactivity of phenols or anilines, allowing more control and avoiding multiple substitutions.

Protect Aniline for Acidic Electrophilic Aromatic Substitution Reactions

Synthesis of Polysubstituted Aromatic TMs: Use of Blocking Groups

The addition of a sulfonic acid group $(-SO_3H)$ is useful for more than preparing sulfonyl-containing target molecules. The sulfonic acid group can also be used as a temporary blocking group, because this group can be removed by treatment with water and heat. Consider the synthesis of *ortho*-chlorophenol via chlorination of phenol. Since the hydroxyl group is an electron-donating group, it is an ortho/para director. The chlorination reaction would produce a mixture of ortho and para products (likely along with some di- and trichlorinated products), with para typically being the major product. While it might be possible to separate these isomers, the desired ortho compound is unlikely to be produced in high yields.

Chlorination of Phenol Gives Small Amount of Ortho Product

A better synthesis would be one that forms a single major product leading to the target molecule, and the use of a blocking group can achieve that goal in this case. Sulfonation of phenol introduces a removable sulfonic acid group to the para position. The presence of this group blocks the para position so chlorination now forces the chlorine into the position ortho to the hydroxyl (the meta-directing $-SO_3H$ group reinforces this regiochemistry). Finally, removal of the blocking group (H_2SO_4, H_2O, heat) gives the desired target molecule.

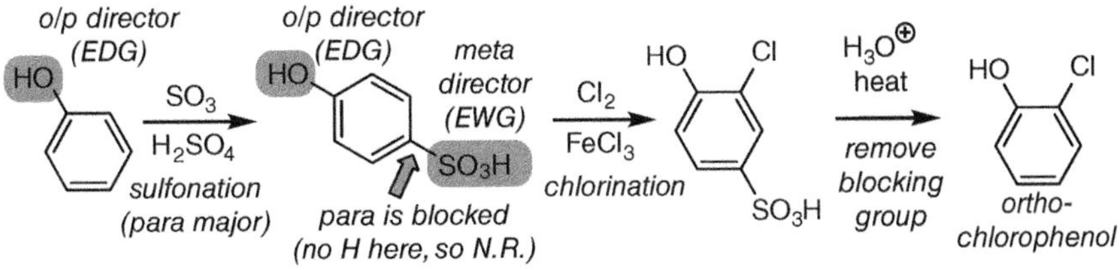

Use of a Sulfonic Acid as a Blocking Group

PRACTICE PROBLEM 5.1: ELECTROPHILIC AROMATIC SUBSTITUTION

5.1 Predict the major product(s) of the following reactions. Do not use abbreviations or condensed formulas in your drawings.

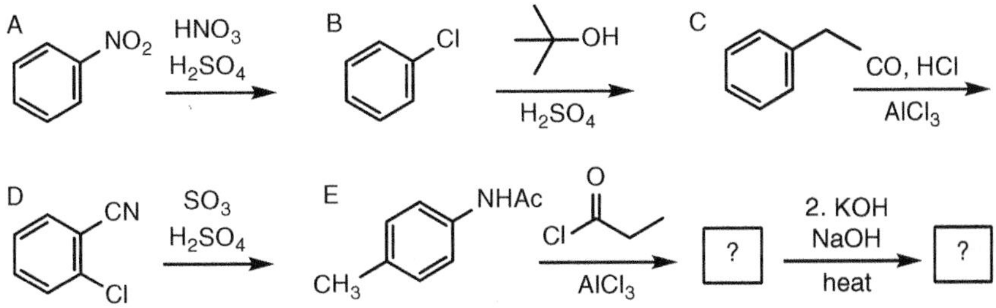

5.1.3 RETROSYNTHESIS OF AROMATIC TMs (ELECTROPHILIC AROMATIC SUBSTITUTION)

Many monosubstituted benzene derivatives can be prepared by electrophilic aromatic substitution. If the substituent is $-Cl, -Br, -NO_2, -SO_3H$, or $-COR$ (acyl), then a possible retrosynthesis involves a disconnection of the bond connecting that substituent to the benzene ring. The aromatic carbon was the nucleophile (simply add a hydrogen here) and the substituent was the electrophile (usually a positively charged atom that is generated by the appropriate set of reagents).

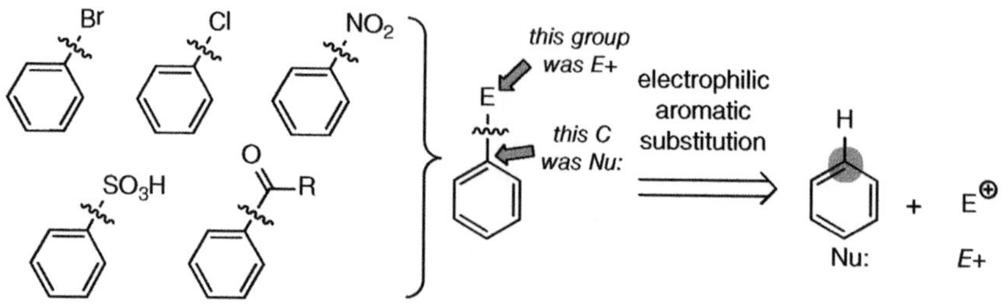

Retrosynthesis of Monofunctional Aromatic TMs

In the case of an alkylbenzene TM (the substituent is —R) in which the alkyl group is 2°, 3°, or allylic, this same disconnection can be applied, resulting in a carbocation electrophile. However, if the alkyl group is one that requires an unstable carbocation that can rearrange, the target molecule cannot be prepared via Friedel–Crafts alkylation. Instead, an FGI retrosynthesis is employed to give an aromatic ketone TM that can be prepared by Friedel–Crafts acylation, followed by reduction (see Section 3.8 for further discussion).

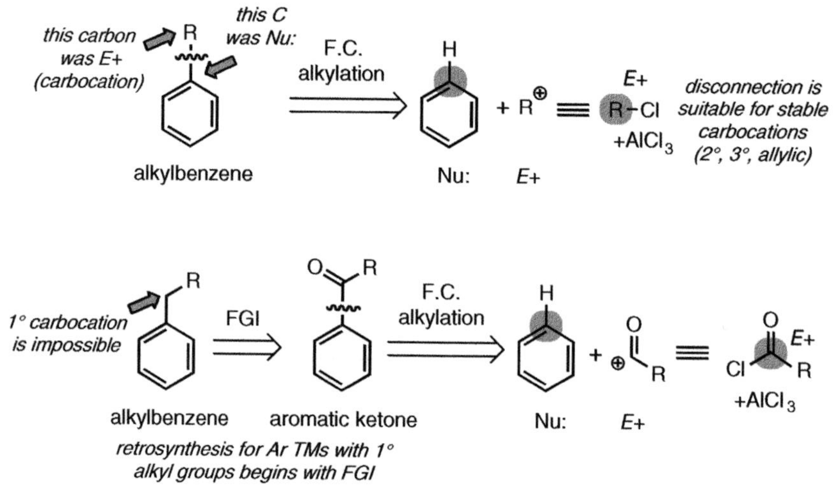

Retrosynthesis of Alkylbenzene TMs

In the case of aromatic target molecules containing more than one substituent, one must determine which substituent to add first (and second, etc.) in order to achieve the desired regiochemistry. For example, the synthesis of a *para*-disubstituted target molecule requires the addition of an ortho/para director first, in order to properly place the second group.

1,3-relationship indicates that the
meta-director must be added first
(disconnected last)

disubstituted aromatic TM

HNO$_3$
H$_2$SO$_4$
nitration

meta
director
(EWG)
NO$_2$

Br$_2$
FeBr$_3$
bromination

Br

NO$_2$

TM

a successful synthesis
adds substituents in the
proper order to give the
required regiochemistry

Br$_2$
FeBr$_3$
bromination

Br   o/p
director
(EDG)

HNO$_3$
H$_2$SO$_4$
nitration

Br

NO$_2$

the wrong sequence fails
to make the desired TM

Retrosynthesis of Difunctional Aromatic TMs

EXAMPLE: AROMATIC TM 1

Synthesize the following target molecule from benzene or toluene as the source of the aromatic rings using readily available reagents.

TM

Both groups on this aromatic target molecule can be added via electrophilic aromatic substitution: chlorination and Friedel–Crafts alkylation with the stable benzyl carbocation. However, both groups are electron-donating and ortho/para directors, so preparing the *meta*-substituted product becomes a challenge. Using the Friedel–Crafts *acylation* to install the benzyl group solves this problem, since the intermediate acyl group is an electron-withdrawing meta director. The required benzoyl chloride is prepared from benzoic acid, which can be synthesized via a Grignard reaction or oxidation of toluene.

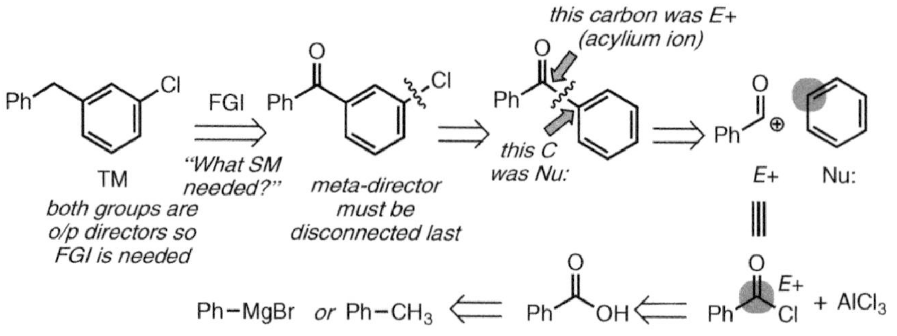

Retrosynthesis of Aromatic TM

Synthesis of the target molecule begins with the preparation of benzoic acid, either by halogenation of benzene, followed by a Grignard reaction with CO_2, or by Jones oxidation of toluene. Conversion to benzoic acid with thionyl chloride ($SOCl_2$) gives the acid chloride needed for a Friedel–Crafts acylation of benzene (addition of $AlCl_3$ generates the required acylium ion electrophile). The resulting aromatic ketone contains an EWG that will direct the incoming chlorine to the meta position (via electrophilic aromatic substitution with Cl_2, $FeCl_3$). Finally, reduction of the carbonyl gives the required benzyl group on the desired target molecule.

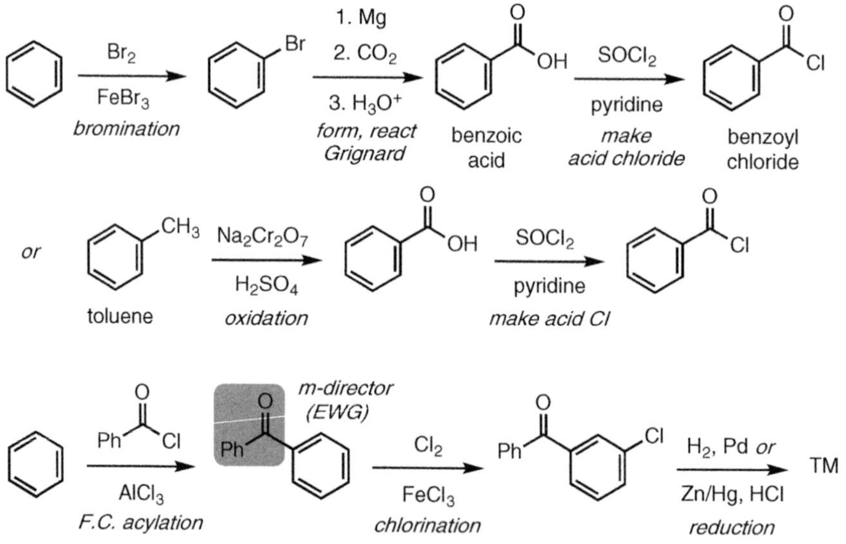

Synthesis of Aromatic TM

SYNTHESIS OF AROMATIC TMs VIA DIAZONIUM SALTS (ArN$_2^+$ + Nu: → ArNu)

A nitro group (—NO$_2$) on a benzene ring can be manipulated further, making possible a wide variety of aromatic synthetic targets. Reduction of the nitro group affords aniline derivatives, and the resulting amino (—NH$_2$) group can be converted to an excellent leaving group that can be replaced by nucleophiles.

5.2.1 PREPARATION OF DIAZONIUM SALTS (ArNH$_2$ → ArN$_2^+$)

When treated with nitrous acid (HONO, prepared *in situ* by reaction of NaNO$_2$ and acid), aniline derivatives (ArNH$_2$) are converted to diazonium salts (ArN$_2^+$). A diazonium salt is a highly reactive species since it has an excellent leaving group, nitrogen gas (N$_2$). Only aromatic diazonium salts are stable enough to be synthetically useful; those derived from alkyl amines (RNH$_2$) decompose quickly to a carbocation and N$_2$, ultimately producing a mixture of substitution and elimination products.

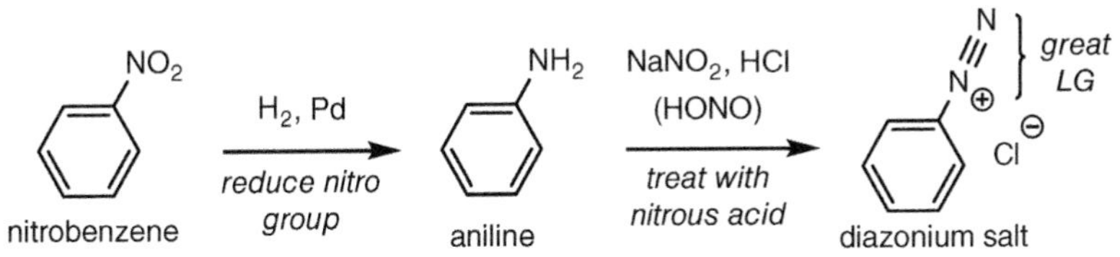

Preparation of Diazonium Salts

Introduction to Strategies for Organic Synthesis, Second Edition. Laurie S. Starkey.
© 2018 John Wiley & Sons, Inc. Published 2018 by John Wiley & Sons, Inc.

5.2.2 USE OF DIAZONIUM SALTS (ArN$_2^+$+Nu:→ArNu)

Reaction of a diazonium salt with a nucleophile results in a substitution reaction in which the N$_2$ leaving group is replaced (called an "ipso" substitution since the incoming nucleophile ultimately occupies the same position as the outgoing leaving group). When copper salts such as CuCN, CuBr, and CuCl are used, the reaction is known as the Sandmeyer reaction (producing ArCN, ArBr, and ArCl, respectively). Other commonly used nucleophiles include KI to make the iodide, HBF$_4$ to prepare the fluoride, H$_2$O to make a phenol (adds −OH), and H$_3$PO$_2$ or NaBH$_4$ to replace the leaving group with hydrogen.

Reaction of Diazonium Salts with Nucleophiles

5.2.3 RETROSYNTHESIS OF AROMATIC TMs (VIA DIAZONIUM SALTS)

Many benzene derivatives can be prepared via diazonium salts. If the target molecule has attached to the benzene any halogen (−X), a cyano (−CN), or a hydroxyl (−OH) functional group, then a possible retrosynthesis involves a functional group interconversion of that substituent to a nitro group (−NO$_2$), since the nitro group is needed to form a diazonium salt. Complex aromatic target molecules may require a combination of electrophilic aromatic substitutions and diazonium salt substitutions in order to install the given substituents with the proper regiochemistry.

Retrosynthesis of Aromatic TMs Involving Diazonium Salts

PRACTICE PROBLEM 5.2: SYNTHESIS OF AROMATIC TMs

5.2 Provide the reagents needed for each of the following transformations (more than one step may be required).

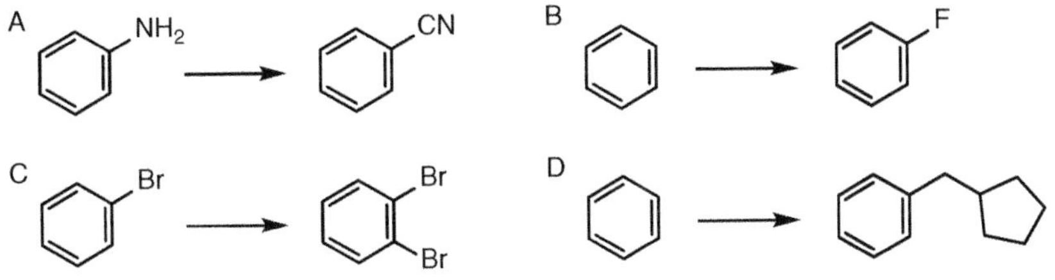

EXAMPLE: AROMATIC TM 2

Synthesize the following target molecule from benzene using readily available reagents.

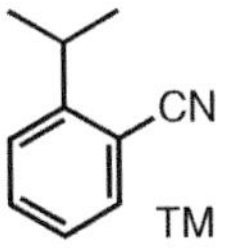

The isopropyl group on this aromatic target molecule can be added via electrophilic aromatic substitution, and the cyano group via a diazonium salt. Since the cyano group is an electron-withdrawing group and a meta director, the isopropyl group (an electron-donating, ortho/para director) must be added first. In order to place the two groups ortho, rather than para to each other, a blocking group must be used.

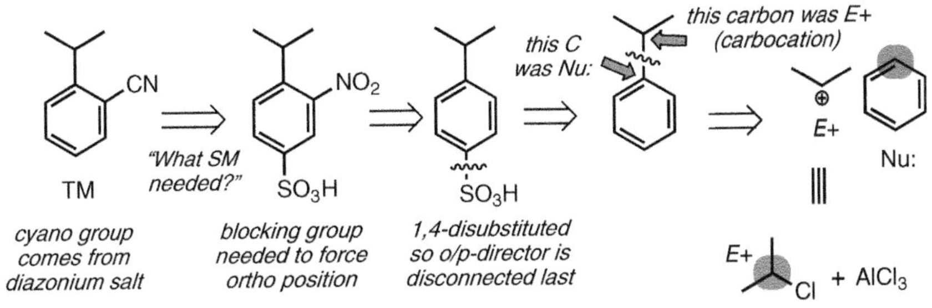

Retrosynthesis of Aromatic TM

Synthesis of the target molecule begins with a Friedel–Crafts alkylation of benzene with isopropyl chloride and $AlCl_3$. The resulting alkylbenzene contains an EDG that will direct the incoming sulfonyl group to the para position (via electrophilic aromatic substitution with fuming H_2SO_4). Nitration while the sulfonyl group is blocking the para position forces the nitro group ortho to the isopropyl; the blocking group is then removed by hydrolysis. Finally, the nitro group is replaced by a cyano group via the diazonium salt.

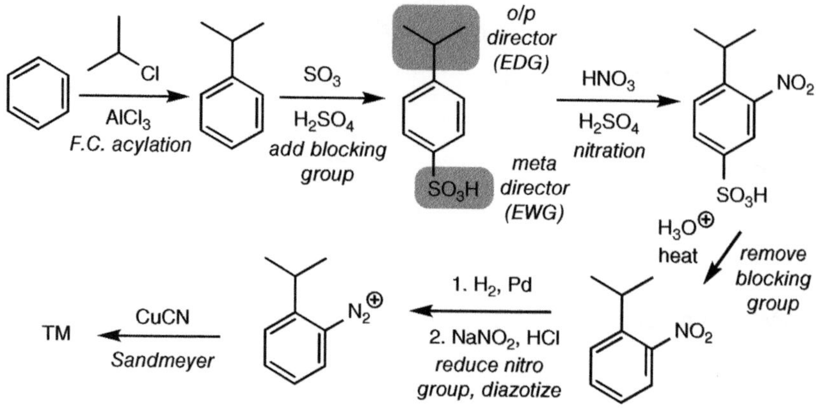

Synthesis of Aromatic TM

NUCLEOPHILIC AROMATIC SUBSTITUTION (ArX + Nu: → ArNu)

5.3.1 MECHANISM OF NUCLEOPHILIC AROMATIC SUBSTITUTION ($S_N Ar$)

Nucleophilic substitution reactions involving certain aryl halides (ArX) are possible. Since the leaving group is on an sp^2-hybridized carbon, the mechanism can be neither backside attack ($S_N 2$) nor via carbocation ($S_N 1$). Instead, a two-step, addition–elimination mechanism may be observed, called nucleophilic aromatic substitution ($S_N Ar$). The nucleophile attacks the carbon bearing the leaving group and breaks the pi bond. While the resulting carbanion intermediate is resonance-stabilized by the two remaining pi bonds, it also requires additional stabilization by electron-withdrawing group(s) suitably placed in the ortho and/or para positions. Rapid reformation of the pi bond by ejection of the leaving group (a mechanism analogous to collapse of a CTI) restores aromaticity and provides the substitution product.

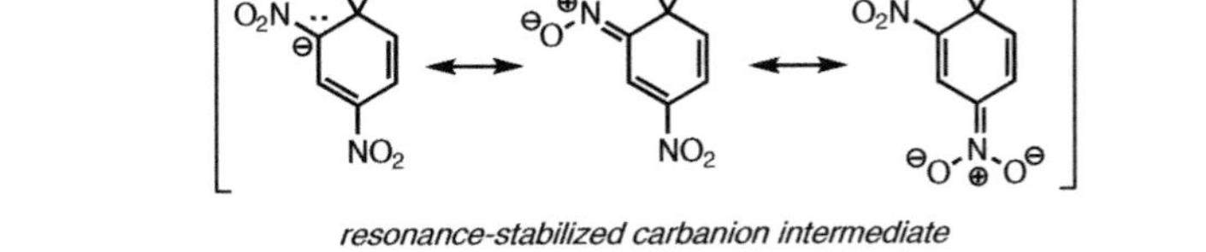

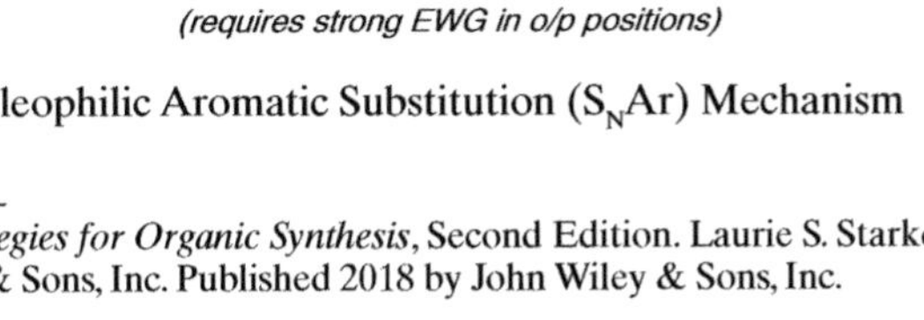

resonance-stabilized carbanion intermediate
(requires strong EWG in o/p positions)

Nucleophilic Aromatic Substitution ($S_N Ar$) Mechanism

Introduction to Strategies for Organic Synthesis, Second Edition. Laurie S. Starkey.
© 2018 John Wiley & Sons, Inc. Published 2018 by John Wiley & Sons, Inc.

Leaving groups for the S_NAr reaction include tosylates and halides, although fluoride affords the fastest substitution reactions. Typical nucleophiles include hydroxide to give phenol products (NaOH + ArX → ArOH), alkoxides to give alkyl aryl ether products (RONa + ArX → ArOR), phenoxides to give diaryl ether products (Ar'ONa + ArX → ArOAr'), and amines to give aryl amine products (RNH$_2$ + ArX → ArNHR or R$_2$NH + ArX → ArNR$_2$).

Nucleophilic Aromatic Substitution (S_NAr) Examples

5.3.2   RETROSYNTHESIS OF AROMATIC TMs (S_NAr)

Certain benzene derivatives are best prepared by nucleophilic aromatic substitution. If the aromatic TM has an −OH, −OR, −OAr, or −NR$_2$, then a possible retrosynthesis involves a disconnection of the bond between that substituent and the benzene ring. The aromatic carbon was the electrophile (simply add a leaving group here, such as a halide) and the substituent was the nucleophile (an amine or a negatively charged oxygen). The S_NAr reaction is facilitated by the presence of EWGs such as nitro groups, so a *para*-nitro group should be added, if there are no EWGs in the TM. The nitro group can be used for the substitution reaction and then either removed (replaced with −H) or converted to another substituent via diazonium salts.

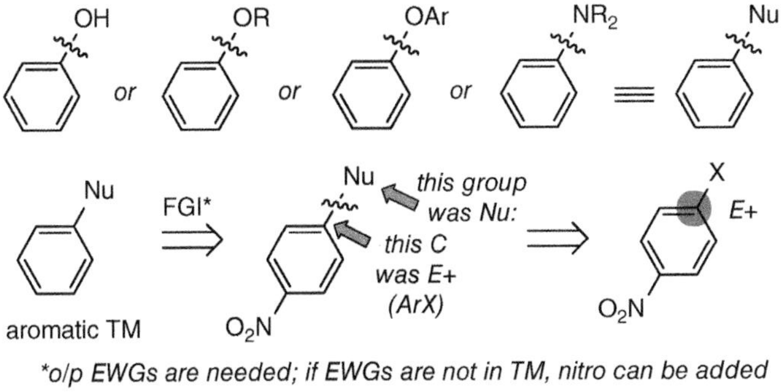

*o/p EWGs are needed; if EWGs are not in TM, nitro can be added

Retrosynthesis via Nucleophilic Aromatic Substitution

EXAMPLE: AROMATIC TM 3

Provide the reagents necessary to transform the given starting material into the desired product. More than one step may be required.

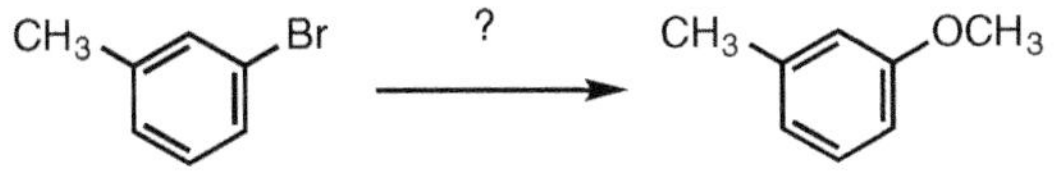

   It appears that a leaving group has been replaced by a nucleophile, but this S_NAr reaction will not occur in the absence of an electron-withdrawing group in the position(s) ortho and/or para to the leaving group. Without appropriately placed EWGs to stabilize the carbanion intermediate, the reaction will be prohibitively slow, and competition with an elimination–addition reaction (involving a benzyne intermediate) arises. After adding a nitro group para to the methoxy group, a disconnection can be made to give the nucleophilic methoxide and electrophilic aryl bromide. Disconnection of the nitro group successfully works backward to the required starting material and completes the retrosynthesis.

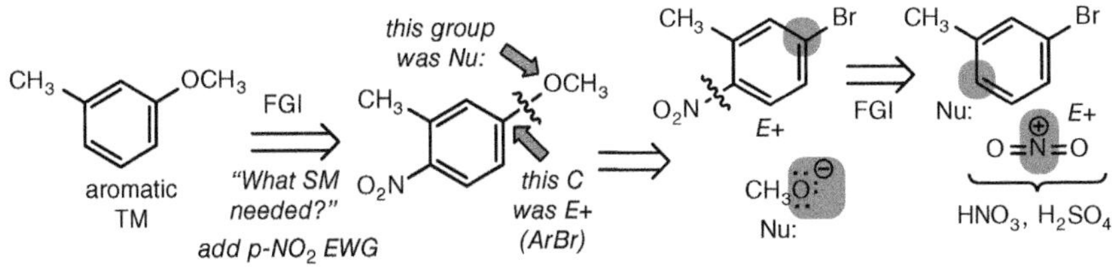

Retrosynthesis of Aromatic TM

Synthesis of the target molecule begins with nitration. Since both the methyl and bromo groups are ortho/para directors, the incoming nitro group will be appropriately placed. To avoid the mixture that is expected for mono-nitration, two nitro groups can be added instead. Treatment with methoxide gives the desired substitution product. To remove the nitro groups, they are converted to diazonium salts and replaced by hydrogen with H_3PO_2.

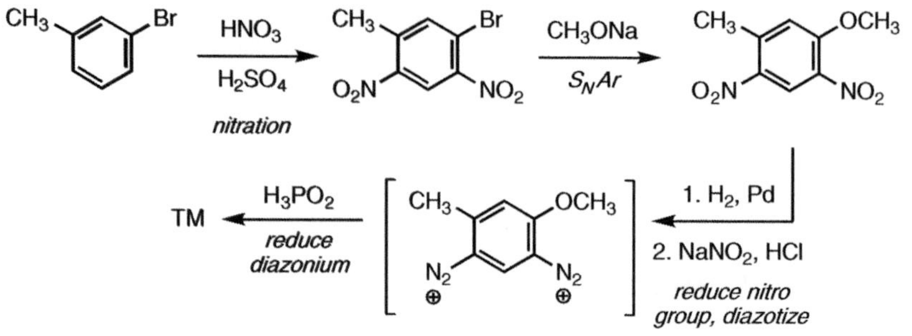

Synthesis of Aromatic TM

AROMATIC TMs

5-1. Synthesize the following target molecules from benzene using readily available reagents. All aromatic rings in the TM must come from benzene.

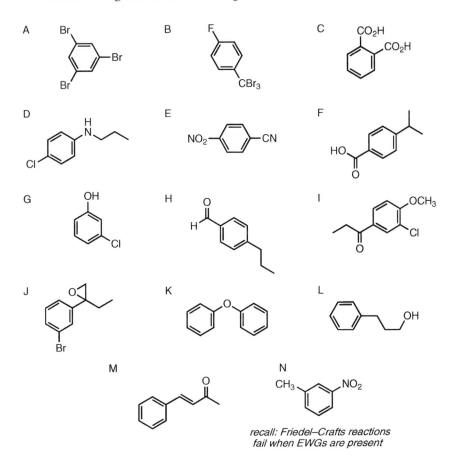

recall: Friedel–Crafts reactions
fail when EWGs are present

Introduction to Strategies for Organic Synthesis, Second Edition. Laurie S. Starkey.
© 2018 John Wiley & Sons, Inc. Published 2018 by John Wiley & Sons, Inc.

5-2. C-14 Synthesis Game Part A. Provide a synthesis for each of the following target molecules. Each synthesis must correctly incorporate the ^{14}C-labeled (*) carbon atom as shown, using the given starting materials as the only sources of carbon. Any commercially available reagents and protective groups may be used, and any previously synthesized molecule can be used as a starting point for another target molecule.

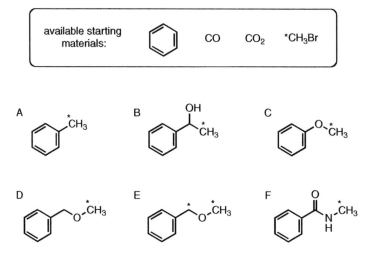

5-3. C-14 Synthesis Game Part B. Provide a synthesis for each of the following target molecules. Each synthesis must correctly incorporate the ^{14}C-labeled (*) carbon atom as shown, using the given starting materials as the only sources of carbon. Any commercially available reagents and protective groups may be used, and any previously synthesized molecule can be used as a starting point for another target molecule.

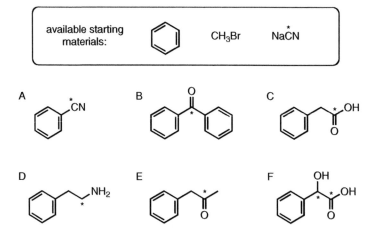

CHAPTER 6

SYNTHESIS OF COMPOUNDS CONTAINING RINGS

Nature chooses five- and six-atom rings far more than any other cyclic patterns. This is likely due to their negligible ring strain and the entropic demands for the two ends to find one another in a ring-forming reaction. Because these are Nature's favorite ring patterns, they are also common synthetic targets and synthetic chemists have therefore developed a wide variety of methods for their synthesis. Many of the Nu:/E+ reactions already presented in this book, such as the S_N2, aldol, and Claisen mechanisms, and the Williamson ether synthesis, can occur in an intramolecular fashion to provide cyclic products. When considering a target molecule containing a five- or six-membered ring, the usual, systematic functional-group analysis developed thus far can be applied, and a disconnection of a carbon–carbon bond within the ring is allowed. Chapter 6 of this book will explore cyclic TMs with particular patterns of functional groups, as well as reactions that have been developed to specifically give cyclic products.

Introduction to Strategies for Organic Synthesis, Second Edition. Laurie S. Starkey.
© 2018 John Wiley & Sons, Inc. Published 2018 by John Wiley & Sons, Inc.

SYNTHESIS OF CYCLOPROPANES

Cyclopropane rings are very reactive due to the large amount of strain contained in a three-membered ring. Such compounds are typically synthesized via carbene addition to an alkene, called a cyclopropanation reaction. Carbenes (R_2C:) are reactive intermediates that can exist in two electronic states. In a singlet carbene, the two unshared electrons have opposite spins and behave as a lone pair of electrons. When a singlet carbene is used (typical in synthetic applications), the stereochemistry of the alkene is retained during the single-step addition mechanism. The nonbonded electrons in a triplet carbene are in two different orbitals and have the same spin. This diradical species can also add to alkenes to give cyclopropane products, but the stepwise, radical mechanism results in a product with mixed stereochemistry.

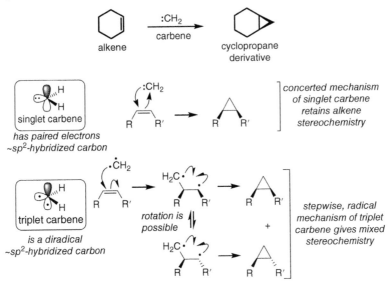

Reaction of Alkenes with Carbenes Gives Cyclopropane Derivatives

Introduction to Strategies for Organic Synthesis, Second Edition. Laurie S. Starkey.
© 2018 John Wiley & Sons, Inc. Published 2018 by John Wiley & Sons, Inc.

Singlet carbenes are formed by treatment of $CHCl_3$ with *t*-BuOK (results in loss of HCl to give $:CCl_2$), by loss of N_2 from diazo compounds such as diazomethane (CH_2N_2), or by reaction of CH_2I_2 with zinc–copper amalgam (cyclopropanation with the resulting $:CH_2$ synthetic equivalent, called a carbenoid, is the Simmons–Smith reaction).

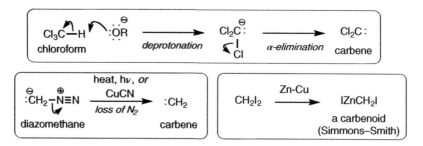

Various Methods Can Be Used to Generate Singlet Carbenes

6.1.1 RETROSYNTHESIS OF CYCLOPROPANE TMs

The retrosynthesis of a target molecule containing a cyclopropane ring begins with the identification of the one-carbon unit that came from a carbene (typically either a CH_2 or a CCl_2 group). Disconnecting both carbon–carbon bonds from that carbon breaks the ring into two pieces: one carbon started as a carbene and the other two carbons started as an alkene. The carbene can be prepared by one of the methods described above, and if the alkene starting material is not a simple, commercially available one, then the retrosynthesis can continue from there (i.e., the alkene can be synthesized from an alkyne or by Wittig, dehydration, E2 elimination, etc.).

Disconnection of a Cyclopropane

SYNTHESIS OF CYCLOBUTANES

Cyclobutane rings are also strained structures that are nearly impossible to form by ordinary ring-closing reactions. The most common method for constructing a four-membered ring is using a [2+2] cycloaddition reaction that involves the interaction of a two-electron π system with another two-electron π system (hence the "2+2" description). This pericyclic reaction of two alkenes is photochemically allowed. The stereochemistry of each alkene is retained during the concerted mechanism, but the regiochemistry may be less predictable, depending on the substituents involved.

Photochemical [2+2] Cycloaddition Forms Cyclobutane Derivatives

6.2.1 RETROSYNTHESIS OF CYCLOBUTANE TMs

The retrosynthesis of a target molecule containing a cyclobutane ring involves a disconnection that breaks the ring into two, two-carbon pieces (both will start as alkenes). Since there are two such possible disconnections, both the regiochemistry and the stereochemistry of the reaction should be taken into consideration. If one of the alkene starting materials is symmetrically substituted, then there can be no question about the regiochemistry of the cycloaddition product, so that would be a reasonable goal in the retrosynthesis of a cyclobutane TM.

Introduction to Strategies for Organic Synthesis, Second Edition. Laurie S. Starkey.
© 2018 John Wiley & Sons, Inc. Published 2018 by John Wiley & Sons, Inc.

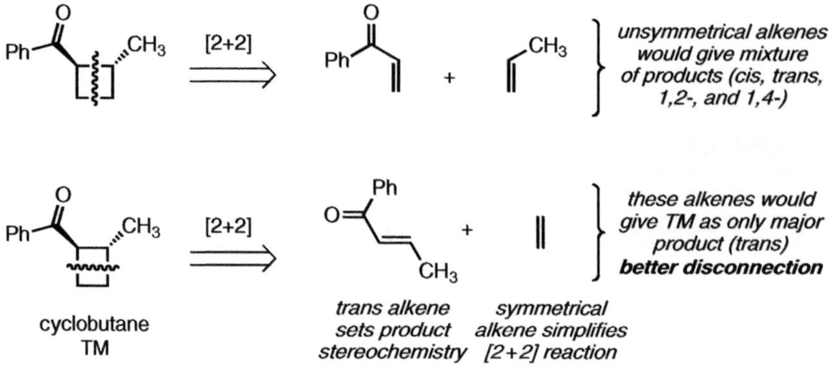

Disconnection of a Cyclobutane

SYNTHESIS OF FIVE-MEMBERED RINGS (RADICAL CYCLIZATION REACTIONS)

Treatment of an alkenyl halide with a tributyltin radical (Bu$_3$Sn, generated by reaction of Bu$_3$SnH with the radical initiator AIBN) causes the abstraction of the halide atom. An alkenyl phenyl selenide can also serve as a precursor to a radical intermediate. The resulting radical can add intramolecularly to an appropriately placed alkene, forming a ring and creating a new radical. Depending on the reaction conditions, this radical can abstract either a hydrogen or a halogen to give a cyclic alkane or alkyl halide product, respectively.

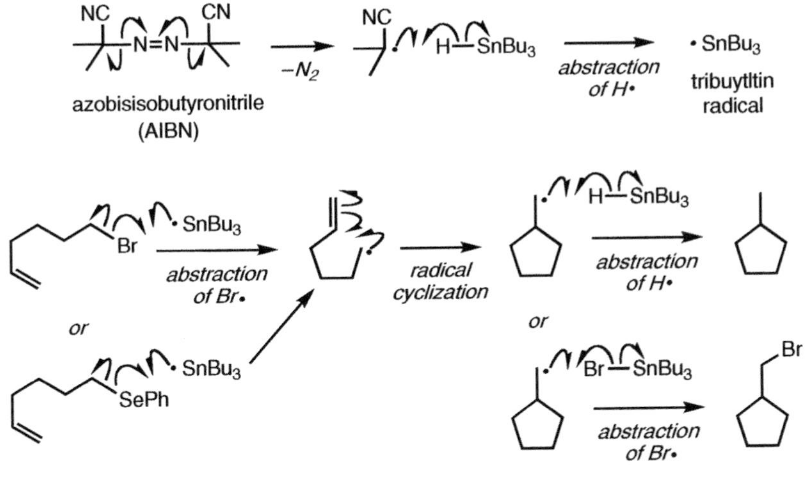

Alkenyl Radical Formation and Cyclization

Introduction to Strategies for Organic Synthesis, Second Edition. Laurie S. Starkey.
© 2018 John Wiley & Sons, Inc. Published 2018 by John Wiley & Sons, Inc.

6.3.1 BALDWIN'S RULES

The size and direction of ring formation are generally described by Baldwin's rules. An exhaustive explanation of Baldwin's rules is beyond the scope of this book, but a brief summary will be presented. A more detailed discussion can be found in a number of advanced organic chemistry books.* Addition of the radical shown above to the trigonal-planar, sp^2 alkene to form a five-membered cyclic intermediate with the radical located outside of the ring (called a "5-*exo*-trig" addition) is the favored mechanism. Less favorable is the formation of a six-membered ring that has the radical on one of the carbons within the ring, called a "6-*endo*-trig" addition. Therefore, radical cyclization mechanisms are suitable for synthesizing methylcyclopentane derivatives.

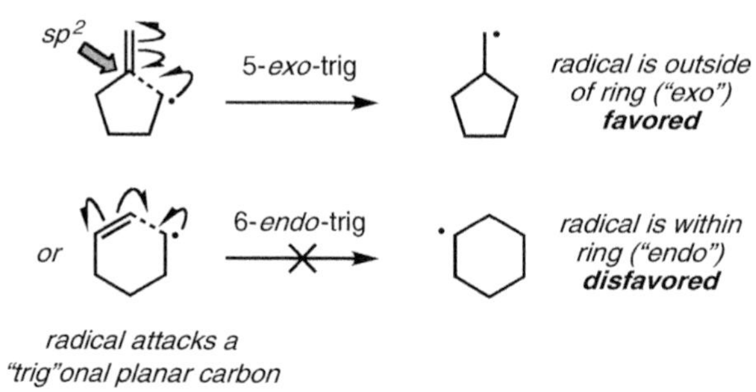

Baldwin's Rules: 5-*exo*-trig is Favored over 6-*endo*-trig

6.3.2 RETROSYNTHESIS OF METHYLCYCLOPENTANE TMs

The retrosynthesis of a target molecule containing a methylcyclopentane ring begins with removal of a hydrogen atom to give a radical. This synthon leads to a disconnection of a carbon–carbon bond to open up the ring at the methyl-substituted carbon. Because the bond will be formed via a radical mechanism, the bond is disconnected *homolytically*. Rather than working backward to a nucleophile and electrophile, the desired starting materials in this case are an alkene and a radical source. This radical will be generated by removal of a halogen, so a halide is placed at this end of the carbon chain. A quick check of the proposed starting material will confirm that formation of a radical, followed by a 5-*exo*-trig cyclization, would give

---

* Michael B. Smith and Jerry March, *March's Advanced Organic Chemistry: Reactions, Mechanisms, and Structure*, 5th ed. (Wiley-Interscience, 2001); Francis A. Carey and Richard J. Sundberg, *Advanced Organic Chemistry, Part A: Structure and Mechanisms*, 5th ed. (Springer, 2007).

the desired TM. The disconnection of a halomethylcyclopentane TM is exactly the same, but with removal of the halogen atom from the TM to give a radical synthon.

Disconnection of a Methylcyclopentane TM

While this cyclization reaction offers a good method for the formation of five-membered rings, the required starting material is nearly as complex as the desired target molecule and is unlikely to be commercially available. Therefore, the use of this synthetic strategy is likely to require the synthesis of the alkenyl halide starting material. A retrosynthetic analysis of the alkenyl halide can begin with either functional group.

EXAMPLE: METHYLCYCLOPENTANE TM

Provide the reagents necessary to transform the given starting material into the desired product.

Recognition of the target molecule as a methylcyclopentane provides a clue as to how the five-membered ring could be formed: radical cyclization. This disconnection leads to a terminal alkene that can be synthesized via a Wittig reaction. The required aldehyde bears some resemblance to the given three-carbon starting material, and a disconnection at the newly formed C—C bond is a logical one, as it involves a carbon beta to a carbonyl (an electrophilic carbon as an enone). The required nucleophile must be one that prefers to undergo conjugate addition to an α,β-unsaturated aldehyde, so a cuprate is considered. Because a cuprate bearing a bromide would not be possible (recall that cuprates react with alkyl halides), the halide is removed retrosynthetically before the cuprate disconnection can be made. One possible starting material for the bromide is an alkene, and such a functional group is well suited as a cuprate ligand, so this is a reasonable FGI to employ. Finally, a disconnection between the alpha and beta carbons leads to a cuprate nucleophile and the required starting material (acrolein).

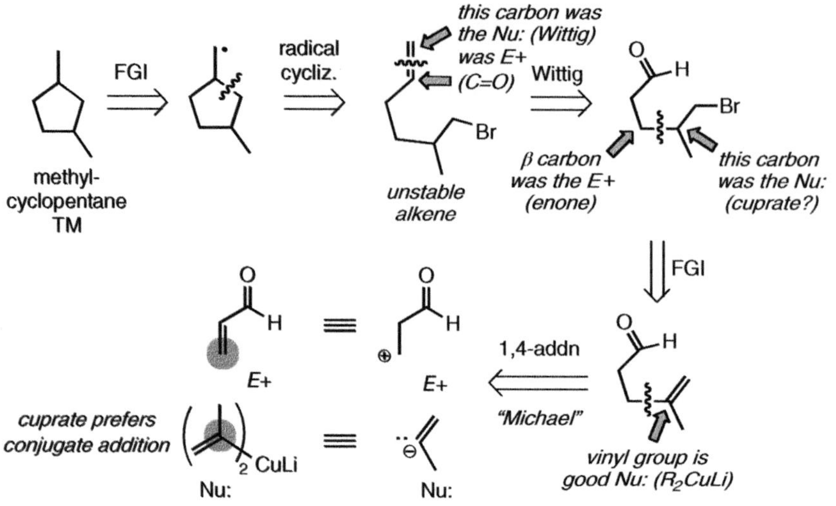

Retrosynthesis of Methylcyclopentane TM

The first step in the synthesis is conjugate addition of the cuprate shown to acrolein. The resulting alkene needs to be converted to a bromide, with *anti*-Markovnikov regiochemistry. Since the peroxide-promoted radical addition of HBr has strongly acidic conditions that may be unfavorable to the aldehyde, a more reliable strategy is hydroboration–oxidation, followed by treatment with PBr$_3$ to achieve the necessary transformation with the desired regiochemistry. Treatment of the bromoaldehyde with the appropriate Wittig reagent gives the bromoalkene needed for the radical cyclization reaction. This reaction is initiated with Bu$_3$SnH and AIBN to give the desired 1,3-dimethylcyclohexane TM.

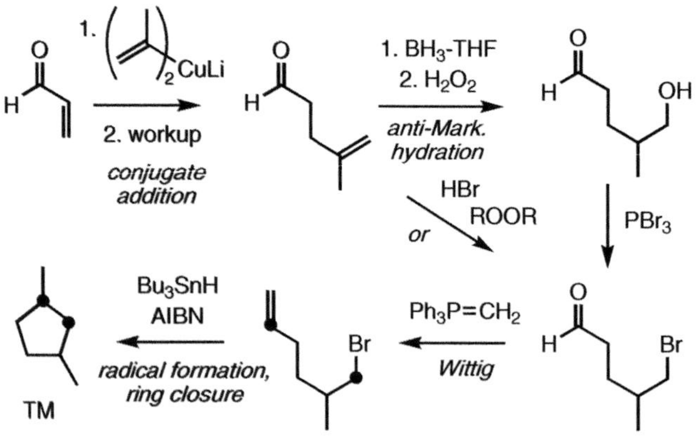

Synthesis of Methylcyclopentane TM

SYNTHESIS OF SIX-MEMBERED RINGS (DIELS–ALDER REACTION)

The Diels–Alder reaction (for which Otto Diels and Kurt Alder were awarded together the Nobel Prize in 1950) involves the reaction of a conjugated diene with another group containing a π bond (referred to as a "dienophile" since it "loves" reacting with dienes). In the presence of heat, a diene and a dienophile will combine to give a cyclohexene product. This concerted mechanism is an example of a pericyclic reaction called a [4+2] cycloaddition since it involves the interaction of a four-electron π system (the diene) with a two-electron π system (the dienophile). While many examples of the Diels–Alder reaction can be easily described as a reaction between a nucleophile and electrophile (the approach to be taken here), the mechanism and the regio- and stereochemistry of the product is best described by frontier molecular orbital theory in which the HOMO of the diene and the LUMO of the dienophile are matched.

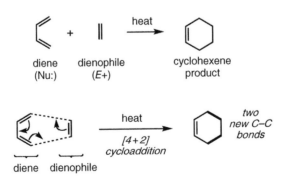

Diels–Alder Reaction

Introduction to Strategies for Organic Synthesis, Second Edition. Laurie S. Starkey.
© 2018 John Wiley & Sons, Inc. Published 2018 by John Wiley & Sons, Inc.

It should be noted that the aza Diels–Alder and oxo Diels–Alder reactions involve nitrogen- and oxygen-containing components, respectively, to give a variety of six-membered heterocyclic compounds. This book will only address the traditional, all-carbon Diels–Alder reaction.

6.4.1 THE DIENOPHILE (*E*+)

In a typical Diels–Alder reaction, the dienophile behaves as the electrophilic species, so the reaction is facilitated by the presence of one or more electron-withdrawing groups (EWG) such as a carbonyl (e.g., an ester $-CO_2R$), a nitro ($-NO_2$), or a cyano ($-C\equiv N$) group. Such groups make the alkene more electron-poor and, therefore, a better electrophile (by lowering the energy of the LUMO). The two pi electrons in the dienophile can come from an alkene or alkyne. In the case of an alkyne, the product formed would be a 1,4-cyclohexadiene derivative.

alkenes or alkynes with one or more electron-withdrawing groups attached make good dienophiles

EWGs

EWG makes pi bond electron-deficient (good *E*+)

diene   alkyne dienophile   heat   cyclohexadiene product

Electron-Deficient Alkenes and Alkynes Make Good Dienophiles

Stereochemistry of Dienophile is Retained

Because the mechanism is a concerted one in which both new carbon–carbon bonds are formed at the same time, the relative stereochemistry of the groups on the dienophile is retained. In other words, two groups that are cis to each other on the dienophile will remain cis to each other in the cyclohexene product (both wedges or both dashes), and trans groups will remain trans. If the product formed is chiral (no internal plane of symmetry), then both enantiomers will be formed, resulting in a racemic mixture. It is important to keep this in mind when drawing the stereochemistry of a Diels–Alder product. You can choose to draw any given group as either a dash or a wedge, and that choice will dictate the *relative* stereochemistry of the remainder of the groups.

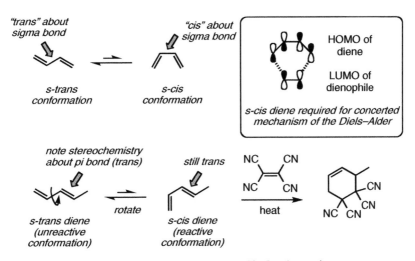

trans dienophile

groups are still trans in product (chiral product is formed as a racemate)

cis dienophile

groups are still cis in product (meso product is achiral, so no enantiomer)

Dienophile Stereochemistry is Retained

6.4.2 THE DIENE (Nu:)

The Diels–Alder reaction requires the diene to have an s-cis conformation, to allow the ends of the diene and dienophile to overlap in a concerted manner. Any substituents on the diene that cause steric strain in the required s-cis conformation will slow down, or even prohibit, the Diels–Alder reaction.

"trans" about sigma bond

s-trans conformation

"cis" about sigma bond

s-cis conformation

HOMO of diene

LUMO of dienophile

s-cis diene required for concerted mechanism of the Diels–Alder

note stereochemistry about pi bond (trans)

s-trans diene (unreactive conformation)

rotate

still trans

s-cis diene (reactive conformation)

heat

Diels–Alder Diene Must Have s-Cis Conformation

Stereochemistry of Bicyclic Diels–Alder Products

Since cyclic dienes are locked in the s-cis conformation, they usually undergo Diels–Alder reactions more readily and are excellent starting materials. Reactions with cyclic dienes give bicyclic products in which "endo" stereochemistry is typically preferred. A Diels–Alder reaction with cyclopentadiene gives a product with a norbornene-type skeleton. If this product is viewed from the side and drawn with the one-carbon bridge pointing "up" (pointing toward the top of the page), then the endo product is the one with the dienophile substituent, typically an EWG, in the "down" position, pointing away from the bridge. If the product is drawn with the cyclohexene ring flat on the page, then this trans relationship is shown using dashes and wedges (e.g., bridge is a wedge and EWG is a dash).

cyclopentadiene

(racemic)
product with endo
EWG is major

(racemic)
product with exo EWG
is typically not favored

1,3-cyclo-   maleic
hexadiene   anhydride

(meso)
endo product

Cyclic Dienes Predominantly Give Products with Endo Stereochemistry

Consideration of Acyclic Diene Stereochemistry

When determining the stereochemistry of substituents on the terminal carbons of an acyclic diene (C_1 and C_4 substituents), it is helpful to be familiar with the endo-rule pattern, even though a bicyclic product is not involved. Just as the one-carbon bridge becomes wedged as the tetrahedral centers are formed to give one enantiomer of the product, the "inside" groups on the diene will both rotate up to a wedged position (or rotate down to a dashed position for the other enantiomer). This is analogous to the "disrotatory" motion observed in electrocyclic mechanisms, with one pi bond of the diene rotating counterclockwise and the other clockwise. For a cycloaddition reaction, this orientation is called "suprafacial" because the formation of each bond takes place on the same face of each of the pi systems; in other words, the two lower p-orbital lobes in the HOMO interact with the two upper p-orbital lobes in the LUMO, as shown in the orbital diagram above. This stereochemistry (supra, supra) is required to conserve orbital symmetry in the Diels–Alder mechanism (true for all six-electron, thermal, pericyclic reactions). Knowing there is a typically a trans relationship between the dienophile's EWG and the bridge (because the endo orientation of the transition state is favored), then if the "inside" groups are drawn as wedges in the product, then the EWG will be drawn as a dash.

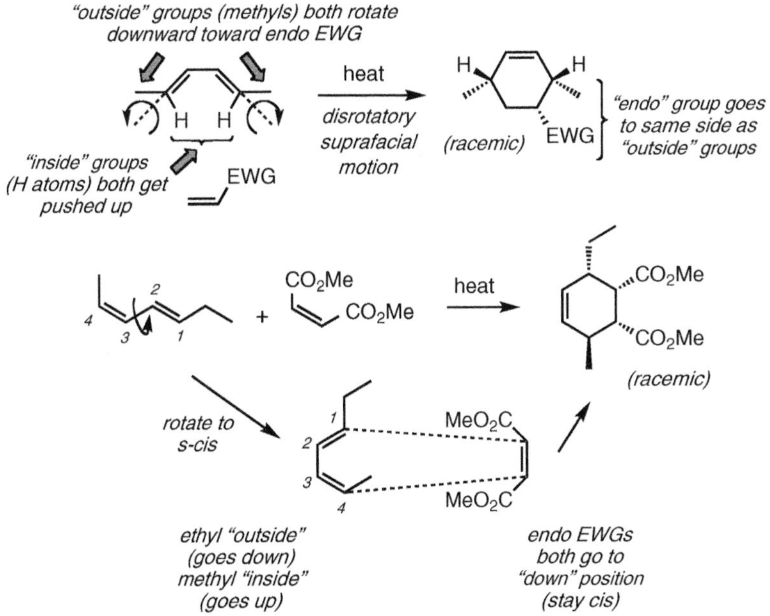

Dienes Undergo Suprafacial Bonding in Diels–Alder

6.4.3 REGIOCHEMISTRY OF THE DIELS–ALDER REACTION

When both the dienophile and the diene are not symmetrically substituted, one must determine the regiochemical alignment of the two components since two different products will be possible. Once again, the observed regioselectivity is most thoroughly explained by MO theory, but a simpler resonance model can provide insight into the expected results. When the dienophile and the diene each contain one substituent, the major product is either the 1,2-disubstituted or the 1,4-disubstituted cyclohexene derivative, rather than the 1,3-disubstituted product. This regioselectivity is sometimes described as favoring the "ortho-like" or "para-like" products because they look similar to benzene derivatives.

1,2-Disubstituted Product is Preferred over 1,3-

If the dienophile contains one EWG substituent (e.g., $-CO_2Et$) and the diene contains an electron-donating substituent on one of its terminal carbons (e.g., $-OCH_3$ on C_1), then the two possible alignments result in a product that is either 1,2-disubstituted or 1,3-disubstituted. The 1,2-product is formed as the major product (note that the stereochemistry of the product has the "outside" methoxy group on the same side as the electron-withdrawing cyano group — both wedges or both dashes).

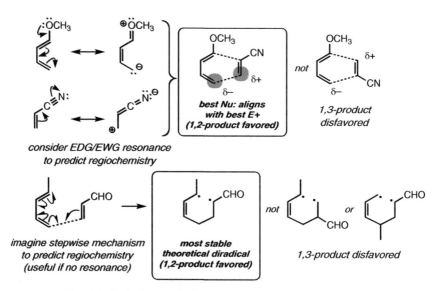

1,2-Product Preferred over 1,3-

From the resonance contributors of the examples given, one can see that the preferred orientation involves a favorable interaction between a large $\delta+$ carbon on the dienophile ($E+$) and a large $\delta-$ carbon on the diene (Nu:). While the best predictor of regioselectivity involves analysis of the participating HOMO and LUMO orbitals, another simple approach is to consider forming one bond at a time and examining the stability of the resulting theoretical diradical intermediate.

Two Methods for Predicting Regioselectivity for 1,2-Product

1,4-Disubstituted Product is Preferred over 1,3-

When the substituent on the diene is instead on one of the internal carbons of the diene (C_2 or C_3), then the two possible Diels–Alder products are the 1,4-disubstituted or 1,3-disubstituted cyclohexenes. Once again, the 1,3-product is disfavored, affording the 1,4-product as the major product. As demonstrated above, resonance and/or diradical methods can be used to explain the formation of the observed major product.

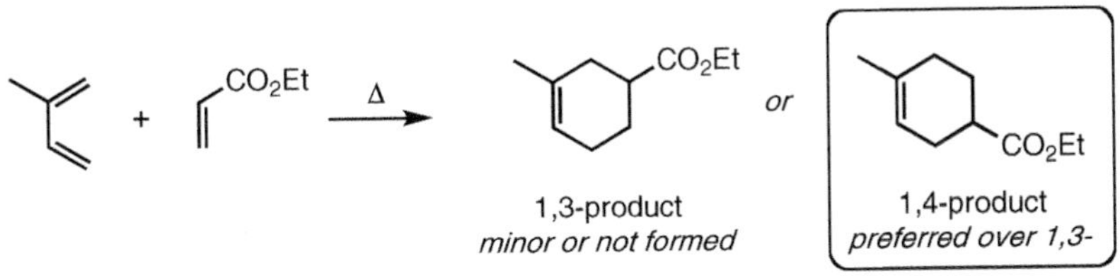

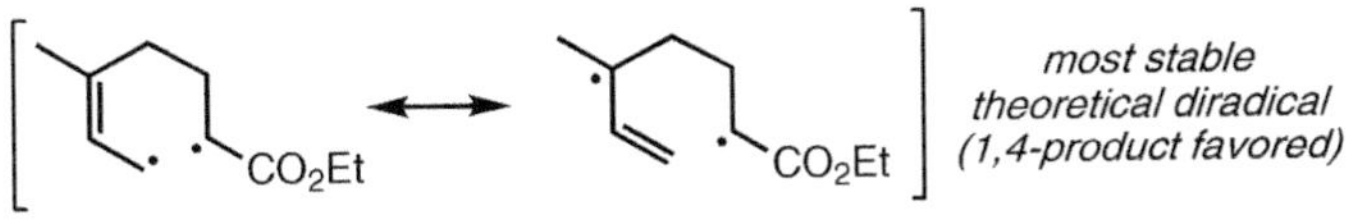

1,3-product
minor or not formed

1,4-product
preferred over 1,3-

most stable
theoretical diradical
(1,4-product favored)

1,4-Product Preferred over 1,3-

PRACTICE PROBLEM 6.4: DIELS–ALDER REACTION

6.4 For each reaction, predict the major product(s) or provide the missing reactants.

A

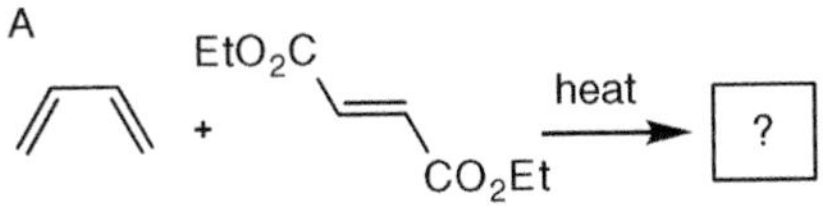

B

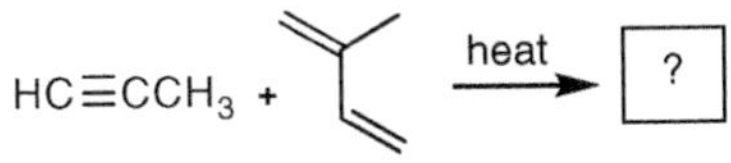

C

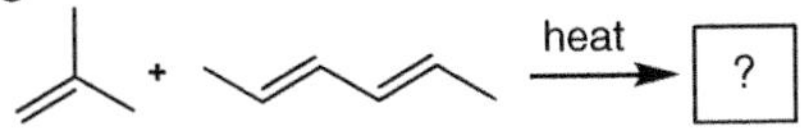

D

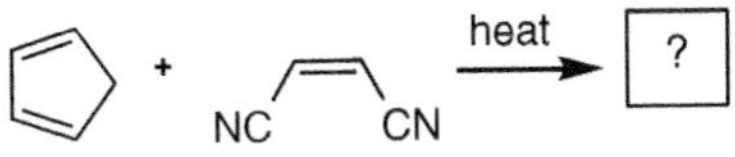

E

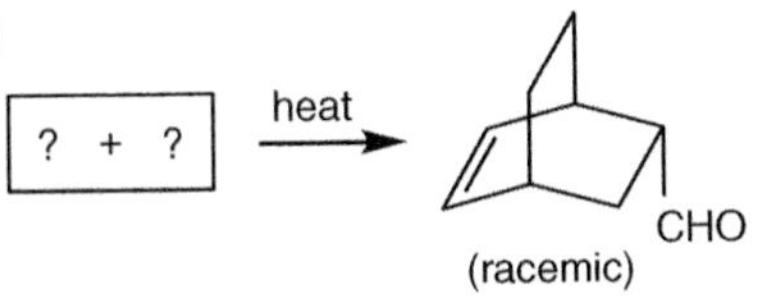

F

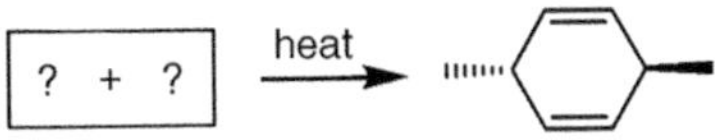

6.4.4 RETROSYNTHESIS OF CYCLOHEXENES (DIELS–ALDER)

The Diels–Alder-based retrosynthesis of a cyclohexene target molecule begins with the identification within the six-membered ring of the four carbons contributed by the diene and the two carbons from the dienophile. The double bond in the cyclohexene TM marks the middle two carbons (C_2 and C_3) of the four-carbon diene. Two disconnections are made in the ring to separate the four-carbon and two-carbon units. It is helpful to show the mechanism for the retro Diels–Alder, since that will not only achieve the required disconnection but also properly locate all of the necessary pi bonds in the starting diene and dienophile.

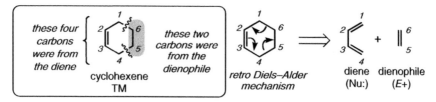

Diels–Alder Retrosynthesis of Cyclohexenes

For bicyclic target molecules, identify the cyclohexene ring and note the size of the carbon bridge to properly guide the retrosynthesis. The stereochemistry of the TM determines where the substituents must be positioned on the starting materials.

Note Stereochemistry of Cyclohexene Substituents

Retrosynthesis of 1,6-Dicarbonyl TMs

The Diels–Alder reaction offers a method to synthesize not only cyclohexene derivatives but also target molecules that can be derived from cyclohexenes. One such TM is a 1,6-dicarbonyl compound, since ozonolysis of a cyclohexene would create this functional group pattern. Any attempt at *disconnecting* a 1,6-dicarbonyl TM results in illogical synthons, but *connecting* the two carbonyl carbons and converting them retrosynthetically into a C=C double bond result in a simple, recognizable starting material (a cyclohexene) that could be prepared in a number of ways, including a Diels–Alder reaction.

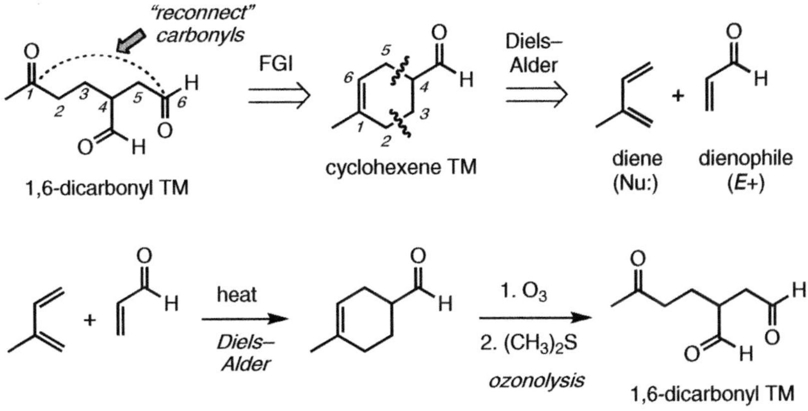

Retrosynthesis of 1,6-Dicarbonyls

6.4.5 RETROSYNTHESIS OF 1,4-CYCLOHEXADIENES

1,4-Cyclohexadiene derivatives can be prepared either by a Diels–Alder reaction with an alkyne or by a Birch reduction of a benzene derivative.

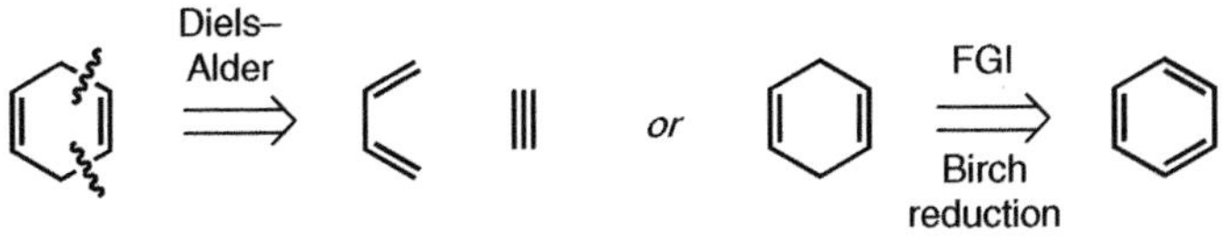

Two Retrosynthetic Strategies for 1,4-Cyclohexadiene TM

The positioning of the substituents on the target molecule will dictate which method is better suited for a particular synthesis. If an electron-withdrawing group is present and is located in an allylic position, then the TM could have come from a Birch reduction since that is the expected regiochemistry for such a reaction. However, if the EWG is on one of the C=C double bonds, then the TM could not have been prepared by a Birch reduction and a Diels–Alder disconnection is required. There are two possible retro Diels–Alders on a cyclohexadiene; the one that places the EWG on the dienophile is the better disconnection (leading to the more favorable Diels–Alder).

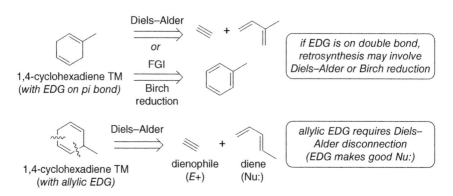

Retrosynthesis of 1,4-Cyclohexadienes with Electron-Withdrawing Group

If a 1,4-cyclohexadiene target molecule contains an electron-donating group, different strategies would be expected. If the electron-donating group is located on one of the C=C double bonds, the TM could have come from a Birch reduction or a Diels–Alder reaction. However, if the EDG is in an allylic position, then a Diels–Alder disconnection that locates the EDG on the diene would be an appropriate retrosynthesis.

Retrosynthesis of 1,4-Cyclohexadienes with Electron-Donating Group

CYCLIC TMs

6-1. Propose a possible disconnection/retrosynthesis for each of the following cyclic target molecules. Include a ring in your disconnection and consider the pattern of functional groups when determining the best bond(s) to break.

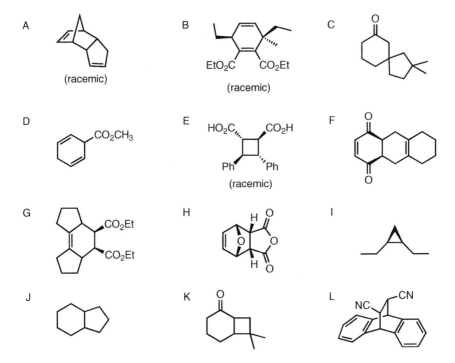

A

(racemic)

B

EtO$_2$C   CO$_2$Et

(racemic)

C

D

CO$_2$CH$_3$

E

HO$_2$C   CO$_2$H

Ph   Ph

(racemic)

F

G

CO$_2$Et

CO$_2$Et

H

I

J

K

L

NC   CN

Introduction to Strategies for Organic Synthesis, Second Edition. Laurie S. Starkey.
© 2018 John Wiley & Sons, Inc. Published 2018 by John Wiley & Sons, Inc.

6-2. Provide a synthesis for each of the following target molecules. Each synthesis must form at least one new carbon–carbon bond. It may help to first do a retrosynthesis of the TM.

A

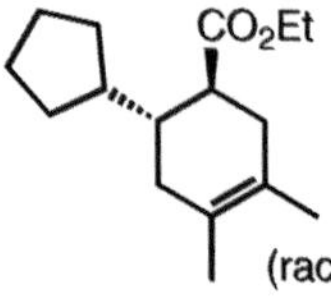

(racemic)

from starting materials with
no more than six carbons

B

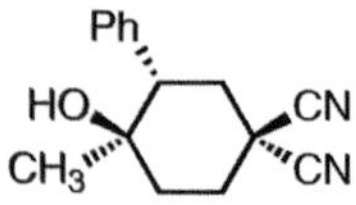

(racemic)

C

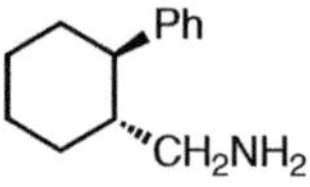

(racemic)

D

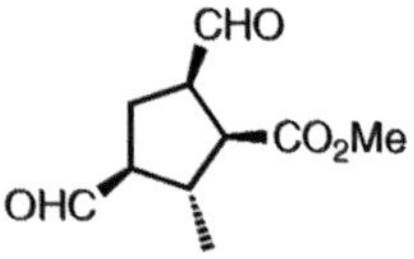

(racemic)

E

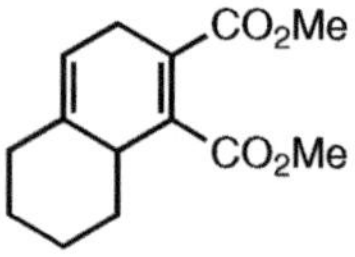

from starting materials with
no more than six carbons

F

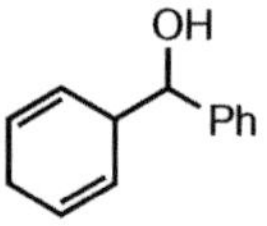

from starting materials with
no more than six carbons

G

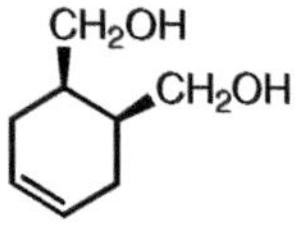

H

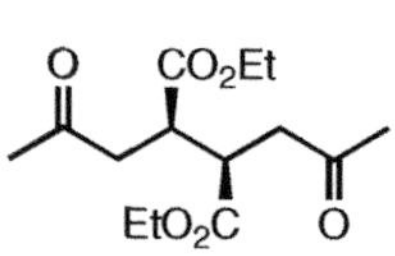

(racemic)

I

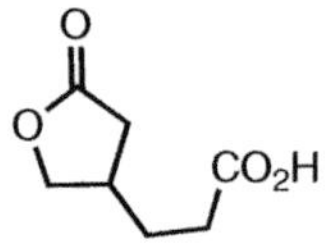

6-3. Predict the product of the following Diels–Alder reaction using Danishefsky's Diene. Select from the possible products shown and provide the correct stereochemistry for the structure. Explain how you determined both the regiochemistry and the stereochemistry.

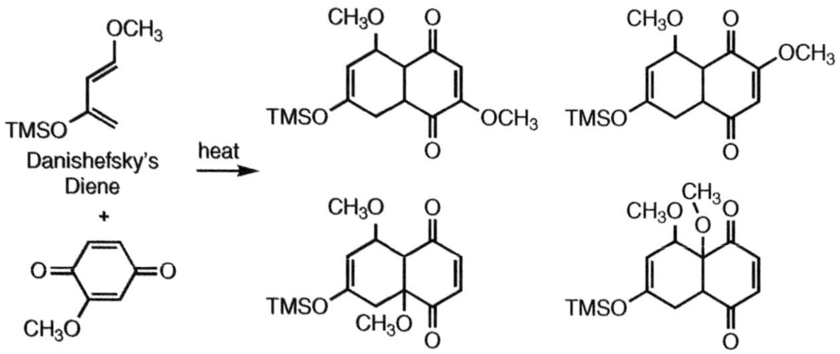

6-4. Each of the following synthetic sequences involves a Diels–Alder reaction followed by a retro Diels–Alder (cycloelimination) reaction. Provide structures for **A** and **B**, and provide a complete mechanism for each reaction sequence.

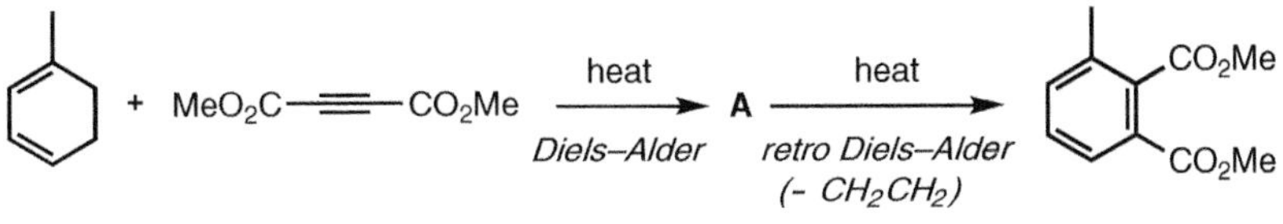

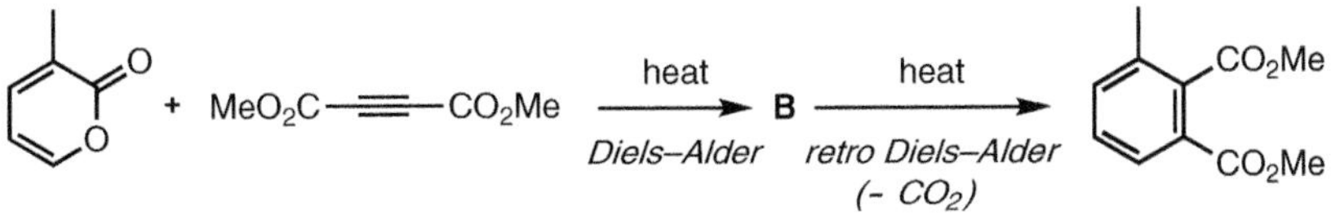

6-5. Each of the following target molecules can be prepared via an intramolecular Diels–Alder reaction. Determine the required disconnection/retrosynthesis for each, paying close attention to the stereochemistry of the TM.

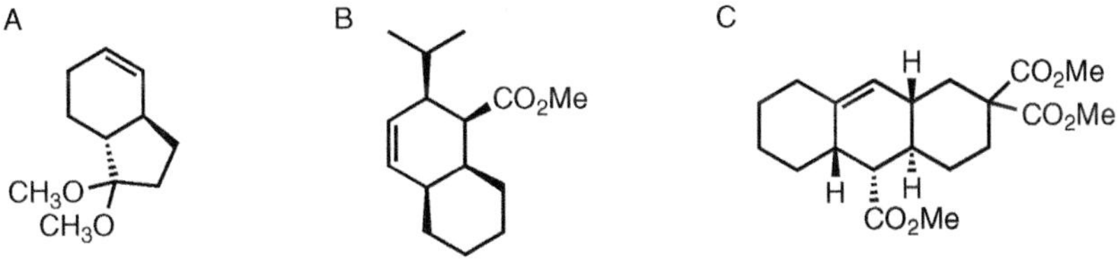

A

B

C

6-6. Provide a complete mechanism for the following transformation.

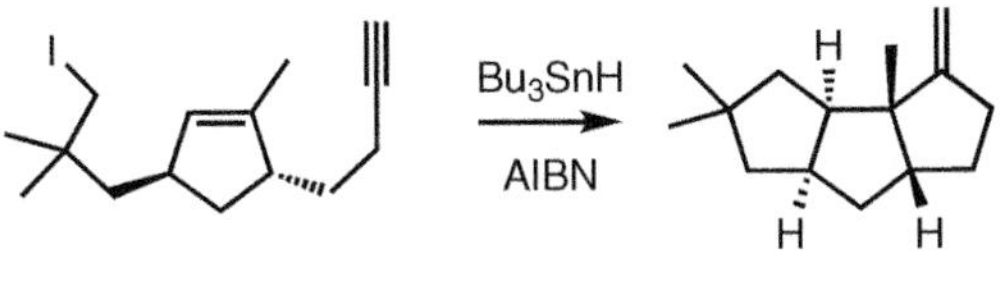

6-7. Provide a complete mechanism for the following transformation.

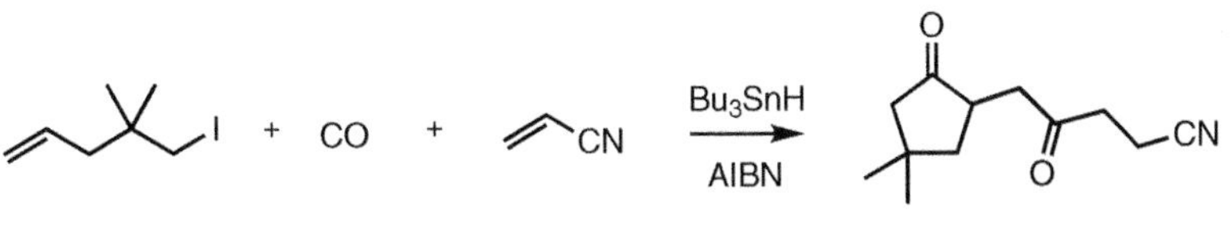

6-8. Provide a complete mechanism for the following transformation and explain the regiochemistry of the hetero Diels–Alder reaction involved.

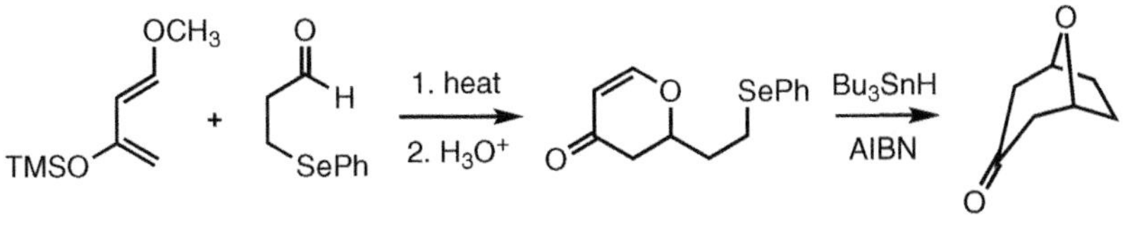

6-9. Determine the starting material (**X**) that is required to form structure **Y** and provide a complete mechanism for its transformation to product **Z**.

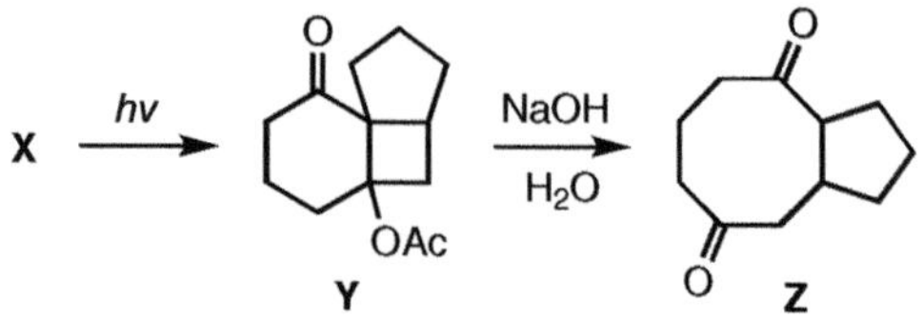

6-10. Provide a complete mechanism for the following transformation.

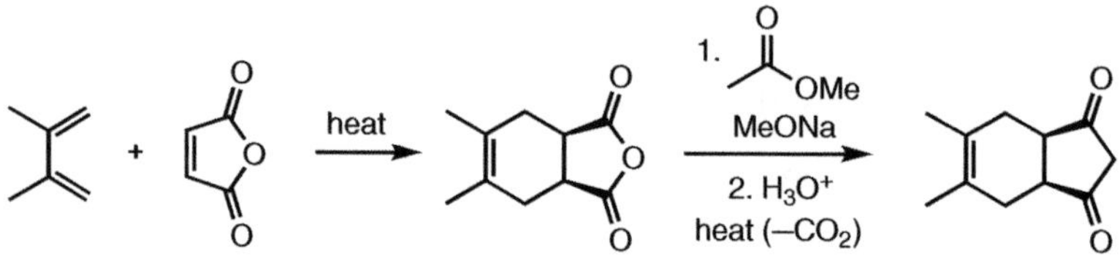

6-11. Provide the missing structures (**A–D**) in the following synthetic sequence (*J. Am. Chem. Soc.*, **1994**, 116, 11275).

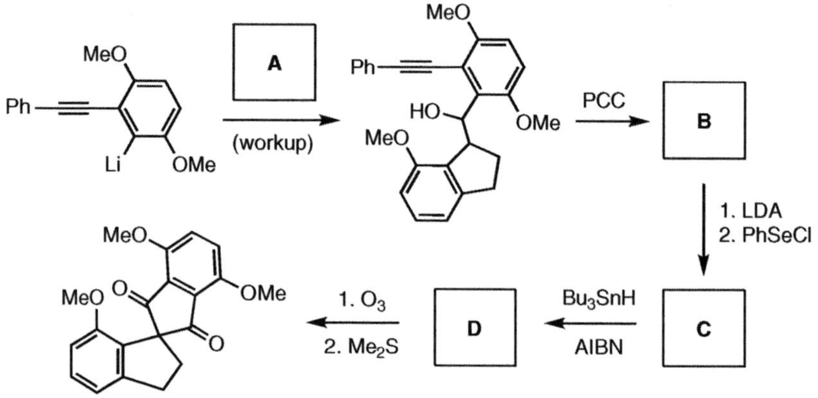

6-12. The following reaction scheme was taken from a journal article reporting the total synthesis of (±)-Przewalskin B, a compound isolated from an ancient folk medicine (*J. Org. Chem.*, **2014**, 79, 2746). Provide the missing reagents.

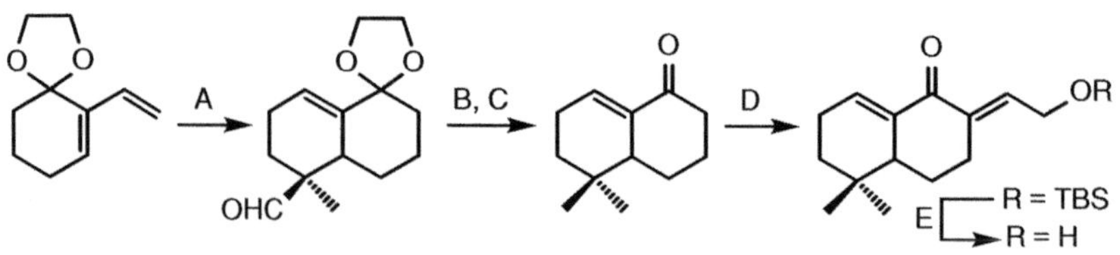

CHAPTER 7

PREDICTING AND CONTROLLING STEREOCHEMISTRY

The synthesis of any target molecule containing one or more stereocenters (e.g., an asymmetric, tetrahedral carbon that can be R or S, or an alkene that can be E or Z) requires careful planning. In the case of many biologically active natural products, hundreds or even thousands of stereoisomers are possible for a given structure, but typically only a single diastereomer is the desired synthetic target. In addition, just a single enantiomer of that compound is typically of interest, so a major goal of modern-day synthesis is the *enantioselective* preparation of chiral molecules. This chapter will provide an overview of the stereochemistry of various organic reactions with which the reader should already be familiar and, after reviewing some basic stereochemical terminology, also give a brief introduction to both diastereoselective and enantioselective reactions.

Introduction to Strategies for Organic Synthesis, Second Edition. Laurie S. Starkey.
© 2018 John Wiley & Sons, Inc. Published 2018 by John Wiley & Sons, Inc.

REACTIONS THAT FORM RACEMATES

Any reaction that forms a single, new chiral carbon center *in an achiral environment* will occur in a racemic fashion (i.e., equal amounts of the *R* and *S* enantiomers will be formed). Having an achiral environment implies that there are no preexisting chiral centers on the molecule and that no chiral reagents are being used. Even if multiple chiral centers are generated in such a situation, a 1:1 mixture of (+/−) enantiomers is still formed, since the transition states leading to those enantiomers (described as enantiomeric transition states) are equal in energy and the products will be formed at equal rates.

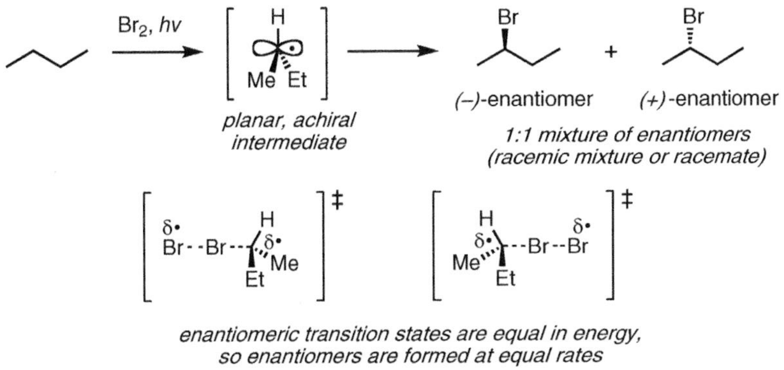

planar, achiral
intermediate

(−)-enantiomer   (+)-enantiomer

1:1 mixture of enantiomers
(racemic mixture or racemate)

enantiomeric transition states are equal in energy,
so enantiomers are formed at equal rates

Enantiomeric Transition States Lead to Racemic Mixtures

Introduction to Strategies for Organic Synthesis, Second Edition. Laurie S. Starkey.
© 2018 John Wiley & Sons, Inc. Published 2018 by John Wiley & Sons, Inc.

7.1.1 FORMATION OF NEW CHIRAL CENTERS

Any reaction that creates a new bond to an achiral carbon atom has the potential to create a new chiral center if that carbon has a tetrahedral geometry in the product. Examples in which the starting carbon was also tetrahedral include free-radical halogenation (replace H· with X·) and α-alkylation of an enolate (replace H⁺ with E+). Additions to alkenes and carbonyls convert a planar, sp²-hybridzed carbon to an sp³-hybridized one, so these reactions frequently create new chiral centers as well. If the starting compound has no chiral centers to begin with, and either a single, new chiral center is being formed, or two chiral centers are generated in a stereoselective mechanism, then these reactions result in the formation of racemic mixtures.

*enantiomeric bromonium ion intermediates
lead to enantiomeric transition states*

Achiral Starting Materials Give Racemic Product Mixtures

However, if two or more chiral centers are being generated in an achiral environment with a non-stereospecific mechanism, then it is possible to form racemic mixtures of various diastereomers. For example, addition of HBr to 1,2-dimethylcyclohexene can give a product with the two methyl groups having either a cis or trans relationship to each other. The cis and trans products are diastereomers and are likely to be formed in *unequal* amounts. Protonation of the alkene from either face results in an enantiomeric pair of carbocations. Attack by bromide on one of these carbocations can occur from the top or bottom face. Because of the new chiral center that has been created in the protonation step, the top and bottom faces of the planar carbocation are no longer equivalent, so attack by bromide is not expected to be equally likely for both faces. As bromide attacks either face, it leads to diastereomeric transition states that are unequal in energy, so the diastereomeric products are formed at different rates and in unequal amounts. Since each of these diastereomers is a chiral product being generated from an achiral starting material, it must be formed as a racemic mixture, so a total of four stereoisomers will be produced. This may seem complex, but it is a common occurrence in chemistry.

Diastereomeric Transition States Lead to Unequal Product Mixtures

When a new chiral center is formed close to an existing chiral center, such a reaction also proceeds through diastereomeric rather than enantiomeric transition states, and diastereoselectivity can be expected in such cases. The chiral center may exert a directing effect on the reagent by being involved in the mechanism, or it might be a source of steric strain that disfavors one transition state over another. This effect, called "asymmetric induction," will be investigated more thoroughly in the upcoming section on carbonyl additions.

oxygen on existing chiral center directs incoming reagent to the same side

mCPBA

not

mCPBA

endo face is more sterically crowded

a single diastereomer is preferred

not

Asymmetric Induction from Existing Chiral Center

7.1.2 LOSS OF A GROUP FROM A CHIRAL CENTER: RACEMIZATION

The term racemization typically refers to the conversion of an enantiopure sample to its racemate, which involves converting the one enantiomer into the other until an equal mixture of both enantiomers exists.

If additional chiral centers exist in the molecule and remain unchanged during this conversion, the pair of compounds with a mixture of R and S configurations at a single site would be related as diastereomers (more specifically, epimers) rather than enantiomers, so such a process is described as an epimerization rather than a racemization. For example, sugars can undergo epimerization at the anomeric carbon via a resonance-stabilized carbocation intermediate.

α-glucopyranose

epimerization of glucose involves inversion at a single center (the anomeric carbon)

inverted stereocenter

β-glucopyranose

Epimerization Produces a Diastereomer Called an Epimer

Chiral centers can be converted to achiral intermediates by loss of a leaving group to give a planar carbocation (as in the S_N1 mechanism). Attempted formation of an organometallic reagent from a chiral alkyl halide often results in significant racemization and is generally not synthetically useful.

Similarly, radical mechanisms result in racemization since a radical intermediate is essentially planar and thus achiral.

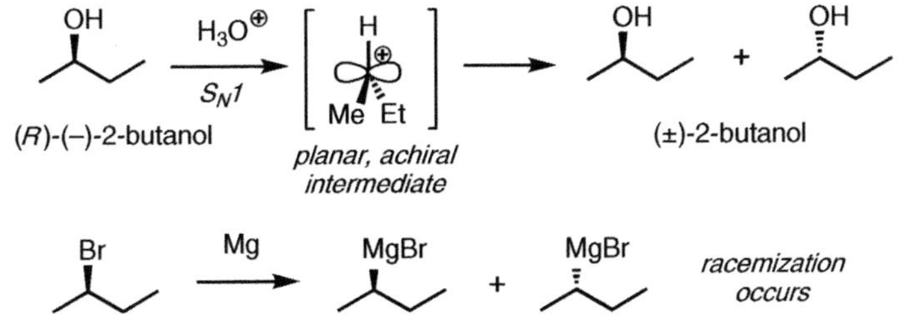

(*R*)-(–)-2-butanol

planar, achiral
intermediate

(±)-2-butanol

racemization
occurs

Racemization by Loss of a Leaving Group

Deprotonation of a carbon alpha to an electron-withdrawing group gives a planar enolate, making chiral centers on such carbons susceptible to racemization. Therefore, care must be taken to minimize deprotonation (or enol formation) when working with these compounds. For example, for peptide synthesis and other reactions involving amino acids, a variety of synthetic methods have been developed to maintain the naturally occurring L stereochemistry throughout various transformations.

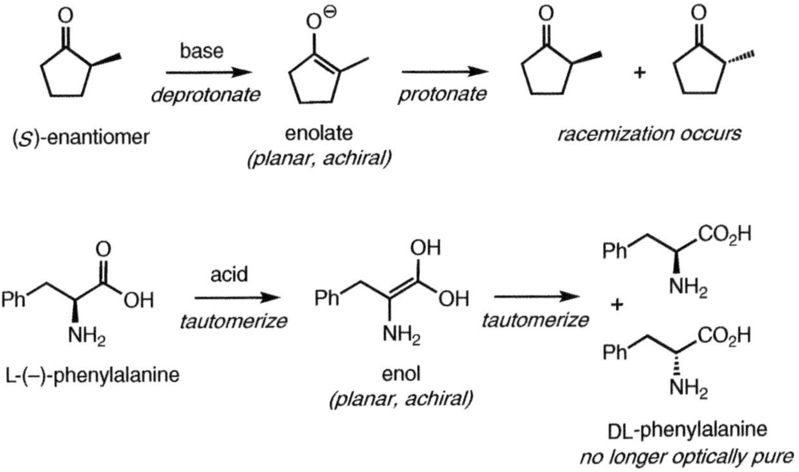

(*S*)-enantiomer

base
deprotonate

enolate
(planar, achiral)

protonate

racemization occurs

L-(–)-phenylalanine

acid
tautomerize

enol
(planar, achiral)

tautomerize

DL-phenylalanine
no longer optically pure

Racemization by Loss of an Alpha Proton (via Enolate or Enol)

S$_N$2 MECHANISM: BACKSIDE ATTACK

The S$_N$2 displacement of a leaving group with a strong nucleophile necessitates a backside approach by the nucleophile. Backside attack can be observed with 2° alkyl halides or tosylates and in epoxide ring-opening reactions. If the S$_N$2 occurs at a chiral center, the nucleophile in the product will be bonded to the side opposite the leaving group in the starting material. Since the nucleophile and leaving group are often both the number one priority group for *R/S* nomenclature purposes, this inversion of stereochemistry (called a Walden inversion) also typically results in an inversion of the configurational label. The S$_N$2 mechanism offers a reliable, predictable way to control stereochemistry for a substitution reaction, although any competition with an S$_N$1 mechanism would result in partial racemization. The use of a polar, aprotic solvent promotes the S$_N$2 mechanism while disfavoring carbocation formation. Racemization via S$_N$2 is possible if the incoming nucleophile matches the outgoing leaving group, as in the case of 2-bromobutane reacting with NaBr, since the product is the enantiomer of the starting material.

Introduction to Strategies for Organic Synthesis, Second Edition. Laurie S. Starkey.
© 2018 John Wiley & Sons, Inc. Published 2018 by John Wiley & Sons, Inc.

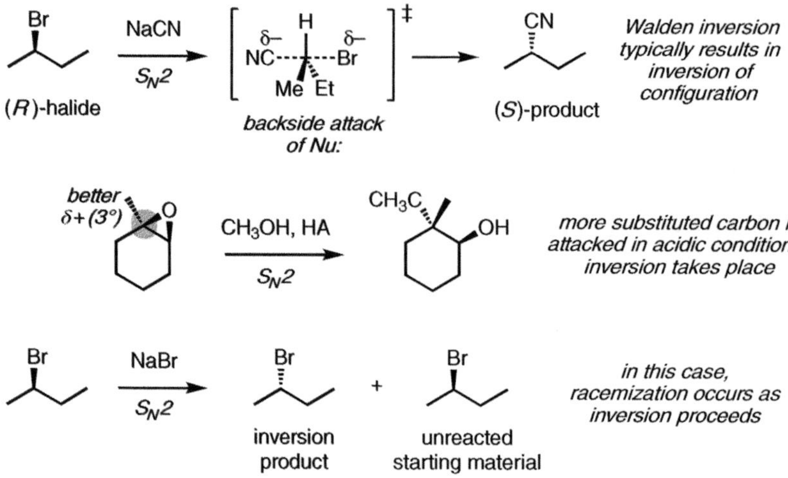

Walden inversion typically results in inversion of configuration

more substituted carbon is attacked in acidic conditions; inversion takes place

in this case, racemization occurs as inversion proceeds

Stereochemistry of S$_N$2 Mechanism: Backside Attack Leads to Inversion

Another interesting example of an inversion reaction is the Mitsunobu reaction, which displaces a hydroxyl with a nucleophile such as a carboxylate. Reaction of a chiral alcohol with triphenylphosphine and diethyl azodicarboxylate (Ph$_3$P, DEAD) in the presence of a carboxylic acid results in the formation of a good leaving group, followed by the S$_N$2 displacement of that group by the carboxylate nucleophile to give an ester product with an inverted stereochemistry. Hydrolysis of the ester produces an alcohol with a configuration that is the opposite of the original alcohol.

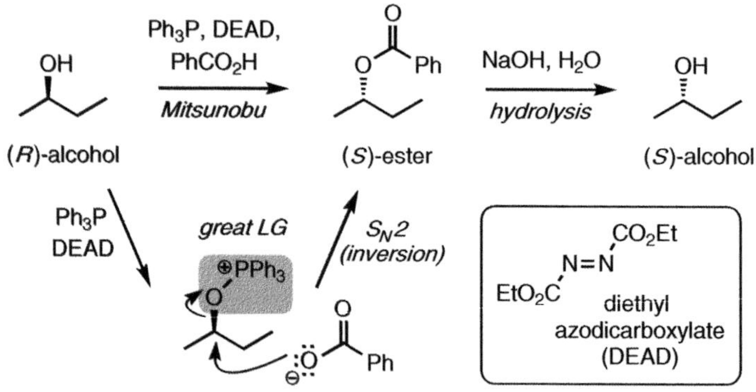

Inversion of Alcohols via the Mitsunobu Reaction

ELIMINATION MECHANISMS

Elimination reactions involve the loss of a leaving group, along with a β-hydrogen. Stereochemical considerations of elimination reactions involve the relationship of these two groups during the mechanism, as well as the relationship of the groups remaining on the alkene product at the end of the reaction. When there is a choice, product stability typically guides the product stereochemistry, such as the less sterically strained trans alkene being favored over the cis. However, the relationship between the β-hydrogen and the leaving group depends on the mechanism of the elimination.

7.3.1   E2 ELIMINATION (ANTI)

Treatment of an alkyl halide with a strong base can result in an E2 elimination reaction. This single-step mechanism usually prefers an *anti*-coplanar relationship between the β-hydrogen and the leaving group (*anti*-periplanar is a better description, since the two groups do not need to be exactly coplanar). The presence of steric interactions in the necessary conformation can slow or even prevent the E2 from occurring. Notable examples of such an effect include the E2 elimination of halocyclohexane derivatives, such as 1-bromo-4-*t*-butylcyclohexane. In the more stable chair conformation, the larger groups prefer to occupy equatorial positions. However, the halide and the β-hydrogen must both be in the axial position (i.e., trans diaxial) in order to be *anti*-coplanar, so a flip to the other chair conformation is required in order for the E2 to occur.

Introduction to Strategies for Organic Synthesis, Second Edition. Laurie S. Starkey.
© 2018 John Wiley & Sons, Inc. Published 2018 by John Wiley & Sons, Inc.

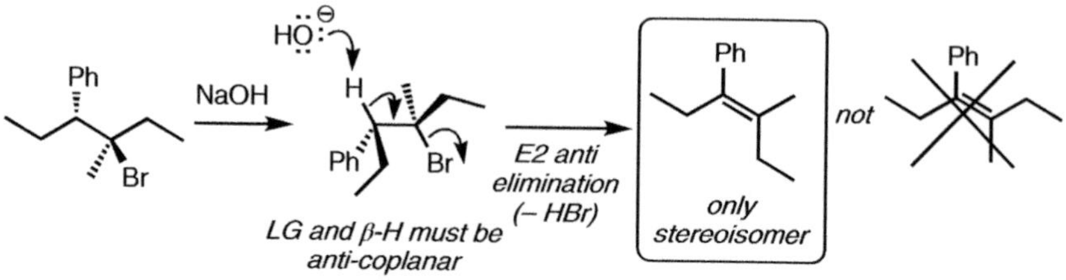

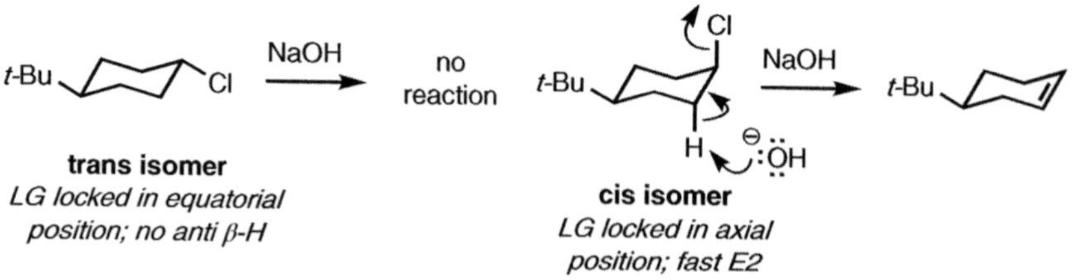

Stereochemistry of E2: *Anti* Elimination

7.3.2 COPE ELIMINATION (SYN)

Oxidation of a tertiary amine with a peroxide (mCPBA) forms an *N*-oxide that undergoes an intramolecular deprotonation with *syn* stereoselectivity upon heating. This reaction is known as the Cope elimination. If given a choice, this deprotonation will occur on a less hindered β-carbon, giving rise to a less substituted alkene product (this regioselectivity, known as the Hofmann's rule, is the opposite predicted by Zaitsev's rule).

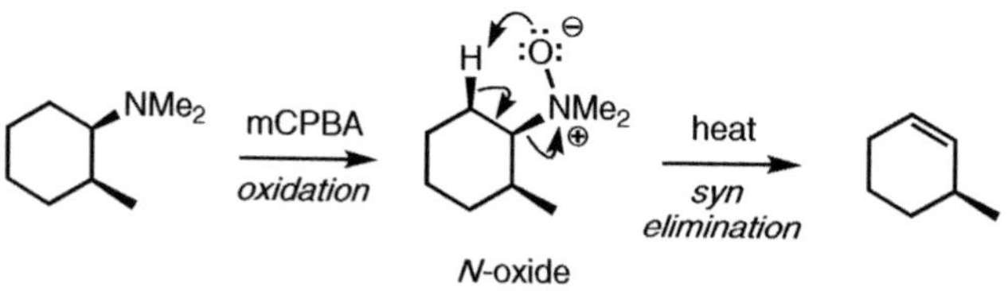

N-oxide

Cope Elimination: *Syn* Elimination Following Hofmann's Rule

ADDITIONS TO ALKENES AND ALKYNES

An addition to an alkene can form up to two new chiral centers, and a reaction that occurs with only a *syn* or only an *anti* addition mechanism will give a product with predictable stereochemistry. Conversion of alkynes to alkenes can also occur with either *syn* or *anti* stereoselectivity. When these alkyne reductions are taken in combination with alkene addition reactions, target molecules with a wide variety of stereochemical relationships can be prepared.

7.4.1 SYN ADDITIONS

When the formation of two new bonds occurs in a concerted fashion (i.e., in a single step with a cyclic transition state), then those two bonds will be added to the same face of the substrate, which is called *syn* addition. Examples of this include dihydroxylation with OsO_4, epoxidation, cyclopropanation, and hydroboration. The mechanism of catalytic hydrogenation of an alkene (or of an alkyne with Lindlar's catalyst) is another example of a *syn* addition, since both hydrogens will be sequentially added to the same face of the pi bond.

Introduction to Strategies for Organic Synthesis, Second Edition. Laurie S. Starkey.
© 2018 John Wiley & Sons, Inc. Published 2018 by John Wiley & Sons, Inc.

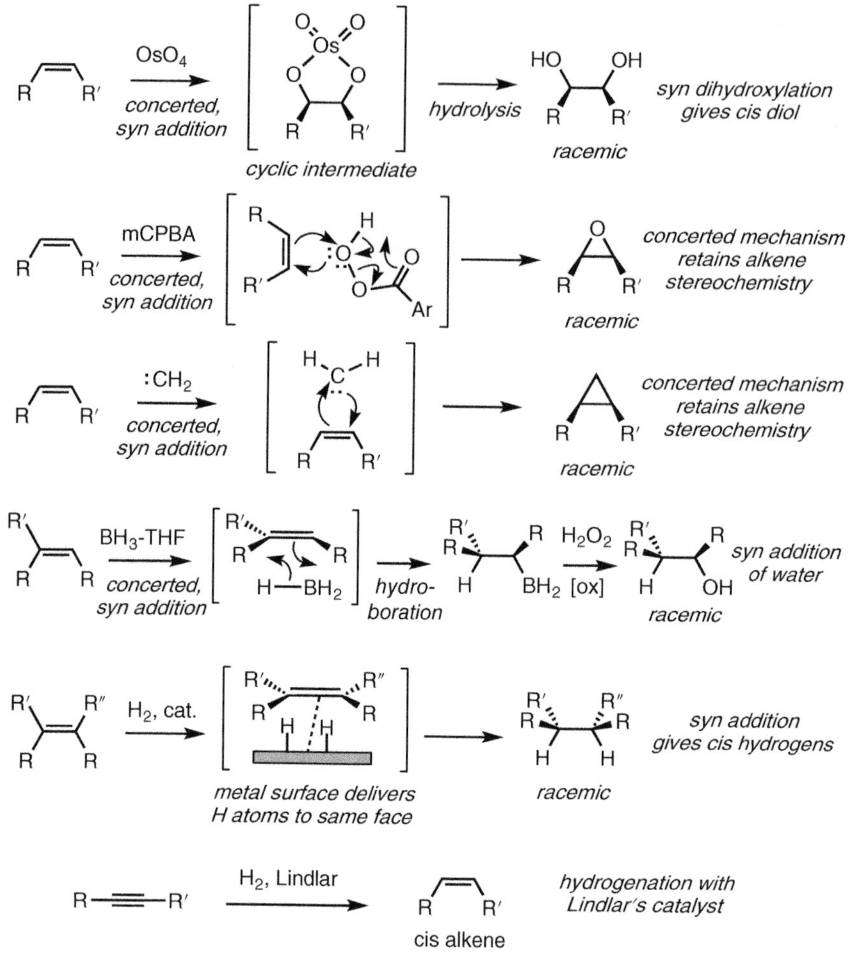

Examples of *Syn* Additions to Alkenes and Alkynes

7.4.2 ANTI ADDITIONS

Anti additions to alkenes typically result from a stepwise mechanism: formation of a cationic cyclic intermediate such as a bromonium ion, followed by back-side attack by a nucleophile to open up the ring. Such is the mechanism for the *anti* addition of Br_2. The bromination reaction results in trans bromine atoms since the second bromine (as Br^-) has to come in from the opposite face as the first bromine (as Br^+) in order to do an S_N2 attack on the bromonium ion. Other mechanisms involving the bromonium ion include reaction of an alkene with Br_2/H_2O (adds —Br and —OH *anti*) and Br_2/ROH (adds —Br and —OR *anti*). Treatment of an alkene with a peroxyacid in water forms an epoxide that undergoes a ring opening *in situ* to give a *trans*-diol product. Trans stereoselectivity is also seen in the dissolving metal reduction of alkynes.

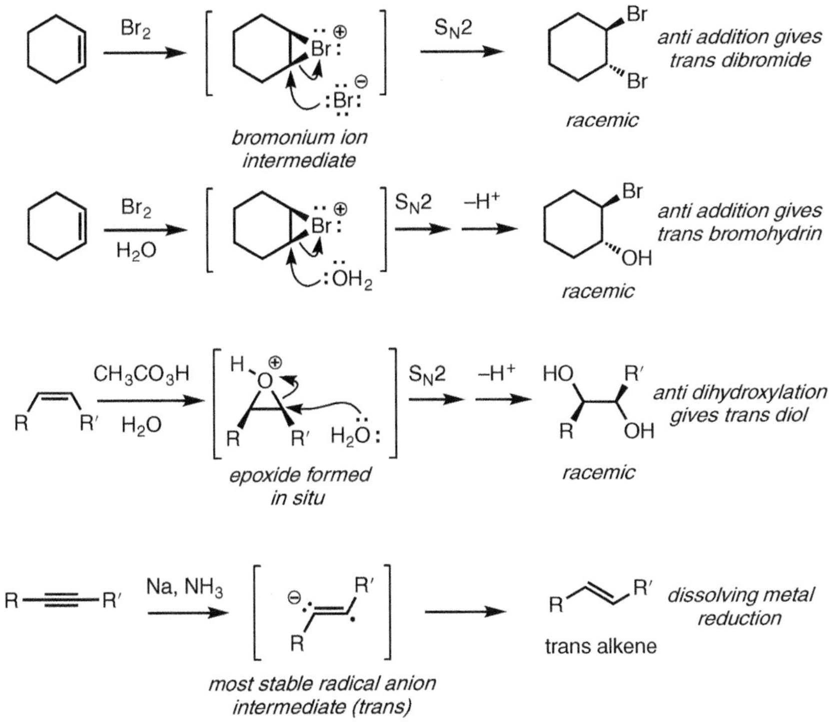

Examples of *Anti*-Additions to Alkenes and Alkynes

PRACTICE PROBLEM 7.4: PREDICTING STEREOCHEMISTRY (SUBSTITUTION, ELIMINATION, AND ADDITION REACTIONS)

7.4 For each of the following reactions, predict the major product(s) or provide the missing reagent(s). Pay close attention to the stereochemistry of each reaction.

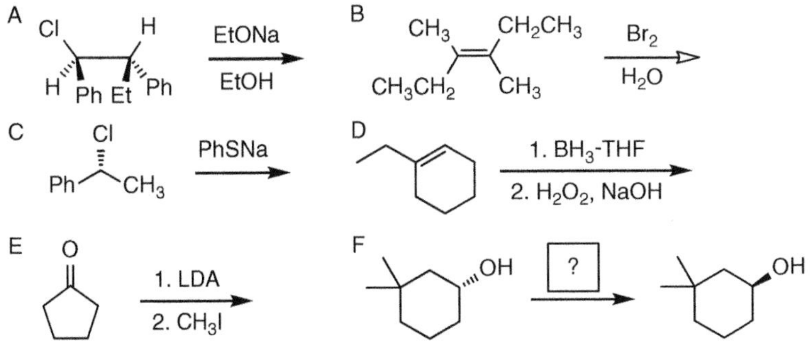

ADDITIONS TO CARBONYLS

Since nucleophilic addition to a carbonyl causes an sp^2-hybridized carbon to become tetrahedral, such reactions can form new chiral centers. For ordinary aldehydes and ketones in an achiral environment, addition of nucleophiles to the enantiotopic faces of the carbonyl results in the usual racemic mixture of alcohols. For carbonyls that already contain chiral centers, however, it is often possible to predict which of the possible diastereomeric products will be preferred, by considering the stabilities of the competing diastereomeric transition states.

If a chiral center exists in close proximity to a carbonyl, it will influence the approach of an incoming nucleophile and one face of the carbonyl is likely to be preferred over the other. This phenomenon is known as asymmetric induction. For example, one face of 2-methylcyclopentanone is partially blocked by the methyl group so nucleophilic attack occurs mostly on the opposite face in order to avoid steric strain in the transition state. The two possible transition states (attack from top face or bottom face) are diastereomeric and thus are not equal in energy. The transition state with the nucleophile approaching away from the methyl group is more stable (i.e., lower in energy), so that product will be formed faster and this is the major product observed.

Nucleophiles Attack the Less Hindered Face of Ketones

Introduction to Strategies for Organic Synthesis, Second Edition. Laurie S. Starkey.
© 2018 John Wiley & Sons, Inc. Published 2018 by John Wiley & Sons, Inc.

7.5.1 DIASTEREOSELECTIVITY IN ACYCLIC SYSTEMS: CRAM'S RULE, FELKIN–AHN MODEL

In an acyclic system, rotations about carbon–carbon bonds can alleviate steric strain. While the existence of many possible conformations makes it more challenging to predict stereochemical preferences, a number of models have been developed to help make such predictions. Cram's rule, along with the updated Felkin–Ahn model, looks at the three groups attached to the carbon alpha to the carbonyl and describes those groups by their size: small (S), medium (M), and large (L). The preferred conformation for nucleophilic attack positions the largest group at a dihedral angle of 90° from the carbonyl. When viewing a Newman projection that places the carbonyl in a vertical position with the oxygen pointing up, the large group L can be pointing either straight out to the left or straight out to the right (both have the necessary 90° dihedral angle). The proper orientation is the one that also places the smallest group S closer to the aldehyde H or the ketone R group attached to the carbonyl carbon, since that is the direction from which the nucleophile will approach, and the smallest group offers the least amount of steric hindrance. When the aldehyde or ketone is in the correct conformation, the nucleophile will predominantly attack the carbonyl from the side *opposite* the largest group (again, to minimize steric strain in the transition state), at an angle approximately 109° from the C=O bond.

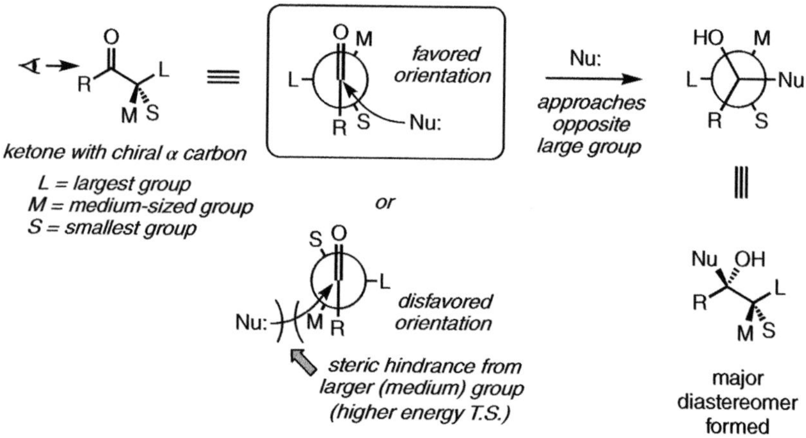

Cram's Rule and Felkin–Ahn Model

If a strongly electron-withdrawing group such as chlorine is attached to the alpha carbon, then electronic effects will position this group at 90° to the carbonyl. As shown below, the smaller of the two remaining groups is then oriented closer to the ketone R group, and the reaction proceeds with the nucleophile attacking from the face opposite the EWG.

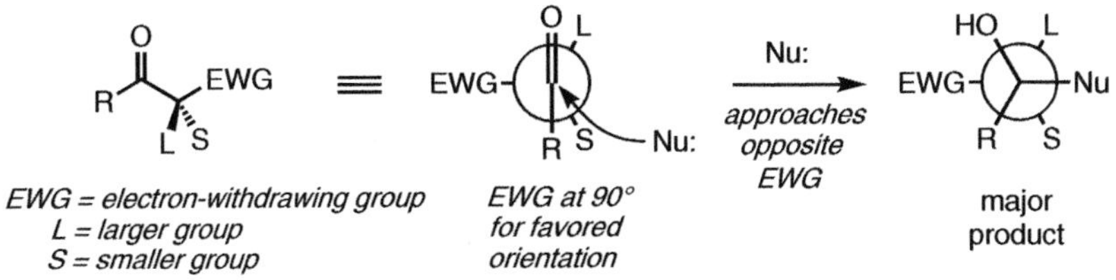

EWG = electron-withdrawing group
L = larger group
S = smaller group

EWG at 90°
for favored
orientation

Nu:
approaches
opposite
EWG

major
product

EXAMPLE: APPLYING CRAM'S RULE

Predict the major product for the following reaction.

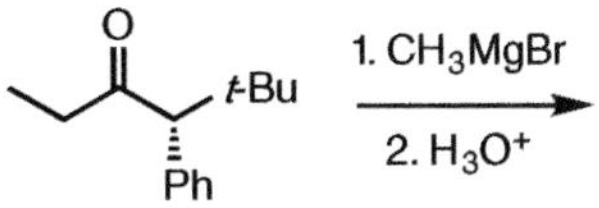

Since there is no EWG in the alpha position, the largest group on the alpha carbon (*t*-butyl) is oriented at 90° to the carbonyl, with the smallest group aligned with the ketone alkyl group (ethyl). Approach by the Grignard nucleophile from the side opposite the *t*-butyl group gives the diastereomer shown as the major product.

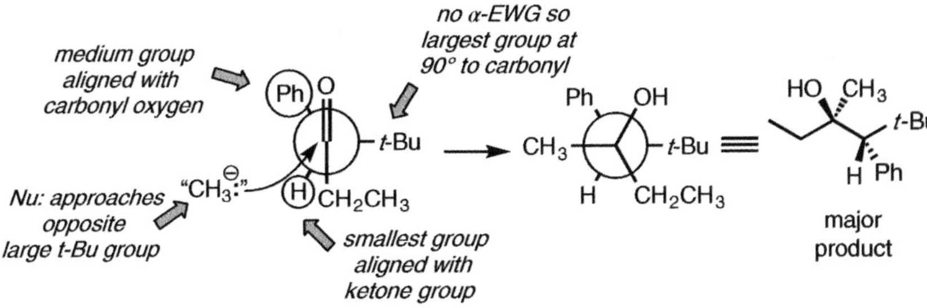

7.5.2 CHELATION CONTROL BY NEIGHBORING GROUPS

If one of the groups on the alpha carbon contains a lone pair and can act as a Lewis base (e.g., $-OH, -OR$, or $-NH_2$), that group can chelate with a Lewis acid or the counterion of the incoming nucleophile. Nucleophilic attack of the resulting cyclic, coordinated species occurs at the less hindered face.

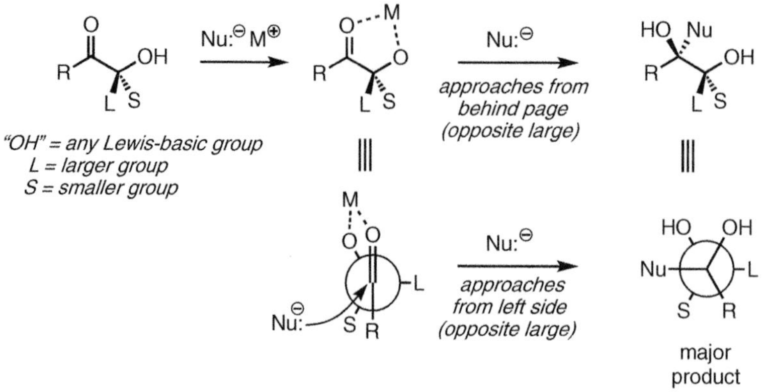

Lewis Basic α-Substituents Can Form Chelates

EXAMPLE: APPLYING CHELATION CONTROL

Predict the major product for the following reaction.

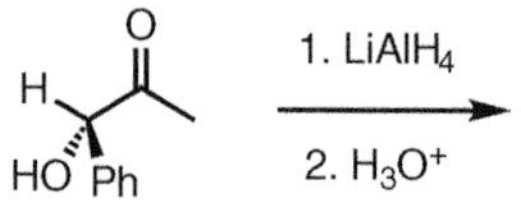

Since there is a hydroxyl in the alpha position, this group will align with the carbonyl oxygen to enable chelation with the aluminum. The less hindered approach by the hydride nucleophile from the side opposite the larger phenyl group gives the diastereomer shown as the major product.

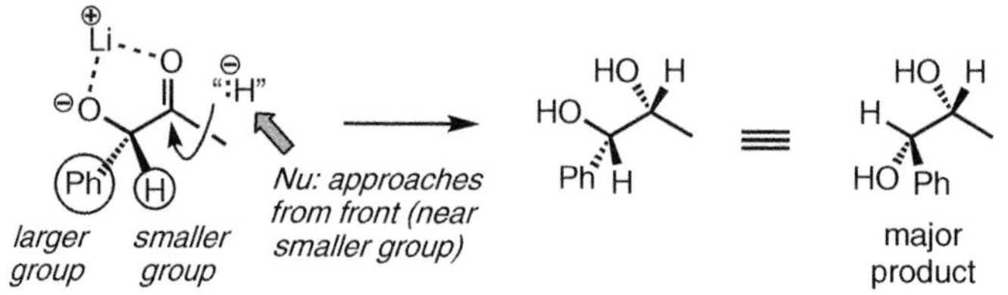

7.5.3 ADDITION TO CYCLOHEXANONES

Cyclohexanone derivatives exist in a chair-like conformation with larger substituents preferring equatorial-like positions. Attack by a nucleophile is described as either an equatorial or an axial approach, depending on whether the nucleophile ends up occupying an equatorial or axial position, respectively, in the product. The direction of the attack is guided by minimizing steric hindrance and depends on the size of the nucleophile. Smaller nucleophiles prefer an axial attack, giving a more stable transition state and resulting in a product with the hydroxyl group in the equatorial position. Larger nucleophiles, however, experience 1,3-diaxial strain with an axial approach, so an equatorial approach is favored in these cases.

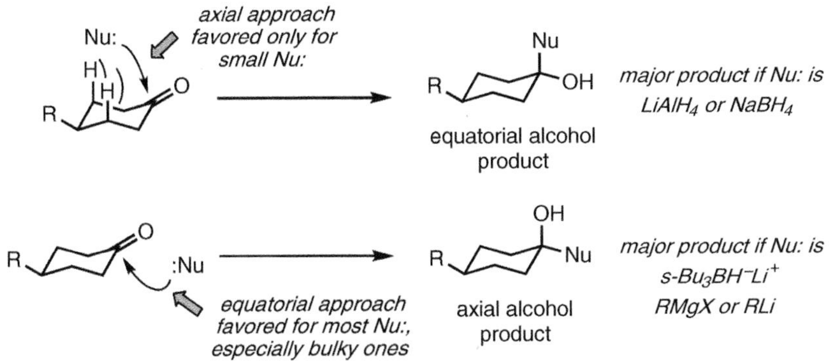

Equatorial Addition to Cyclohexanones is Usually Favored

ADDITIONS TO ENOLATES: ALDOL STEREOCHEMISTRY

The aldol reaction, the reaction of an enolate nucleophile with a carbonyl electrophile to give a β-hydroxy carbonyl product, has the potential to form two new chirality centers. It is possible to both predict and control the stereochemical relationship between a group on the alpha carbon and the newly formed hydroxyl group on the beta carbon. Since this relationship depends on the stereochemistry of the enolate involved, formation of the enolate must first be discussed.

Aldol Product Stereochemistry Depends on Enolate Stereochemistry

7.6.1 FORMATION OF (*E*)- AND (*Z*)-ENOLATES

The stereochemistry of an enolate is described by examining the relationship between the groups attached to the double bond of the enolate. If the two higher priority groups are on the same side, then it is described as a (*Z*) (or cis) enolate. In the (*Z*)-enolate, the alkyl group on the alpha carbon is on the same side as the oxygen (−OLi). Since this arrangement minimizes the steric strain by placing the R groups trans to each other across the double bond, the more stable (*Z*)-enolate is typically favored for ketones, especially when an additive such as hexamethylphosphoramide (HMPA) is used. Such

Introduction to Strategies for Organic Synthesis, Second Edition. Laurie S. Starkey.
© 2018 John Wiley & Sons, Inc. Published 2018 by John Wiley & Sons, Inc.

an additive forms a complex with the metal cation, allowing it to dissociate from the O⁻ anion. The resulting open transition state allows equilibration, favoring formation of the (Z)-enolate.

If a bulkier base is used, such as lithium tetramethylpiperidide (LTMP), or in the case of ester enolate formation in which the "R_1" group (actually "OR") is smaller, the (E)-enolate is major. The Ireland model uses the possible six-membered, chair-like, cyclic transition states to help rationalize the stereoselectivity of enolate formation.

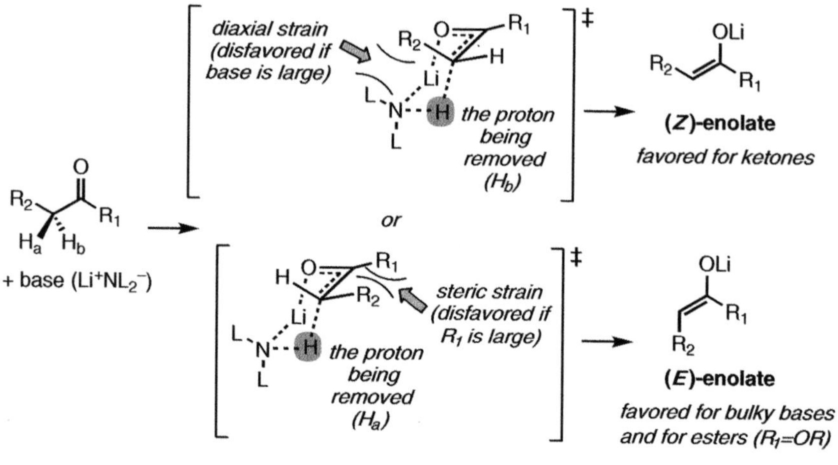

Predicting the Formation of (Z)- or (E)-Enolates

7.6.2 ALDOL REACTION WITH (E)- AND (Z)-ENOLATES

In the aldol reaction, the stereochemistry of the product depends on the stereochemistry of the enolate starting material. Reaction of a (Z)-enolate with an aldehyde gives a *syn* aldol product in which both the OH and the aldehyde R group are on the same side of the plane of the zigzag line drawing (both are wedges in one enantiomer and both are dashes in the other enantiomer).

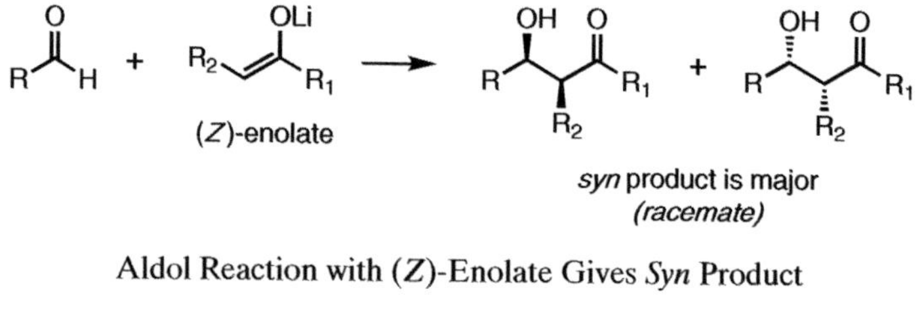

Aldol Reaction with (Z)-Enolate Gives *Syn* Product

On the other hand, reaction of an (E)-enolate with an aldehyde gives the other diastereomer, an *anti* aldol product with one group as a wedge and the other as a dash.

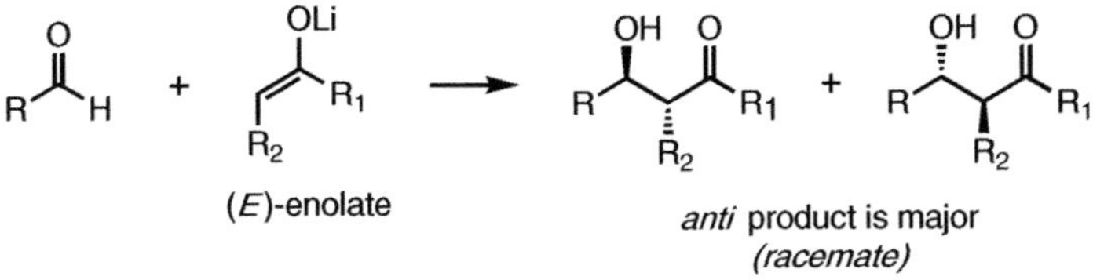

Aldol Reaction with (*E*)-Enolate Gives Anti Product

Once again, consideration of the chelated, cyclic transition state, known as the Zimmerman–Traxler model, provides the rationale for this diastereoselectivity. In the most favorable chair-like transition state, the aldehyde R group is in an equatorial position. This preferred orientation produces the *syn* product from the (*Z*)-enolate and the *anti* product from the (*E*)-enolate. Each transition state shown is forming a single enantiomer product; attack by the enolate to the opposite face of the aldehyde would give rise to the other enantiomer.

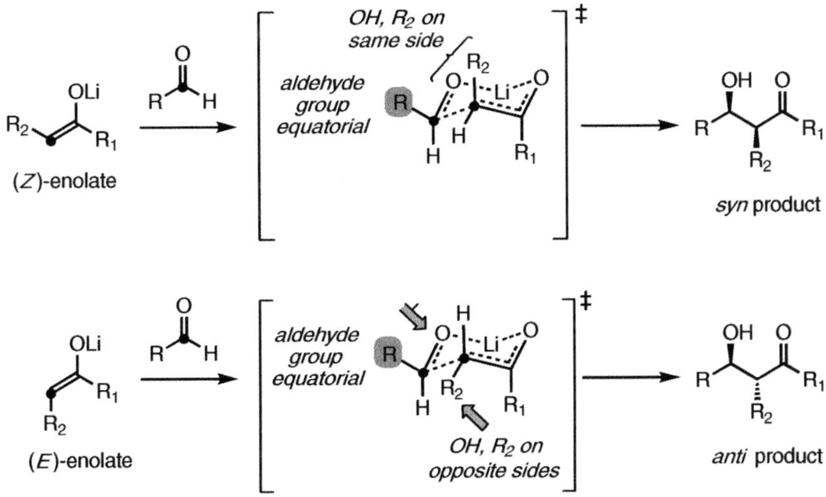

Zimmerman–Traxler Transition States Explain Aldol Stereoselectivity

EXAMPLES: PREDICTING ALDOL STEREOCHEMISTRY

Predict the major product for each of the following reactions.

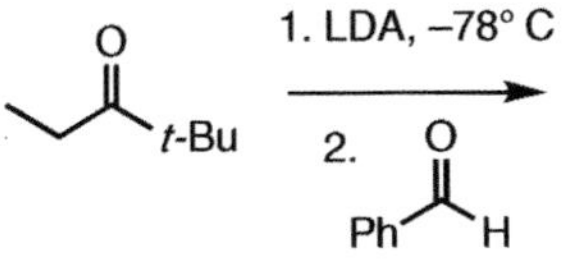

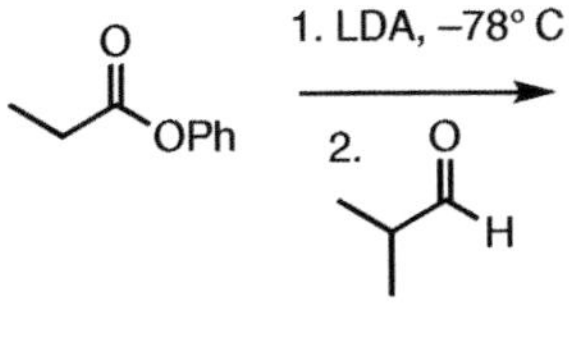

The bulky *t*-butyl group gives rise to a (*Z*)-enolate and *syn* aldol product. Esters predominantly form (*E*)-enolates, so the second example produces an *anti* aldol product.

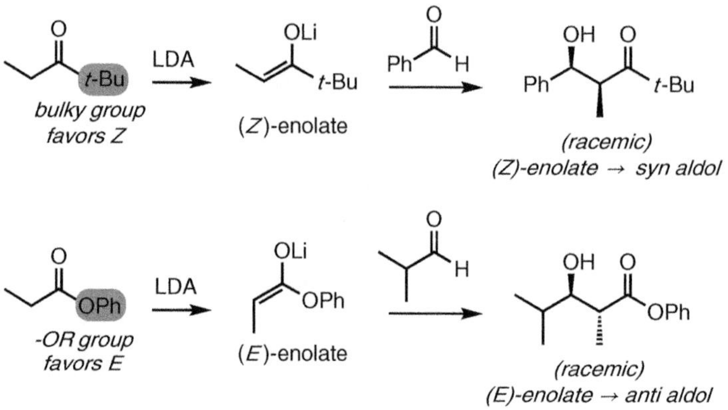

(*Z*)-enolate

(racemic)
(*Z*)-enolate → syn aldol

bulky group
favors *Z*

-*OR* group
favors *E*

(*E*)-enolate

(racemic)
(*E*)-enolate → anti aldol

ENANTIOSELECTIVITY AND ASYMMETRIC SYNTHESES

Any reaction that forms a new chiral carbon center in an achiral environment will occur in a racemic fashion (equal amounts of the R and S enantiomers are formed). Because so many naturally occurring synthetic targets are chiral, a multitude of different reaction strategies have been developed to preferentially produce one enantiomer over the other. Beginning a synthesis with a naturally occurring or commercially available chiral molecule is one such strategy; sugars, amino acids, and terpenes are examples of chiral starting materials. The presence of one or more chiral centers in a starting structure can induce chirality in a newly formed, neighboring chiral center to give predominantly a single diastereomer as a major product (this effect is called asymmetric induction). Another option is to start with an achiral starting material and use a chiral reagent or a chiral catalyst to enable formation of a single enantiomer product.

7.7.1 PROCHIRAL ENVIRONMENTS

A prochiral atom or molecule is one that has the potential to become chiral by making a single change to the structure. For example, a tetrahedral carbon atom with two identical groups attached to it is not chiral, but if one of these groups was changed to something new then the carbon center would be asymmetric and chiral (i.e., a carbon atom with four different groups on it that can be described as R or S). Such a carbon atom is called a prochiral carbon. The two groups are described as being either enantiotopic,

Introduction to Strategies for Organic Synthesis, Second Edition. Laurie S. Starkey.
© 2018 John Wiley & Sons, Inc. Published 2018 by John Wiley & Sons, Inc.

if the replacement of either group with something different would result in a pair of enantiomers, or diastereotopic, if replacement would lead to a pair of diastereomers (i.e., if the molecule already contains another chirality center). Trigonal planar atoms that have three different groups attached are also described as being prochiral, since addition of a fourth unique group would result in a tetrahedral chiral center. In this case, the two faces of the plane are described as either enantiotopic or diastereotopic.

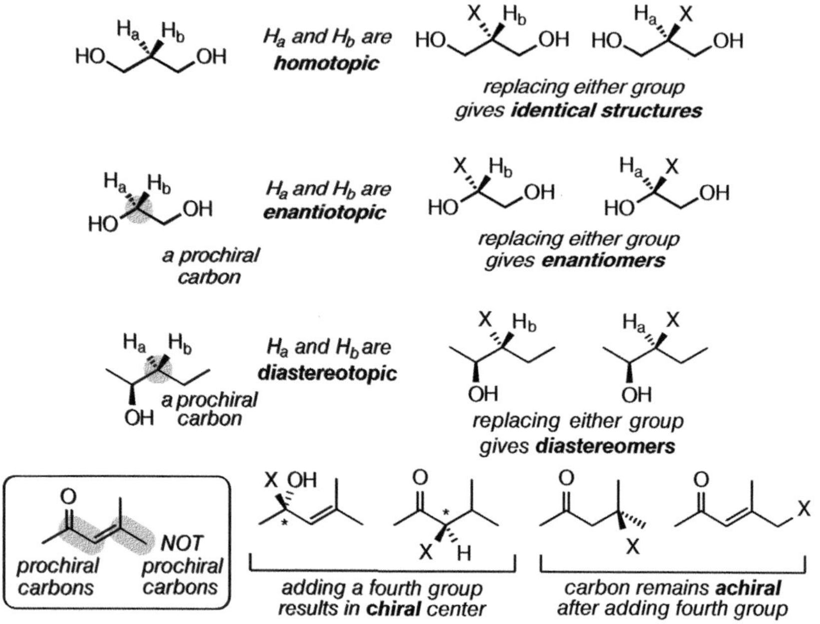

Prochiral Atoms are either enantiotopic or Diastereotopic

Since an asymmetric synthesis is one that preferentially attacks only one prochiral face of a planar substrate or reacts with only one prochiral group on a tetrahedral atom, some terminology is needed to unambiguously describe such species.

Pro-R and Pro-S Groups

The two otherwise identical groups on a tetrahedral prochiral carbon are described as "*pro-R*" and "*pro-S.*" These labels are determined by mentally replacing the first atom in one of the identical groups with a heavier isotope, making it a slightly higher priority (based on Cahn–Ingold–Prelog rules), and allowing all four groups to be ranked. If the new, imaginary structure has the *R* configuration, then the group that was replaced is described as being *pro-R*. The *Pro-S* group is the one whose replacement (and resulting increase in rank) would lead to an *S* configuration.

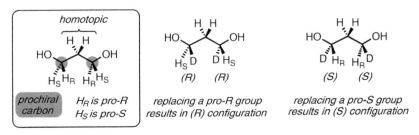

Tetrahedral Prochiral Carbons Have a *Pro-R* Group and a *Pro-S* Group

PRACTICE PROBLEM 7.7: PROCHIRAL GROUPS

7.7 Identify the labeled groups as homotopic, enantiotopic, or diastereo-topic and determine whether the indicated hydrogen is *pro-R* or *pro-S* (or neither).

Re and Si Faces of a Trigonal Planar Atom

The two faces of a prochiral trigonal planar, sp^2-hybridized atom are described as "*re*" and "*si*." Assigning these names begins with prioritizing the three groups around the trigonal planar carbon atom (using Cahn–Ingold–Prelog rules). When viewing the *re* face, a *clockwise* motion is observed when going from #1 → #2 → #3 (analogous to the *R* configuration of a tetrahedral carbon). When viewing the structure from the opposite *si* face, going from #1 → #2 → #3 is a *counterclockwise* motion.

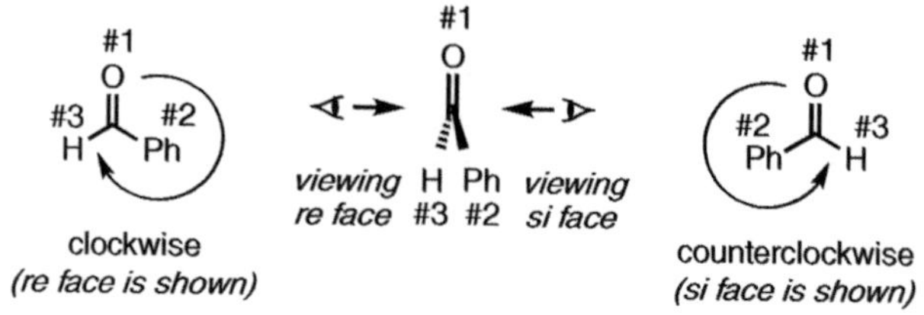

Trigonal Planar Prochiral Carbons Have a *Re* Face and a *Si* Face

7.7.2 ENANTIOSELECTIVE TECHNIQUES

In all of the reactions explored to this point, chiral products formed from achiral starting materials were always generated as racemic mixtures. Modern organic synthesis demands a higher goal: to produce just a single enantiomer of a chiral target molecule. One way to do this is to synthesize a racemic mixture and then perform an optical resolution to separate the two enantiomers, using a chiral resolving agent or perhaps by chiral chromatography. While this method is ideal if it would be useful to study small amounts of both enantiomers of a given TM, it is an otherwise costly process in which half of the final product is discarded. As a general rule to minimize material usage and waste, the earlier in a synthesis that the resolution is performed, the better. A highly desirable strategy is a synthesis that produces just a single enantiomer, either by starting with a single enantiomer of a chiral starting material or by conducting an asymmetric reaction that uses a chiral reagent, a chiral auxiliary, or perhaps an enzyme to impart enantioselectivity. A few examples of these techniques will be shown here.

Separation of Enantiomers via Resolution of a Racemate

While it might be possible to separate a mixture of enantiomers using chromatography with a chiral stationary phase (CSP), such methods are expensive and not suitable for all substrates. The separation of a racemic mixture more commonly relies on the use of a chiral resolving agent that will undergo a temporary, reversible reaction with both enantiomers, such as formation of a salt. For example, a racemic mixture of a carboxylic acid (RCO_2H) could be resolved using a single enantiomer of a chiral amine ($R*NH_2$). The two resulting salt products (RCO_2^- RNH_3^+) are related as diastereomers and, therefore, can potentially be separated in a number of ways, such as by crystallization. Removal of the chiral resolving agent (by neutralization, in this case) generates the neutral carboxylic acid in an enantiopure form.

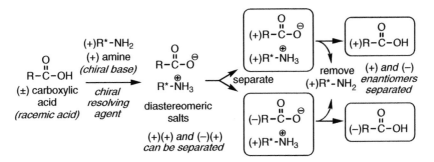

Racemic Mixtures Can Be Resolved to Separate Enantiomers

In a similar manner, a racemic amine could be resolved using a chiral carboxylic acid, such as the naturally occurring tartaric acid.

Resolution of compounds containing functional groups that are not acidic or basic requires a reversible functional group interconversion involving a chiral resolving agent. For example, reaction of a racemic alcohol with a chiral carboxylic acid results in a mixture of diastereomeric esters that can be separated. After separation, hydrolysis of the ester regenerates the desired alcohol. To resolve a racemic ketone or aldehyde, a reaction with a chiral diol or a chiral amine could be employed to form diastereomeric acetals or imines, respectively.

Asymmetric Synthesis: Sharpless Epoxidation

Asymmetric synthesis represents a robust field of research, with entire journals and books devoted to the topic. The current literature is full of examples of new and improved chiral reagents, catalysts, and auxiliaries that can be used for enantioselective syntheses. In every case, a chiral environment is produced that favors reaction with one prochiral face of a planar center, or one prochiral group of a tetrahedral center. The ongoing research seeks not only to develop new asymmetric reactions but to improve those already known, perhaps by making them higher yielding, more selective (higher ee or de), more environmentally friendly (less toxic, greener solvents, etc.), or more cost effective (more efficient, catalytic amounts, etc.).

The Sharpless asymmetric epoxidation of allylic alcohols (one of the reactions that helped K. Barry Sharpless earn his part of the 2001 Nobel Prize in Chemistry) offers a good example of an enantioselective technique that can be used to create either enantiomer of an epoxide product. This reaction uses a diester of tartaric acid, such as diethyl tartrate (DET) or diisopropyl tartrate (DIPT), as the source of chirality. The dialkyl tartrate coordinates with the titanium tetraisopropoxide [Ti(O*i*-Pr)$_4$] catalyst and *t*-butyl hydroperoxide (*t*-BuOOH) to make a chiral oxidizing agent. Since both enantiomers of tartaric acid are commercially available, and each enantiomer will direct the reaction to a different prochiral face of the alkene, both enantiomers of an epoxide can be synthesized.

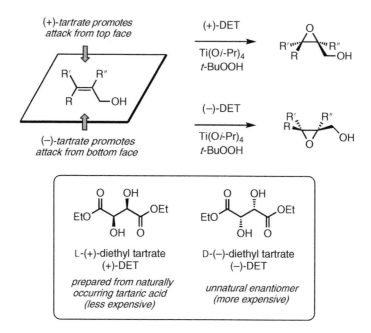

Sharpless Asymmetric Epoxidation of Allylic Alcohols

Enzymatic Transformations: Biocatalysis

Asymmetric syntheses are ubiquitous in nature, and some of Nature's catalysts have been exploited by chemists for use in the laboratory. Enzymes are proteins that can be isolated from natural sources and can be used to catalyze certain chemical transformations. Often, one enantiotopic group on a starting material will react preferentially. Alternatively, only one enantiomer of a racemic starting material will undergo a reaction (called a kinetic resolution), leaving the other enantiomer as unreacted starting material. Advantages to the use of biocatalysts include low cost and high stereoselectivity, but disadvantages include the catalysts' sensitivity to reaction conditions and the limited number of chemical transformations that are possible. Most enzymes used in synthesis catalyze an oxidation, a reduction, or the formation/hydrolysis of an ester. The pursuit of designer enzymes—those created specifically for use as a synthetic catalyst—is inspired by these challenges. Examples of biocatalysts include yeast for reduction reactions and pig liver esterase (PLE) for ester hydrolysis reactions.

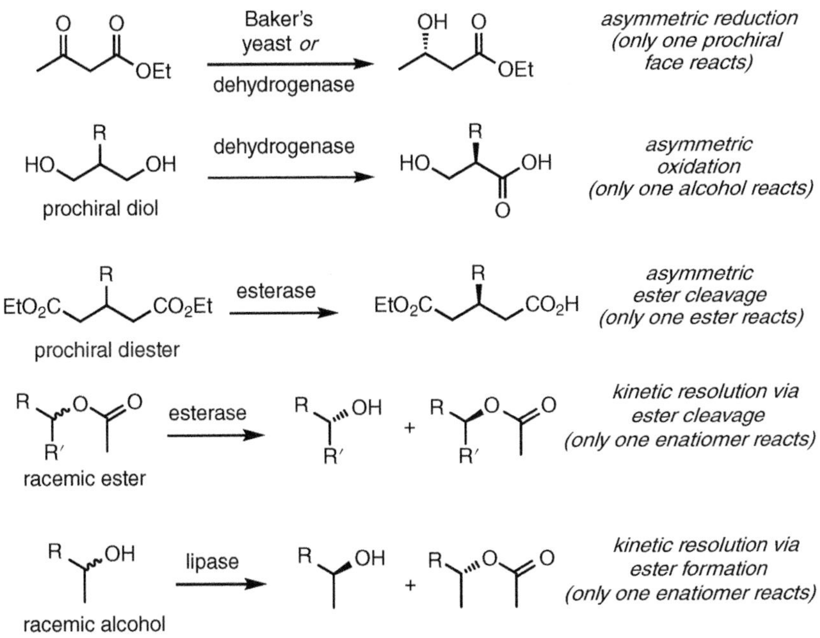

Asymmetric Transformations Using Enzymes

STEREOCHEMISTRY

7-1. Predict the major product(s), paying close attention to stereochemistry of each reaction product.

A
$$CH_3CO_3H$$
$$H_2O$$

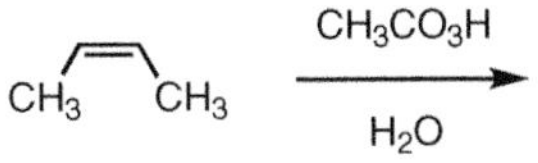

B
1. mCPBA
2. heat

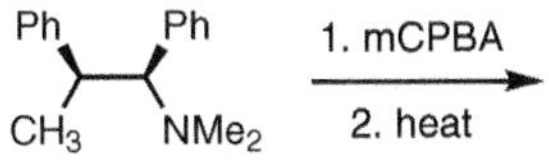

C
t-BuOK
t-BuOH

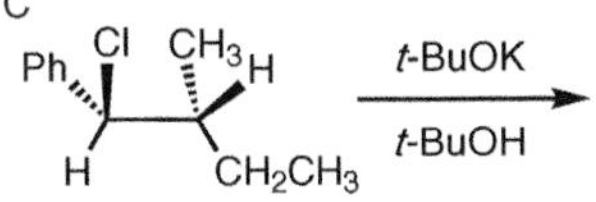

D
NaCN

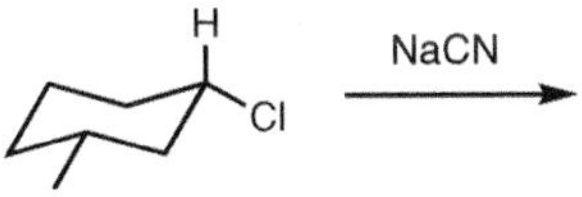

E
heat

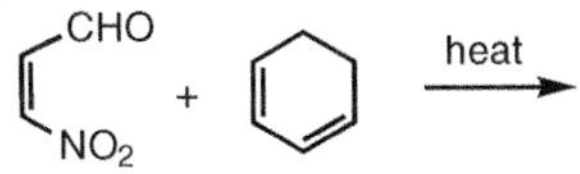

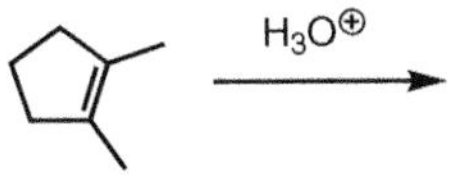

F
H_2 (excess)
Pd

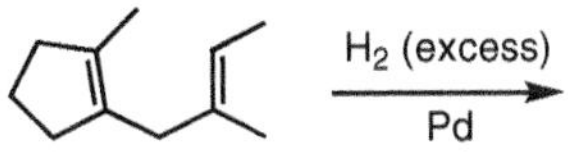

G
$H_3O^\oplus$

H
$KMnO_4$

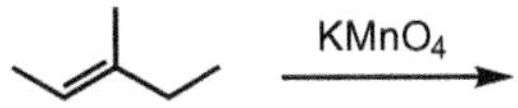

I
H_2O

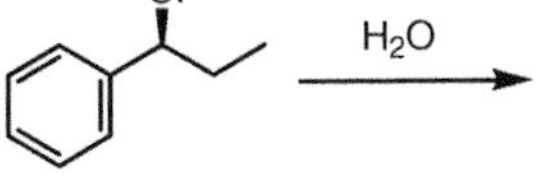

J
t-BuOK
t-BuOH

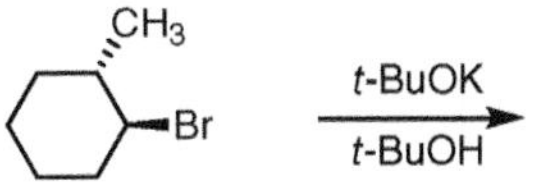

K
heat

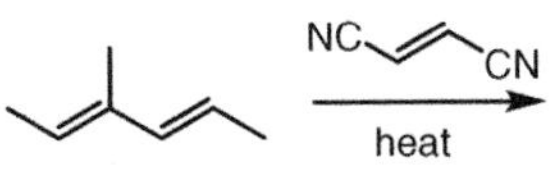

L
Br_2

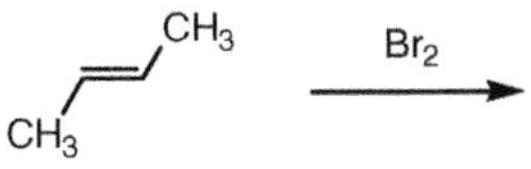

7-2. For each of the following compounds, predict the possible monochlorination products (imagine reacting each with Cl_2, $h\nu$). Consider all possible regioisomers and stereoisomers.

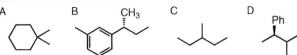

A   B   C   D

Introduction to Strategies for Organic Synthesis, Second Edition. Laurie S. Starkey.
© 2018 John Wiley & Sons, Inc. Published 2018 by John Wiley & Sons, Inc.

7-3. Predict the major product(s), paying close attention to stereochemistry.

A

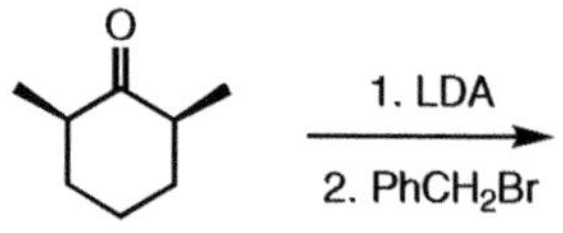

O
1. LDA
2. $PhCH_2Br$

B

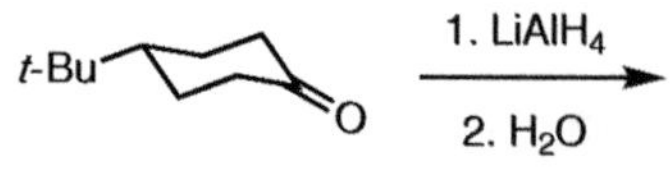

t-Bu
O
1. $LiAlH_4$
2. H_2O

C

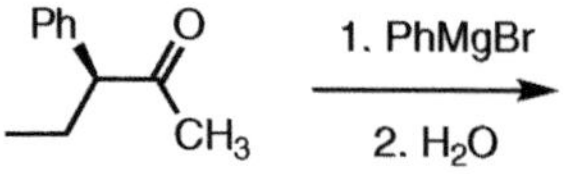

Ph
O
CH_3
1. PhMgBr
2. H_2O

D

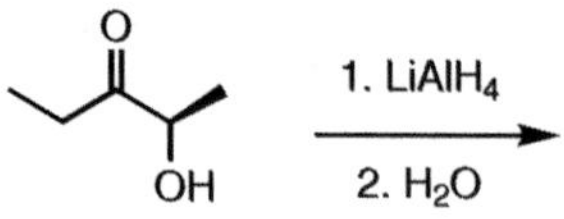

O
OH
1. $LiAlH_4$
2. H_2O

E

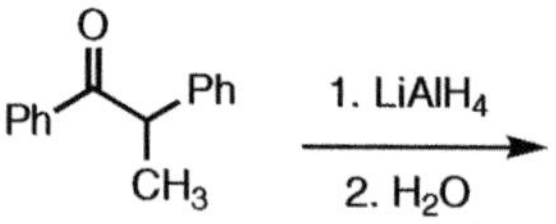

O
Ph
Ph
CH_3
1. $LiAlH_4$
2. H_2O

(racemic)

F

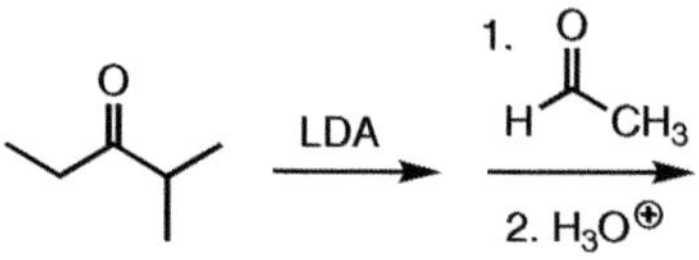

O
LDA
1. O
H
CH_3
2. $H_3O^{\oplus}$

7-4. For each of the following reactions, do you expect to form as the major product the (*E*)- or (*Z*)-enolate?

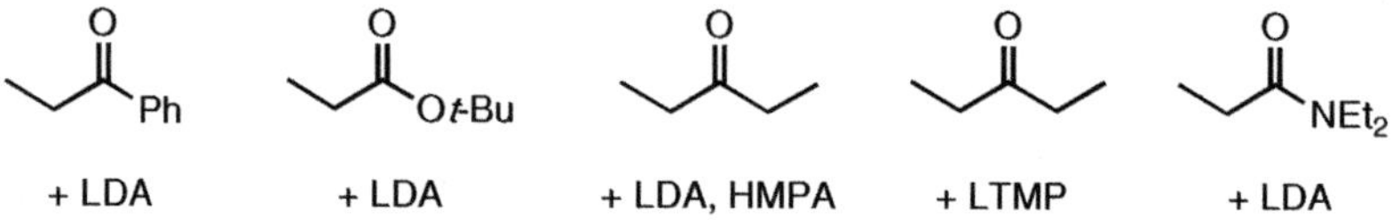

Ph
+ LDA
O*t*-Bu
+ LDA
+ LDA, HMPA
+ LTMP
NEt_2
+ LDA

7-5. Which would be the appropriate starting material to synthesize the given target molecule with the stereochemistry shown? Explain.

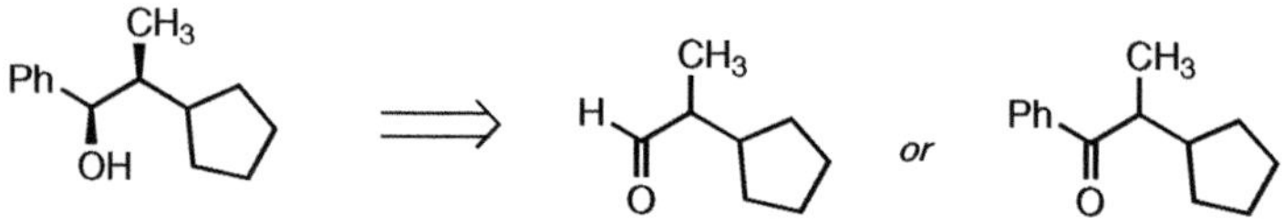

CH_3
Ph
OH
CH_3
H
O
or
CH_3
Ph
O

+ enantiomer

7-6. Briefly describe how the following compound can undergo racemization. Would you expect the following compound to undergo racemization faster under **acidic** or **basic** conditions? Explain using drawings as appropriate.

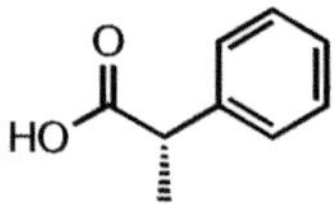

O
HO

7-7. 3-Methylcyclohexanol undergoes epimerization upon treatment with acid. Which chiral center is involved in this reaction? Propose a mechanism. Which diastereomer is predicted to be the major one formed at equilibrium? Explain.

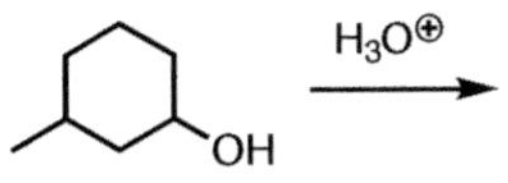

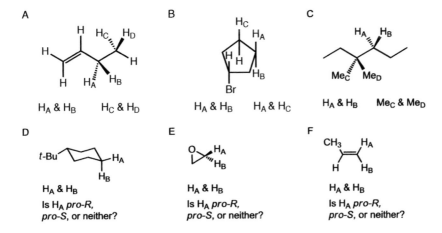

7-8. Draw the two products that will be formed in the following bromination reaction. Do you expect these two products to be formed in equal amounts? Explain using an energy diagram and appropriate drawings to support your answer.

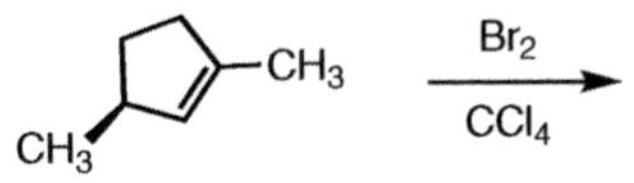

7-9. Identify the labeled groups as homotopic, enantiotopic, or diastereotopic and determine whether the indicated hydrogens are pro-R or pro-S (or neither).

A   B   C

H_A & H_B H_C & H_D H_A & H_B H_A & H_C H_A & H_B Me_C & Me_D

D   E   F

H_A & H_B H_A & H_B H_A & H_B

Is H_A pro-R,   Is H_A pro-R,   Is H_A pro-R,
pro-S, or neither?   pro-S, or neither?   pro-S, or neither?

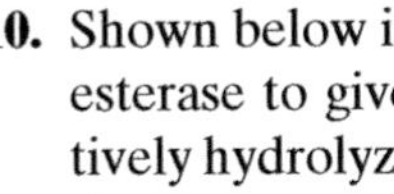

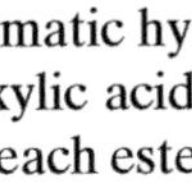

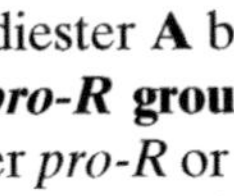

7-10. Shown below is the enzymatic hydrolysis of the diester **A** by pig liver esterase to give a carboxylic acid product. The ***pro-R* group** is selectively hydrolyzed. Label each ester group as either *pro-R* or *pro-S* and draw a structure of the product (including stereochemistry).

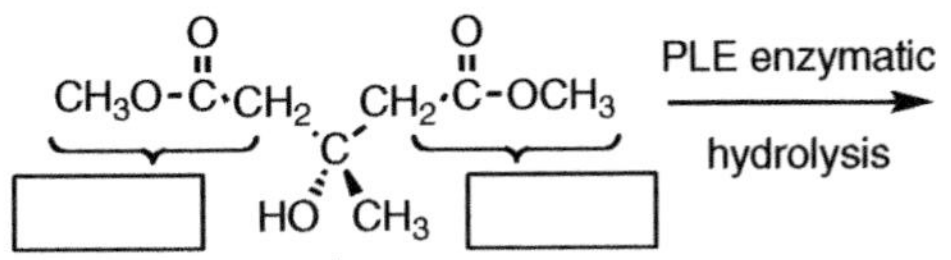

$CH_3O-\overset{\overset{O}{\|}}{C}\cdot CH_2 \quad CH_2\cdot\overset{\overset{O}{\|}}{C}-OCH_3$ PLE enzymatic

HO CH_3

A   hydrolysis

7-11. Enzymatic oxidation of naphthalene by bacteria proceeds to give the *cis*-diol shown. Which prochiral faces of C-1 and C-2 of naphthalene are hydroxylated in this process? (Label each as *re* or *si*.)

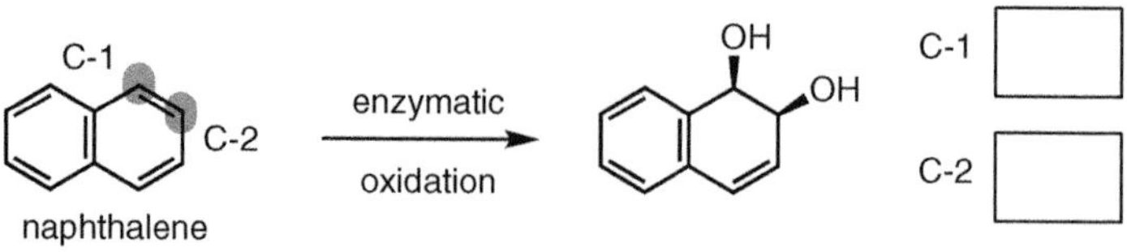

naphthalene   enzymatic oxidation

7-12. The compound given below can be prepared via a Sharpless asymmetric epoxidation. Provide the structure of the required starting material and determine if the (+) or (−) enantiomer of diethyl tartrate (DET) is required to produce the enantiomer shown.

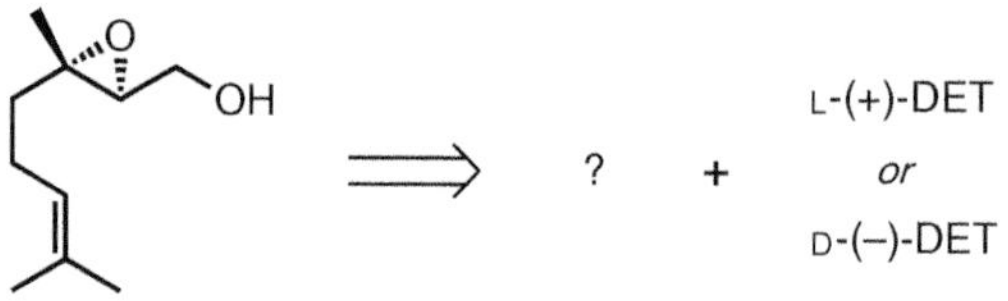

⟹   ?   +   L-(+)-DET
                    or
              D-(−)-DET

7-13. You need to separate the enantiomers in a racemic mixture of 1-amino-2-cyclohexene. Choose an appropriate resolving agent from the compounds shown below and outline the complete procedure to accomplish the resolution.

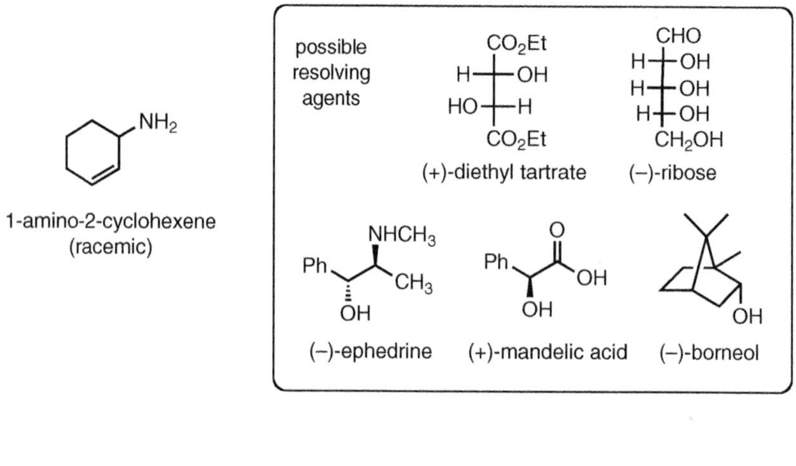

1-amino-2-cyclohexene
(racemic)

possible resolving agents

CO₂Et
H——OH
HO——H
CO₂Et
(+)-diethyl tartrate

CHO
H——OH
H——OH
H——OH
CH₂OH
(−)-ribose

(−)-ephedrine

(+)-mandelic acid

(−)-borneol

CHAPTER 8

TRANSITION METAL-MEDIATED CARBON–CARBON BOND FORMATION

As the field of organic synthesis moved from the 20th to the 21st century, the continuous development of new tools opened doors to an ever-growing list of new synthetic strategies. While the identification of suitable nucleophiles and electrophiles continues to be critical to the understanding of the *logic* of planning an organic synthesis, the fact is that the majority of new carbon–carbon bonds in modern synthetic processes are formed via reactions mediated by transition metals. The study of transition metals lies at the crossroads of inorganic, physical, and organic chemistry, and a thorough treatment of the topic is beyond the scope of this textbook.* The goal of this chapter is to introduce some basic skills for understanding the structure and reactivity of transition metal complexes, and their application to a select sample of common organic transformations. After working with this chapter you will learn to see certain functional groups and disconnections with new eyes, and you will be better prepared to follow many of the synthetic schemes encountered in the modern literature.

---

* For an excellent introduction to synthetic applications of transition metals, please see L. Hegedus and B. Soderberg, *Transition Metals in the Synthesis of Complex Organic Molecules*, 3rd ed. (Sausalito, CA: University of Science Books, 2009).

Introduction to Strategies for Organic Synthesis, Second Edition. Laurie S. Starkey.
© 2018 John Wiley & Sons, Inc. Published 2018 by John Wiley & Sons, Inc.

TRANSITION METAL COORDINATION COMPLEXES

Most transition metals serve as effective and versatile homogeneous catalysts because they can exist in multiple oxidation states and most are capable of reversibly bonding with up to six or more groups (atoms or molecules). The neutral or cationic metal serves as the central atom that is surrounded by electron-sharing groups called ligands, which coordinate (bond) to the metal center. The resulting complex, described by the ligand coordination sphere, acts like an amazing factory that brings in a variety of starting materials, holds them just tightly enough to manipulate them, and then releases the resulting products into solution. The central metal undergoes a sequence of changes throughout the process, but the metal complex ultimately must restore itself to its original configuration so it is ready to resume the cycle as an effective catalyst. Therefore, although these metals may be expensive they are ideally needed in only catalytic amounts, so these microscopic factories are, in fact, affordable work horses that can orchestrate transformations that are otherwise difficult to achieve.

8.1.1 COUNTING ELECTRONS: THE 18-ELECTRON RULE

The stability and reactivity of Main Group elements, including carbon, are driven by the octet rule; a total of eight valence electrons around a carbon atom corresponds to a stable configuration with filled valence orbitals (one s orbital + three p orbitals = four orbitals total, needing eight electrons to be full for stability). Transition metals, however, have additional orbitals (five d orbitals), so they can accommodate a total of 18 valence electrons, and such metal complexes with 18 electrons are especially stable. The electron count under consideration is the total of the metal's d electrons and the electrons donated by each ligand. It is important to note that according to the formalism

Introduction to Strategies for Organic Synthesis, Second Edition. Laurie S. Starkey.
© 2018 John Wiley & Sons, Inc. Published 2018 by John Wiley & Sons, Inc.

used, the valence electrons that are normally placed in an s orbital when predicting an atomic configuration of an isolated metal are included in the count of d electrons in the metal complex. This formalism accounts for the fact that the presence of the ligands affects the electronic structure of the metal, and it agrees best with what is typically observed experimentally. As usual, the number of d electrons is reduced by one for every charge on a cationic metal. Some select neutral and cationic transition metals are given below, along with the number of electrons needed from ligands to satisfy the 18-electron rule.

Examples of Transition Metals Used in Synthesis

Metal	Sc	Ti Zr	V	Cr Mo W	Mn	Fe Ru Os	Co(I) Rh(I) Ir(I)	Ni Pd Pt	Cu(I)	Zn(II)
"d" electrons	3	4	5	6	7	8	8	10	10	10
Electrons needed	15	14	13	12	11	10	10	8	8	8

Transition metal complexes with fewer than 18 electrons are described as being coordinatively unsaturated, because they are able to accommodate additional ligands. To determine the oxidation state of the metal using the ionic model, the ligands are detached and the bonding electrons leave with the ligand. Any resulting formal charge on the ligand is combined with the "charge" on the metal to give the overall charge on the complex. Let us take a look at some examples. The first complex, $RhCl(PPh_3)_3$, known as Wilkinson's catalyst, is a neutral compound that is used for hydrogenation reactions. Since there is only one anionic ligand (Cl^-), and the complex has no overall charge, the rhodium must have a +1 oxidation state. Rh(I) has 8 d electrons and the ligands provide an additional 8 electrons, for a total of 16, so Wilkinson's catalyst is coordinatively unsaturated.

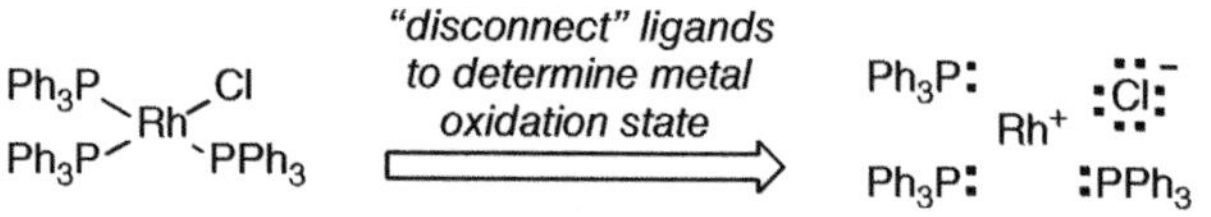

Net charge: $+1(Rh^+) + (-1)(Cl^-) = 0$

Electron count: $Rh(I)d^8 + 2e^-(Cl) + 6e^-(PPh_3 \times 3) = 16e^-$

Ionic Model for Determining Metal's Oxidation State

While a solvent molecule (which can coordinate using a lone pair of electrons) is likely to occupy the vacant coordination site in order to stabilize the metal complex, it is easily displaced when another ligand is introduced, such as an alkene. Let us take a look at the structure of Wilkinson's catalyst as it is facilitating the hydrogenation of an alkene, as shown below. Here, we see an alkene ligand that is using pi bond electrons (rather than a lone pair) to coordinate to the metal, and two commonly encountered notations are provided. At this stage, the catalyst has a Rh(III) metal center and is coordinatively saturated.

Net charge: $+3(Rh^{3+}) + (-1)(Cl^-) + (-1)(H^-) + (-1)(H^-) = 0$

Electron count: $Rh(III)d^6 + 2e^- (Cl) + 4e^-(H \times 2) + 4e^-(PPh_3 \times 2) + 2e^- (alkene) = 18e^-$

Hydrogenation Metal Complex (Wilkinson's Catalyst)

The next example is a salt, tetraamminedichlorocobalt(III) chloride $[Co(NH_3)_4Cl_2]Cl$. Normally, the coordination complex is given in brackets, with the counterion(s) listed before or after, as appropriate (typically the cation is listed first, as in NaCl). Please note that this compound name does not contain a typographical error—amine describes a class of organic compounds, while ammine refers to an ammonia (NH_3) ligand. This nomenclature twist reminds us that we are indeed on an interdisciplinary adventure as we explore organometallic chemistry!

Net charge: $+3(Co^{3+}) + (-1)(Cl^-) + (-1)(Cl^-) = +1$

Electron count: $Co(III)d^6 + 4e^-(Cl \times 2) + 8e^-(NH_3 \times 4) = 18e^-$

Transition Metal Complexes may be Charged

Finally, let us look at another interesting type of ligand as demonstrated in the complex given below, known as a metal carbene. The exact structure of the carbene, a carbon atom with two bonds and two nonbonded electrons, can affect the bonding relationship of the ligand with the metal. The "singlet" carbene shown is a neutral ligand, as are the carbon monoxide (CO) ligands. A carbon monoxide ligand is named as a carbonyl group.

Net charge: $0(Cr^0) + 0(CO \times 4) + 0(carbene) = 0$

Electron count: $Cr(0)d^6 + 8e^-(CO \times 4) + 2e^-(carbene) = 16e^-$

Carbene Ligands

Polydentate Ligands

In all of the preceding examples, each ligand donated one pair of electrons (from a lone pair or a pi bond) and occupied one position in the coordination sphere of the metal. Ligands that bind at two sites are described as bidentate ("two-toothed") and are said to chelate (grab like a claw) with the metal as they occupy two coordination sites. As with the monodentate ligands, bidentate ligands can be neutral or negatively charged. Some examples include the oxalate dianion (the salt of oxalic acid, named the oxalato ligand), as illustrated below in the complex $K_3[Fe(C_2O_4)_3]$, and ethylenediamine (abbreviated as "en"), as shown in $[CoCl_2(en)_2]Cl$.

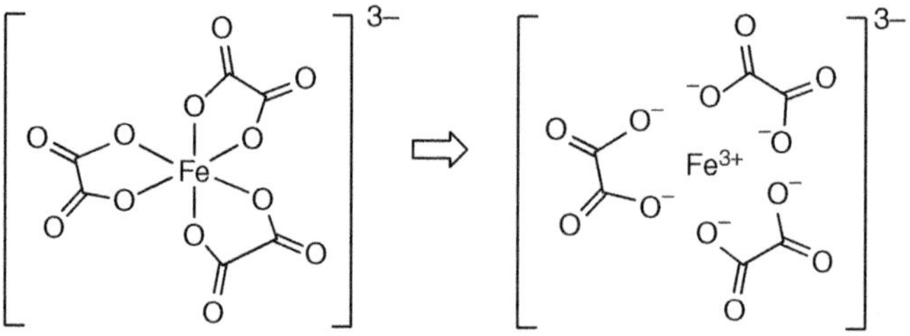

Net charge: $+3(Fe^{3+}) + (-6)(C_2O_4{}^{2-} \times 3) = -3$
Electron count: $Fe(III)d^5 + 12e^-(C_2O_4{}^{2-} \times 3) = 17e^-$

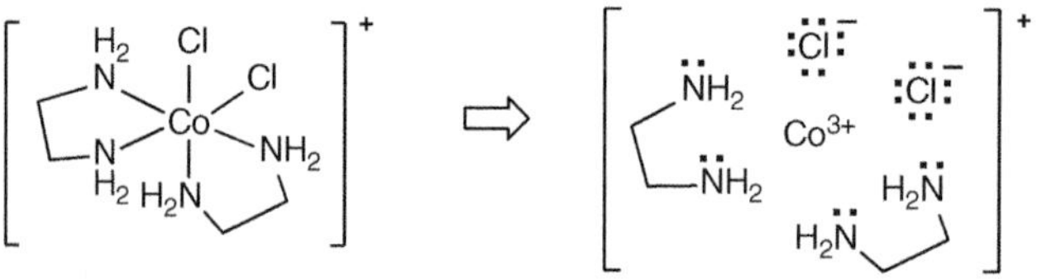

Net charge: $+3(Co^{3+}) + (-2)(Cl^- \times 2) + 0(en \times 2) = +1$
Electron count: $Co(III)d^6 + 4e^-(Cl \times 2) + 8e^-(en \times 2) = 18e^-$

Examples of Bidentate Ligands

Some ligands are capable of donating more than two pairs of electrons, such as benzene and the cyclopentadienyl anion (both occupy three coordination sites and donate six electrons in the ionic model). The well-known ferrocene complex contains an Fe(II) atom "sandwiched" between two cyclopentadienyl anions. Recall that the cyclopentadienyl anion (Cp⁻) is aromatic, so the delocalized pi electrons can be represented with a circle inside the five-membered ring. Ferrocene is named bis(η^5-cyclopentadienyl) iron—the η^5 in the name indicates that the p orbitals on all five atoms of the ring are involved in the bonding.

Net charge: $+2(Fe^{2+}) + (-2)(Cp^- \times 2) = 0$
Electron count: $Fe(II)d^6 + 12e^-(Cp \times 2) = 18e^-$

Cyclopentadienyl Ligands (Ferrocene)

Finally, let us see if we can take what we have learned so far and make sense of a transition metal complex that is fairly complex! Allylpalladium chloride, $(\eta^3\text{-}C_3H_5)_2Pd_2Cl_2$, is a dimer that involves two bridged chloro ligands and the resonance-delocalized propenyl anion (the bidentate allyl ligand). Two different representations of this "allyl complex" are given below, as well as an analysis of the oxidation state of the metal. Note that even as a dimer this compound remains coordinatively unsaturated.

allyl group is a bidentate ligand

Cl is neutral
2e⁻ donor
to one Pd

Cl is anionic
2e⁻ donor
to other Pd

Net charge: $+2(Pd^{2+}) + (-1)(allyl^-) + (-1)(Cl^-) = 0$
e⁻ count: $Pd(II)d^8 + 4e^-(Cl \times 2) + 4e^-(allyl) = 16e^-$

Allyl Ligands

8.1.2 PALLADIUM CATALYSTS

Many of the carbon–carbon bond forming reactions found in modern syntheses are catalyzed by Pd(0) complexes. A Pd(0) catalyst can be used directly, or one can be prepared by *in situ* reduction of a Pd(II) catalyst. A wide variety of these palladium catalysts exist, so it can sometimes be overwhelming when it seems like no two articles in the literature are using the same experimental procedures! Each catalyst has its own advantages, and reagent choice depends on a number of factors including reactivity, stability, and cost. Ligand selection is especially important as ligands influence the metal's reactivity due to steric and electronic effects (each ligand has varying abilities to donate electrons to, and in some cases receive electrons from, the metal center). Many organometallic reagents are sensitive to air (they can be oxidized by O_2 and/or may react with water), so they may need to be freshly prepared. Air-sensitive complexes are typically handled under an inert atmosphere in a glove box or a glove bag or in a fume hood fitted with specialized vacuum/gas manifolds called Schlenk lines. Using this equipment, one can achieve an oxygen-free environment by using a vacuum pump to evacuate the working space and then filling the chamber with an inert gas, such as nitrogen or argon. Shown below are two commonly used Pd(0) catalysts, tetrakis(triphenylphosphine)palladium, $Pd(PPh_3)_4$, and bis(dibenzylideneacetone)palladium, $Pd(dba)_2$.

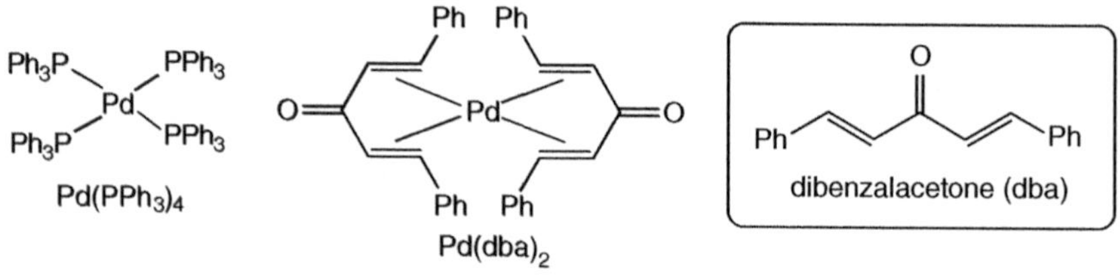

Palladium(0) Catalysts

Palladium(II) complexes are readily reduced to Pd(0) by many different compounds, including carbon monoxide, alcohols, tertiary amines, alkenes, and phosphines. It may seem confusing when a Pd(II) reagent is used for a Pd(0)-catalyzed reaction, but if you look closely at the reaction conditions you will find an available reducing agent. The details of the reduction are beyond the scope of this textbook, but it is important to recognize that a mechanism involving a Pd(0) species may be possible when *any* palladium reagent is given in a reaction. Shown below are some commonly encountered Pd(II) catalysts, bis(acetonitrile)dichloropalladium(II), bis(triphenylphosphine)palladium(II) chloride, lithium tetrachloropalladate(II), and palladium(II) acetate.

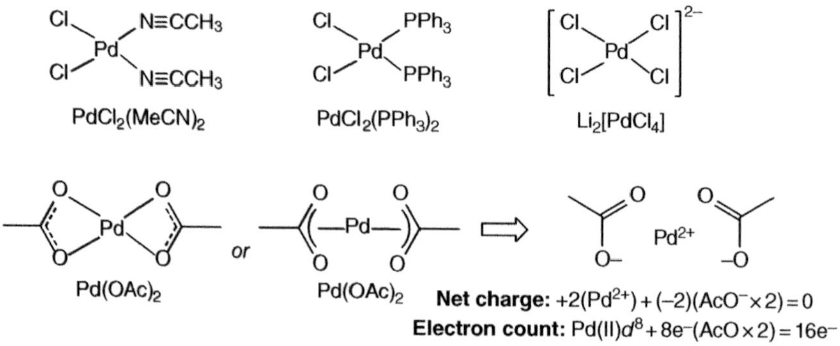

Net charge: $+2(Pd^{2+}) + (-2)(AcO^- \times 2) = 0$
Electron count: $Pd(II)d^8 + 8e^-(AcO \times 2) = 16e^-$

Palladium(II) Catalysts

PRACTICE PROBLEM 8.1: TRANSITION METAL COMPLEXES

8.1 For each of the following organometallic compounds, give the electron count, the formal oxidation state of the metal, and the metal's d^n configuration.

A

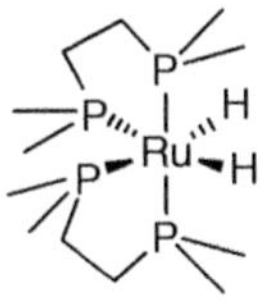

Ru(dmpe)$_2$H$_2$ (dmpe,
1,2-Bis(dimethylphosphino)ethane)

B

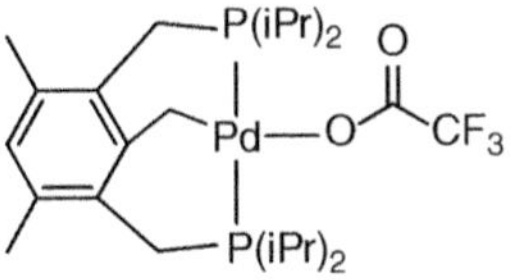

C

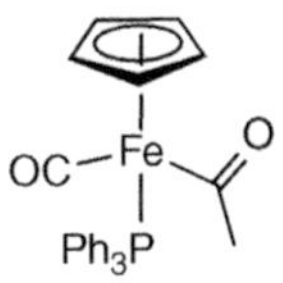

D

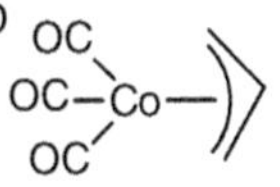

E

[Rh(NH$_3$)$_5$Br]Br$_2$

ORGANOMETALLIC REACTION MECHANISMS

As the microscopic factory known as a coordination complex does its work, ligands are continuously coming and going at the metal center. Throughout the course of a catalytic cycle, there are a number of common mechanisms employed. The details of these mechanisms (e.g., concerted vs. stepwise, direction of electron flow "arrow pushing") are beyond the scope of this book. Instead, we will be looking at a sampling of organometallic complexes in catalytic processes.

8.2.1 LIGAND SUBSTITUTION

Ligand substitution (or ligand exchange), the swapping of one ligand for another, is a substitution reaction that has no effect on the oxidation state of the metal. Because any given ligand may be tightly or loosely bound, some ligand exchanges are facile while others are unlikely to occur. If a metal has a similar affinity for two ligands, the direction of the reversible substitution is determined by Le Chatelier's principle. Many transition metal complexes are colored, so ligand exchange reactions may be accompanied by a color change. Such is the case when the pink-colored hydrated cobalt(II) complex $[Co(H_2O)_6]^{2+}$ is treated with excess hydrochloric acid to give the blue $[CoCl_4]^{2-}$ product. This reaction has a fairly complicated mechanism, but the overall transformation results in a ligand substitution. Because chloride ions are larger than water molecules, only four chloride ions are coordinated to the metal after exchanging with six water ligands. As always, the metal stays at the same formal oxidation state, +2 in this case.

Introduction to Strategies for Organic Synthesis, Second Edition. Laurie S. Starkey.
© 2018 John Wiley & Sons, Inc. Published 2018 by John Wiley & Sons, Inc.

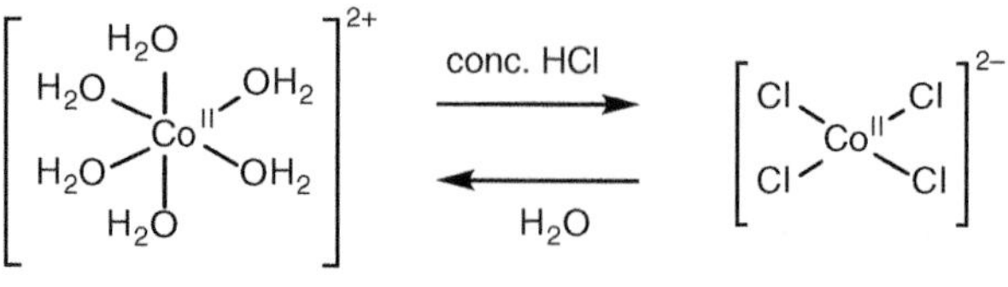

hexaaquacobalt(II) ion
pink solution

$[CoCl_4]^{2-}$ ion
blue solution

Ligand Substitution

The manufacturers of Drierite™ desiccant have found a practical use for a similar color-changing reaction. "Indicating" Drierite is made from the colorless drying agent calcium sulfate, $CaSO_4$, mixed with a small amount of anhydrous cobalt(II) chloride, $CoCl_2$, which has a blue color. As the desiccant does its job adsorbing water, the cobalt(II) hydrate is formed and the Drierite begins to turn purple and will eventually turn pink. The drying agent can be regenerated to its anhydrous, blue state by heating to drive off the adsorbed water.

8.2.2 MECHANISMS THAT CHANGE THE METAL'S OXIDATION STATE

Oxidative Addition

When a transition metal atom is inserted into a sigma bond, the result is the addition of two new ligands on the metal center. This process is called an oxidative addition because it results in an *increase* in the oxidation state of the metal by two (i.e., the metal has been oxidized). In order to accommodate two new groups, the metal must be either coordinatively unsaturated or bonded to ligand(s) that can be easily exchanged. In the example shown below, a palladium(0) complex becomes Pd(II) when it adds the bromo and phenyl groups from bromobenzene.

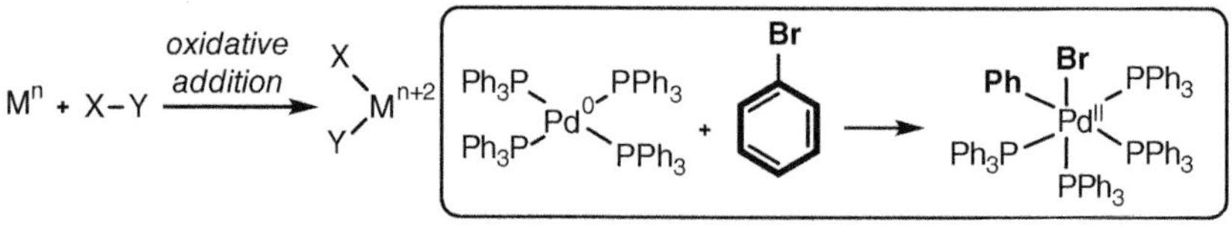

Oxidative Addition Results in +2 Change in Oxidation Number

Reductive Elimination

The reverse of an oxidative addition process is called a reductive elimination. Here, two adjacent ligands leave the metal and become joined together with a sigma bond. This process is described as a reductive elimination because it results in a *decrease* in the oxidation state of the metal by two (i.e., the metal has been reduced). In the example shown, the methyl and acyl groups are eliminated to produce a ketone, and the Rh(III) is reduced to Rh(I) in the process.

Reductive Elimination Results in −2 Change in Oxidation Number

The formation of a new carbon–carbon bond in the example given above illustrates the synthetic utility of transition metal-catalyzed reactions. At first glance, oxidative addition and reductive elimination may simply appear to be the forward and reverse of the same reaction, but countless possibilities unfold when the two groups being eliminated are different from the two groups that were initially added. This is the strategy employed for the "cross-coupling" reactions that will be explored in this chapter. (The significance of such reactions was recognized with the 2010 Nobel Prize in Chemistry, which was awarded to Richard F. Heck, Ei-ichi Negishi, and Akira Suzuki for their pioneering work in this field.)

Palladium-Catalyzed Cross-Coupling Reaction

8.2.3 MECHANISMS THAT RETAIN THE METAL'S OXIDATION STATE

Coordination of a transition metal to the pi bond of an unsaturated species, such as an alkene or carbon monoxide, does not change the metal's formal oxidation state. The next two mechanisms involve the interaction of a C or H ligand with an adjacent unsaturated ligand.

Migratory Insertion

The insertion of an unsaturated group such as an alkene into a M–H or M–R (R = alkyl, vinyl or aryl) bond is described as a migratory insertion (more specifically, this is described as a 1,2-migratory insertion). The hydride or carbon atom brings with it two electrons and adds to one end of the pi bond, and the other end of the pi bond coordinates to the metal in the position that was just vacated by the migrating group. There is no change to the oxidation state of the metal, but it will be coordinatively less saturated because two ligands have merged to become one group.

1,2-Migratory Insertion: Hydrometalation of an Alkene

From the alkene's point of view, a "hydrometalation" addition reaction has occurred where a metal and a hydride have been added across the double bond. The migratory insertion mechanism forms two new sigma bonds in a stereospecific manner (*syn* addition). If the migrating group is alkyl, vinyl, or aryl, then a new carbon–carbon bond has been formed!

Beta Elimination

The reverse of a 1,2-migratory insertion process is described as beta elimination. The most commonly encountered examples are beta-hydride eliminations, because they occur most readily. Here, a hydride that is adjacent to the metal bond is transferred to the metal with the simultaneous formation of a pi bond.

β-Hydride Elimination to Form an Alkene

As with a sequence of oxidative addition and reductive elimination reactions, an insertion/elimination sequence can also lead to a coupling reaction (see the Heck Reaction). Another application for this sequence is the migration of a pi bond that results when the hydrogen removed in the beta-elimination step is in a different position than the hydrogen that was added in the insertion reaction. The following metal-catalyzed isomerization of an alkene illustrates such a combination of migratory insertion and beta-hydride elimination mechanisms. To follow along in the catalytic loop provided below, begin at the rhodium complex at the top and continue in a clockwise direction. The loop gives the structure of the metal complex throughout the catalytic cycle, with the final step of the mechanism involving the regeneration of the original catalyst (please note that simplified versions of the metal complexes are shown in this chapter's mechanisms, and no geometry is implied). As you can see by tracking the highlighted carbon atom (C2) below, a new hydride is added to the pi bond (at C1) and a different hydride is removed (from C3) in the beta elimination.

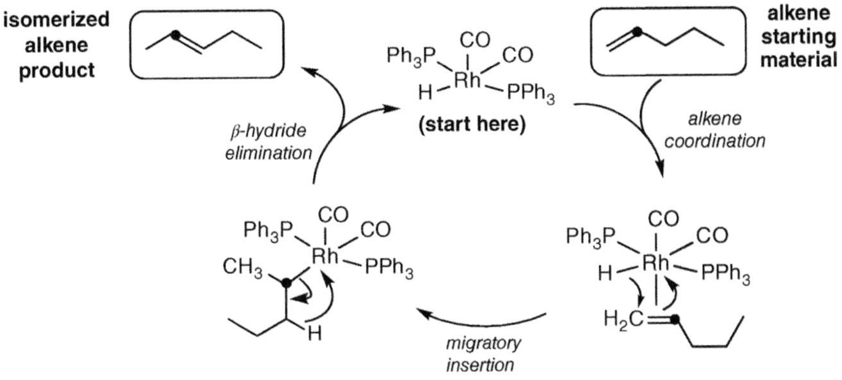

Catalytic Cycle of Alkene Isomerization

8.2.4 TRANSMETALATION

When the metal coordinated to a carbon atom is exchanged for different metal, the process is described as transmetalation. The equilibrium of this exchange can be driven by a variety of factors, including the electronegativities of the metals involved.

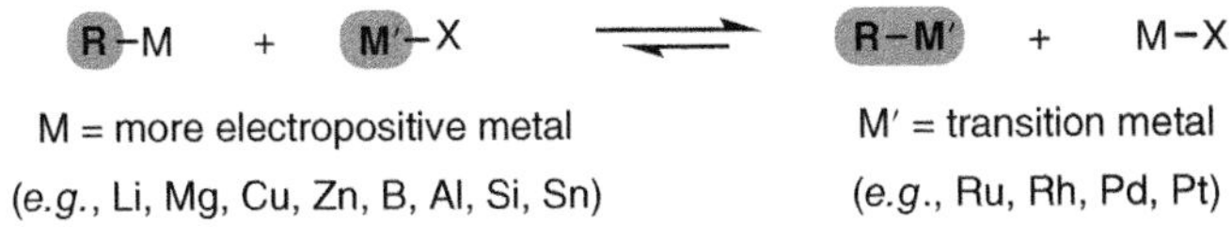

M = more electropositive metal

(e.g., Li, Mg, Cu, Zn, B, Al, Si, Sn)

M′ = transition metal

(e.g., Ru, Rh, Pd, Pt)

Transmetalation Exchanges Coordinated Metal Atoms

In the following example, one of the carbon groups of an organozinc compound displaces a halogen on a palladium complex; this reaction is thermodynamically favorable because the halogen moves to a more electropositive metal. The methyl group has undergone a transmetalation and as a result the palladium complex has a new coordinated alkyl group.

Transmetalation Driven by Differences in Electronegativities

Transmetalation plays a prominent role in a variety of carbon–carbon bond-forming reactions that are important in organic synthesis, such as Stille (Sn), Suzuki–Miyaura (B), Kumada (Mg), Negishi (Zn), Sonogashira (Cu), and Hiyama (Si) coupling reactions.

EXAMPLE OF A TYPICAL CATALYTIC CYCLE: ALKENE HYDROGENATION WITH WILKINSON'S CATALYST

Now that we have an understanding of the basic types of organometallic reactions, we can make sense of each step in a typical catalytic cycle. Recall that the hydrogenation of an alkene requires a catalyst. Catalysts that do not dissolve in the reaction solvent are described as *heterogeneous*. Heterogeneous catalysts, such as Pd, Pt, and Ni metals, are convenient because they are commercially available and they can be easily removed at the end of the reaction by simple filtration. If organic ligands are added to a metal, the resulting organometallic complex is generally soluble in organic solvents. Such *homogenous* catalysts are synthetically useful because they may allow for milder reaction conditions, they may be more tolerant of other functional groups, and they may be more selective (by preferentially reducing just one double bond in a diene, for example).

$$H_2C=CH_2 \xrightarrow[\textbf{catalyst}]{H_2} CH_3CH_3$$

heterogeneous catalyst	**homogeneous catalyst**
Pd, Pt, Ni	organometallic reagent such as Wilkinson's catalyst, RhCl(PPh₃)₃

Heterogeneous and Homogeneous Catalysts for Hydrogenation

Let's take a look at a proposed mechanism for the catalytic hydrogenation of an alkene using Wilkinson's catalyst, tris(triphenylphosphine)rhodium(I) chloride.

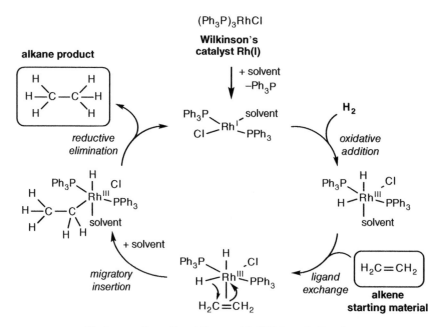

Hydrogenation of an Alkene with Wilkinson's Catalyst

Wilkinson's catalyst is coordinatively unsaturated, so after ligand exchange to replace one of the PPh_3 groups it readily undergoes oxidative addition with H_2 gas. The alkene starting material coordinates to the resulting Rh(III) atom by exchanging with a solvent ligand, and the first hydrogen atom is added to the alkene via migratory insertion. Finally, reductive elimination forms the fully hydrogenated alkane product and regenerates the Rh(I) complex so the catalytic cycle can continue. Notice the cycle involves Rh(I)-Rh(III) oxidation state changes, with two electrons involved in each bond-breaking and bond-forming step.

CARBONYLATION AND DECARBONYLATION

Migratory insertion of a CO ligand to a metal-carbon bond results in the addition of a carbonyl group, so metal carbonyls can be useful catalysts for the synthesis of aldehydes, ketones, and carboxylic acid derivatives. Insertion of CO into metal–carbon sigma bonds (described as a 1,1-migratory insertion) occurs readily, and the migrating group R can be alkyl, vinyl, or aryl.

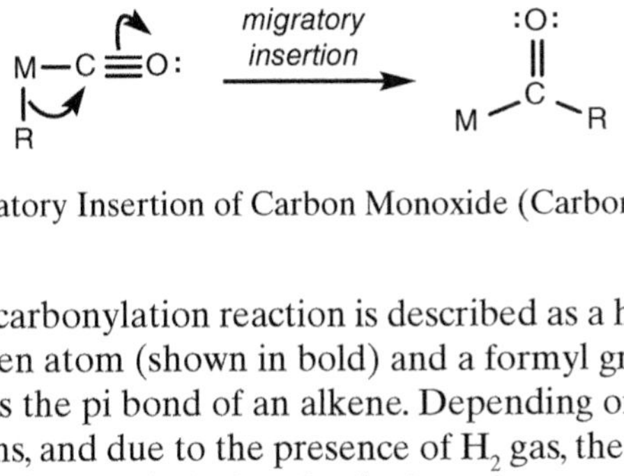

1,1-Migratory Insertion of Carbon Monoxide (Carbonylation)

The following carbonylation reaction is described as a hydroformylation, because a hydrogen atom (shown in bold) and a formyl group (CHO) have been added across the pi bond of an alkene. Depending on the catalyst and reaction conditions, and due to the presence of H_2 gas, the aldehyde may be reduced *in situ* to give an alcohol as the final product.

Introduction to Strategies for Organic Synthesis, Second Edition. Laurie S. Starkey.
© 2018 John Wiley & Sons, Inc. Published 2018 by John Wiley & Sons, Inc.

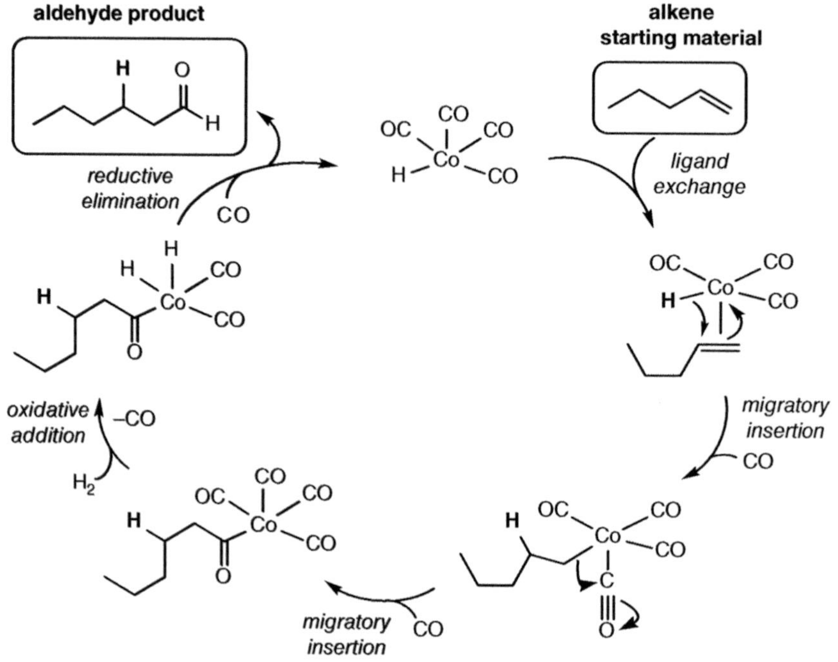

Hydroformylation of an Alkene

The CO insertion reaction is reversible, so transition metal catalysts also mediate decarbonylation reactions. The mechanism is described as α-elimination and can be thought of as a migratory deinsertion.

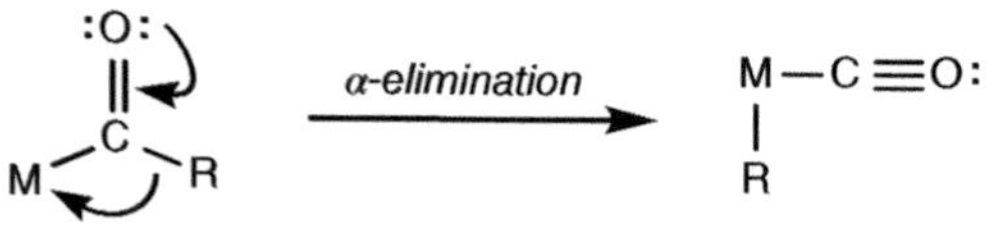

α-Elimination of Carbon Monoxide (Decarbonylation)

Decarbonylation can be used for a variety of transformations, including converting aldehydes to alkanes and acid halides to alkyl halides. A proposed mechanism is shown below for the removal of CO from an aldehyde using Wilkinson's catalyst [RhCl(PPh$_3$)$_3$]. To simplify the mechanism, the abbreviation L$_n$ is used to represent an unspecified number of ligands coordinated to the metal (Ph$_3$P and/or solvent molecules in this case). Note that stereochemistry is retained for both carbonylation and decarbonylation reactions.

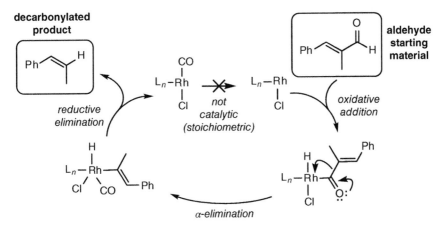

Decarbonylation of an Aldehyde with Wilkinson's Catalyst

THE HECK REACTION
(ArX + ALKENE → Ar-ALKENE)

The Heck reaction is one of the earliest Pd-catalyzed carbon–carbon bond forming reactions, dating back to 1972. Recall that Richard Heck was one of the chemists who shared the 2010 Nobel Prize in Chemistry for this work.

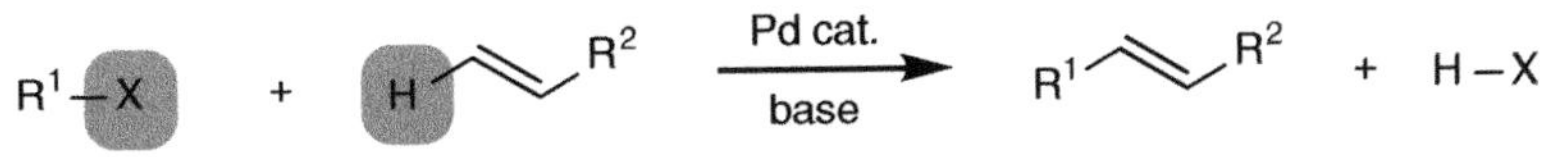

R^1 = aryl or vinyl; R^2 = EWG or EDG (Ar, CO_2R, CN, R, H); X = I, Br, OTf

Heck Reaction

Using an alkene and an organohalide (or triflate), the Heck reaction joins two sp^2-hybridized carbons and can be described as the arylation or alkenylation of a monosubstituted alkene. The Heck reaction offers excellent stereoselectivity, typically giving a trans alkene as the major product, and the reaction is tolerant of a wide variety of functional groups.

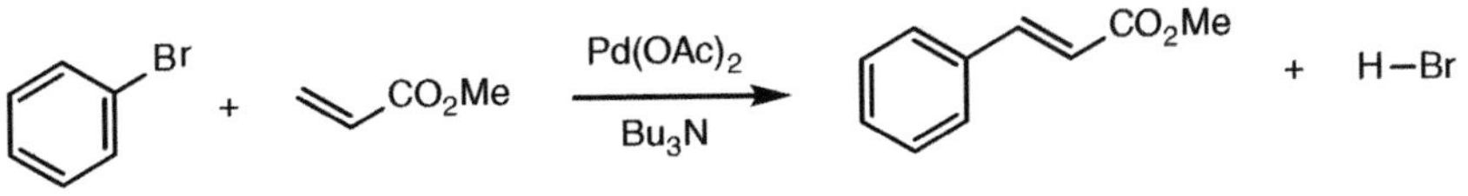

Example of the Heck Reaction

After *in situ* reduction of the Pd(II) catalyst, the Heck reaction mechanism involves a Pd(0)–Pd(II) catalytic cycle, and a stoichiometric amount of base is required to neutralize the acid byproduct that is generated.

Introduction to Strategies for Organic Synthesis, Second Edition. Laurie S. Starkey.
© 2018 John Wiley & Sons, Inc. Published 2018 by John Wiley & Sons, Inc.

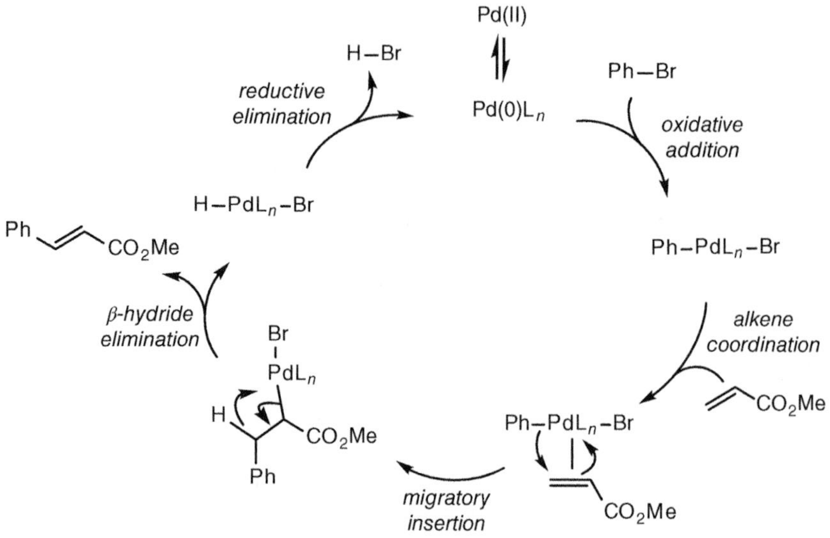

Heck Reaction Mechanism

PRACTICE PROBLEM 8.4: THE HECK REACTION

8.4 Predict the product or provide the missing reagent in the following Heck reactions.

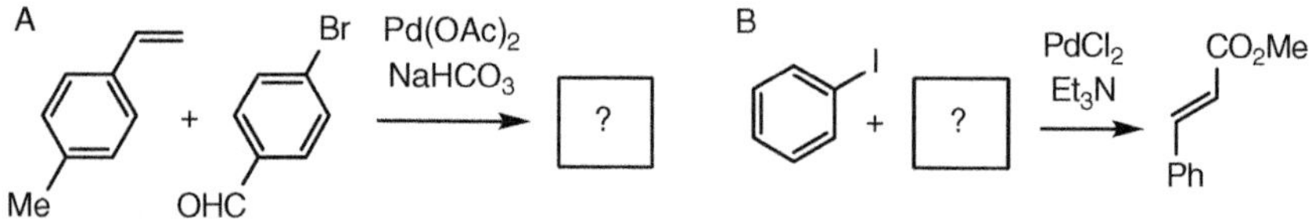

PALLADIUM-CATALYZED CROSS-COUPLING REACTIONS (RX + R′M → R-R′)

General Cross-Coupling Reaction

A large number of carbon–carbon bond forming reactions have been developed that use palladium to catalyze the coupling of two components: an organohalide and an organometallic reagent. These cross-coupling reactions are typically named after the chemists who developed the methodology, and a sampling of such "name reactions" is given in the table below.

Reaction	Metal (M)
Hiyama	Si
Suzuki–Miyaura	B
Stille	Sn
Negishi	Zn
Sonogashira	Cu
Kumada	Mg

The general catalytic cycle for a cross-coupling reaction is given below. The mechanism includes a key transmetalation step to transfer an organic group (that can be aryl, akenyl, alkynyl, or possibly alkyl) to the palladium complex.

Introduction to Strategies for Organic Synthesis, Second Edition. Laurie S. Starkey.
© 2018 John Wiley & Sons, Inc. Published 2018 by John Wiley & Sons, Inc.

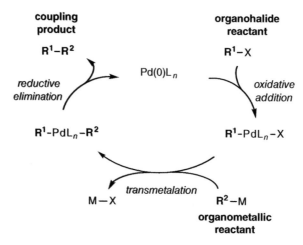

General Cross-Coupling Reaction Catalytic Cycle

8.5.1 STILLE REACTION

The Stille cross-coupling reaction combines an organotin compound with a halide or pseudohalide (e.g., triflate) electrophile. Its wide use can be attributed to mild reaction conditions that tolerate a wide variety of functional groups and the stability of organostannanes to both air and water. In the presence of carbon monoxide, the Stille coupling involves carbonylation as well.

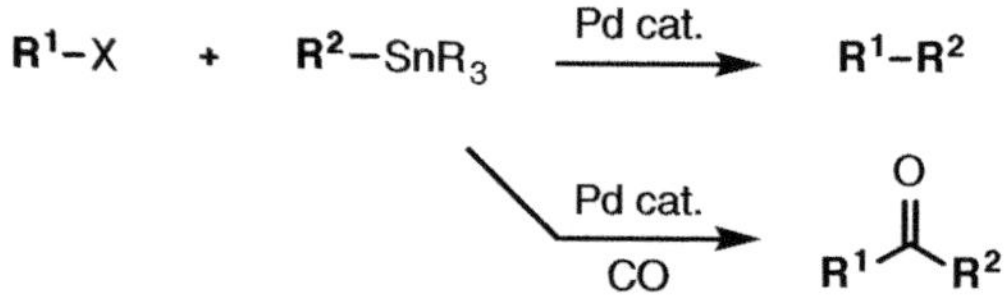

Stille Coupling and Carbonylative Stille Reaction

Stille coupling reactions can be applied to a variety of substrates, yielding a diverse array of target molecules. A sampling of potential reactants is shown below.

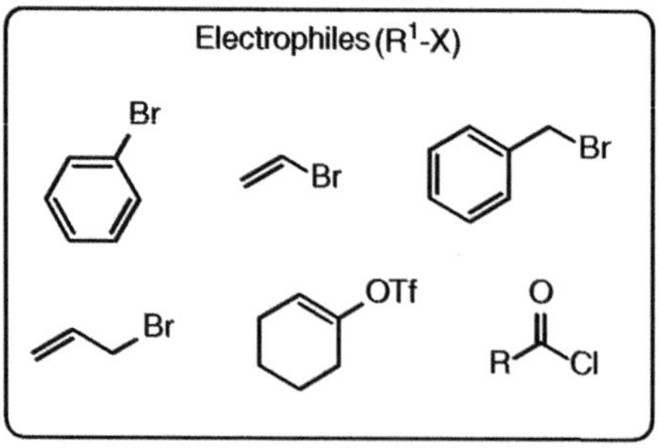

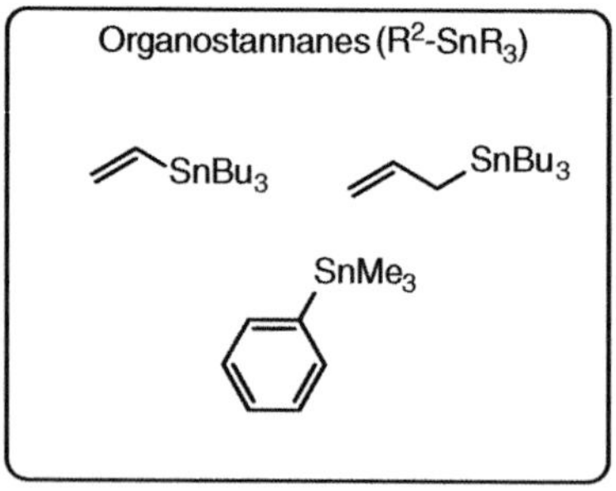

When substituted vinyl halides are used, the stereochemistry of the alkene is typically retained. Let us take a look at an example of a Stille cross-coupling reaction and explore its mechanism.

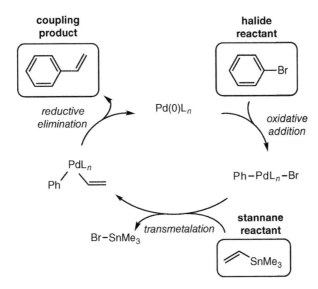

Example of Stille Coupling

The catalytic cycle begins with oxidative addition of the aryl halide to the Pd(0) complex. The transmetalation step, in which a second organic group is transferred to the palladium, is typically the rate-determining step. Because alkyl groups are very slow to be transferred in a transmetalation step, the three alkyl groups on the organostannane (such as methyl or butyl groups) will remain on the tin while the fourth organic group (typically an sp^2 or sp-hybridized carbon) is transferred to palladium. Finally, the two carbon ligands are coupled together by a reductive elimination step that also regenerates the Pd(0) catalyst.

Stille Coupling Reaction Mechanism

8.5.2 SUZUKI–MIYAURA REACTION

In the Suzuki–Miyaura reaction, an organoboron compound undergoes a palladium-catalyzed coupling reaction with an aryl, vinyl, or alkynyl halide.

$$\boxed{R^1}-X \quad + \quad \boxed{R^2}-B(OH)_2 \quad \xrightarrow[\text{NaOH}]{\text{Pd cat.}} \quad \boxed{R^1}-\boxed{R^2}$$

R^1 = aryl, vinyl, alkynyl; R^2 = aryl, vinyl, allyl, alkyl

$R^2-B\overset{OH}{\underset{OH}{\diagup}}$	$R^2-B\overset{OR'}{\underset{OR'}{\diagup}}$	$R^2-B\overset{R'}{\underset{R'}{\diagup}}$
boronic acid	boronic ester	organoborane

Suzuki–Miyaura Coupling Reaction

Thanks to multiple preparation methods, a variety of organoboron compounds are readily available; the Suzuki–Miyaura coupling employs organoboronic acids, as well as organoboronic esters and organoborane compounds. Boron also offers the advantage of being environmentally friendly (boron is safer than tin, which is toxic, and Suzuki–Miyaura reactions can be run with water as a solvent or cosolvent). Akira Suzuki was recognized for his work on this reaction as one of the chemists who shared the 2010 Nobel Prize in Chemistry. (It is reasonable to wonder why John Stille was not also included in this award, but he died in 1989 and the Nobel Prize is not awarded posthumously.)

A variety of methods exist to synthesize the organoborane compounds needed for a Suzuki–Miyaura reaction. Borohydride reagents such as disiamylborane, 9-BBN, or catecholborane add regiospecifically to alkenes. Recall that hydroboration reactions proceed with a concerted mechanism that adds the larger boron atom to the less sterically hindered carbon of the alkene, giving *anti*-Markovnikov regiochemistry. *Syn* addition of a borohydride reagent to a terminal alkyne gives a trans alkenyl borane, with the boron atom adding to the terminal carbon. Organoboronic esters and acids can also be prepared from Grignard reagents and organolithium compounds via transmetalation.

Boron Reagents and Organoboron Compound Preparation

The mechanism of the Suzuki–Miyaura reaction is similar to the Stille coupling, beginning with an oxidative addition of the organohalide to a Pd(0) catalyst.

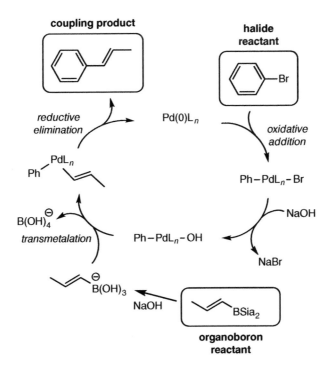

Example of Suzuki–Miyaura Coupling

The added base, such as NaOH, EtONa, or K_2CO_3, is required to facilitate the transmetalation step, by activating both the Pd complex and the organoboron substrate. Finally, a reductive elimination couples the two organic ligands.

Suzuki–Miyaura Reaction Mechanism

The Suzuki–Miyaura reaction offers excellent stereoselectivity, and it is useful for preparing a wide variety of target molecules, including alkenes, conjugated dienes and trienes, and biaryl compounds.

PRACTICE PROBLEM 8.5—CROSS-COUPLING REACTIONS

8.5 Predict the product(s) and provide the name of the coupling reaction.

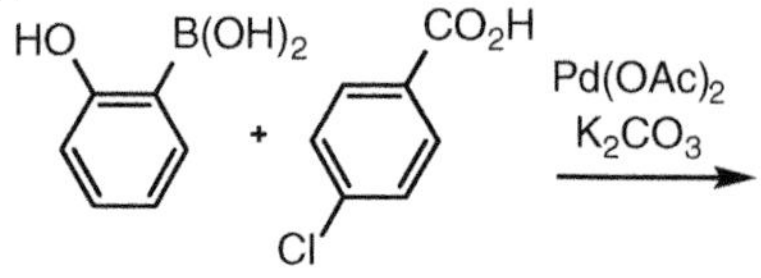

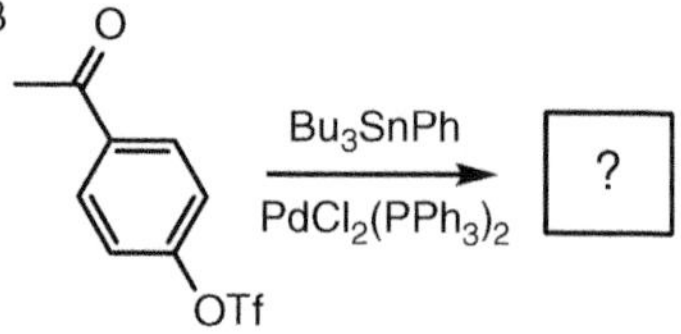

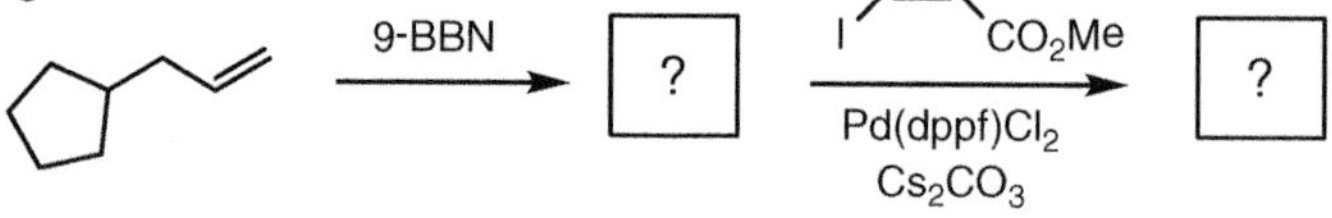

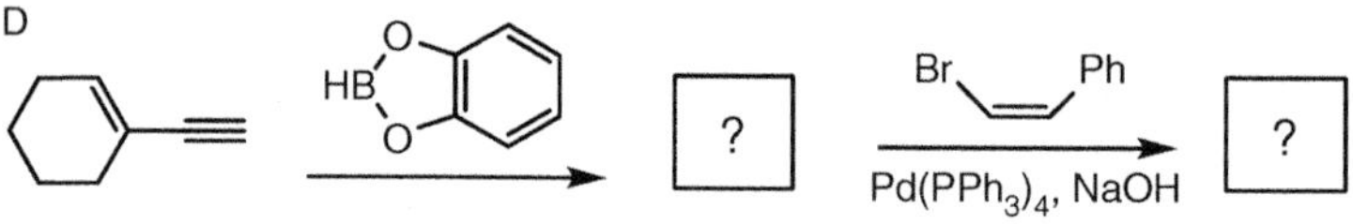

OLEFIN METATHESIS REACTIONS

Two alkenes (olefins) can be joined together in a transition metal-mediated transformation known as an olefin cross-metathesis reaction. These reactions have many important applications, and the 2005 Nobel Prize in Chemistry was awarded to Yves Chauvin, Robert Grubbs, and Richard Schrock for their olefin metathesis research.

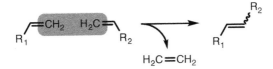

General Olefin Metathesis Reaction

In the olefin metathesis catalytic cycle, an alkylidene metal complex undergoes a [2+2] cycloaddition with an alkene, and a subsequent retro [2+2] mechanism releases a different alkene (the formation of a volatile product such as ethylene or propylene helps drive the reaction forward). When the sequence is repeated with a second alkene, the two olefin starting materials are ultimately combined as a single alkene product.

Introduction to Strategies for Organic Synthesis, Second Edition. Laurie S. Starkey.
© 2018 John Wiley & Sons, Inc. Published 2018 by John Wiley & Sons, Inc.

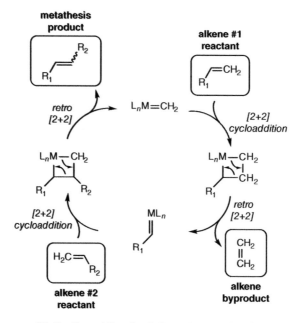

Olefin Cross-Metathesis Reaction Mechanism

8.6.1 RING-CLOSING METATHESIS (RCM)

When a *diene* starting material undergoes an alkene metathesis reaction, a ring is formed. Such conversions, described as ring-closing metathesis (RCM) reactions, have found widespread use in synthesis since the 1990s to form rings of nearly any size. Examples of RCM reactions range from the formation of a five-membered ring to a macrocycle containing 100 atoms!* If the cycloalkene product is a large ring, it may be formed with E or Z stereochemistry.

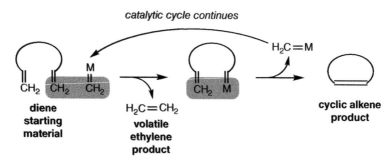

General Ring-Closing Metathesis (RCM) Reaction

* Y. Cao, L. Wang, M. Bolte, M. O. Vysotsky, and V. Bohmer, *Chem. Commun.*, **2005**, 3132–3134.

Many RCM catalysts have been developed. Grubbs' ruthenium-based catalysts are stable to air and tolerate a variety of functional groups. Schrock's molybdenum-based catalysts are also widely used, but they are sensitive to water and oxygen, and they are incompatible with protic functional groups.

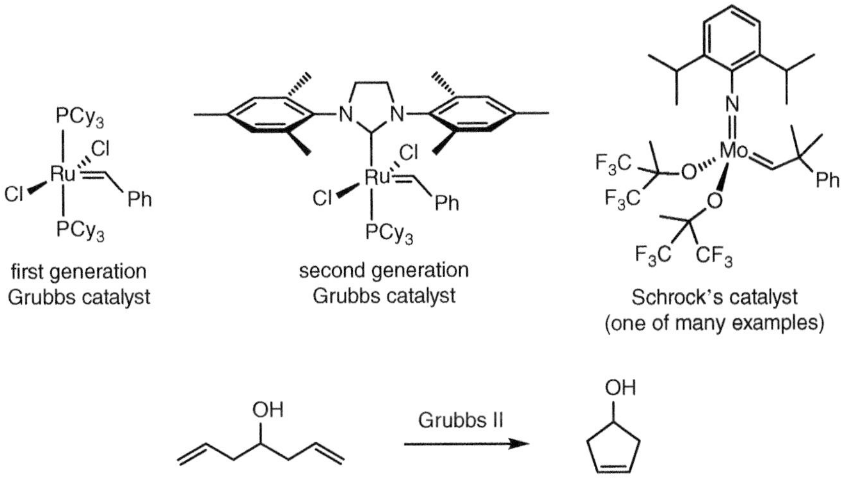

first generation
Grubbs catalyst

second generation
Grubbs catalyst

Schrock's catalyst
(one of many examples)

Grubbs II

Ring-Closing Metathesis Catalysts

RCM reactions are regularly used in natural product synthesis to create rings that are otherwise difficult to make. For example, shown below is the ring-closing step in the synthesis* of the cytotoxic compound clonostachydiol 1, which has been isolated from marine algae.

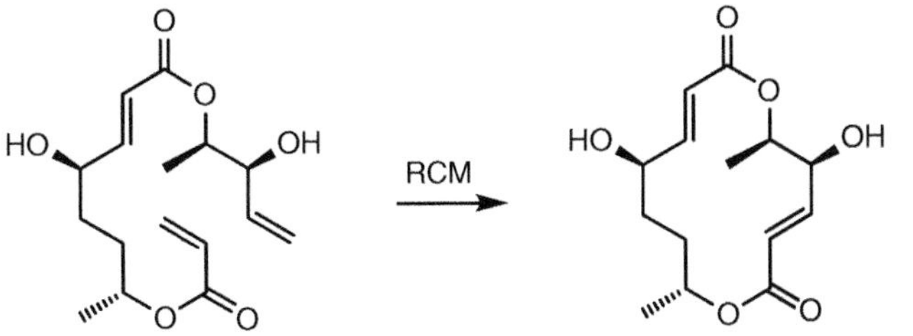

RCM

Clonostachydiol 1

Natural Product Synthesis via RCM

* U. Ramulu et al., *Tetrahedron: Asymmetry*, **2012**, *23*, 117–123.

RETROSYNTHESIS: DISCONNECTIONS BASED ON METAL-MEDIATED REACTIONS

The reactions covered in this chapter deviate from the typical combination of nucleophilic and electrophilic species, so the usual strategy of using those reactivities to identify logical disconnections cannot be applied here. Transition metal-mediated reactions form carbon–carbon single bonds where the carbon atoms involved are typically sp^2 or sp-hybridized (although sp^3-hybridized alkyl groups can also be involved in coupling reactions). To illustrate retrosynthetic strategies using the selected reactions covered in this chapter, we will focus on the vinyl and aryl regions of the TM when looking for potential bonds to disconnect.

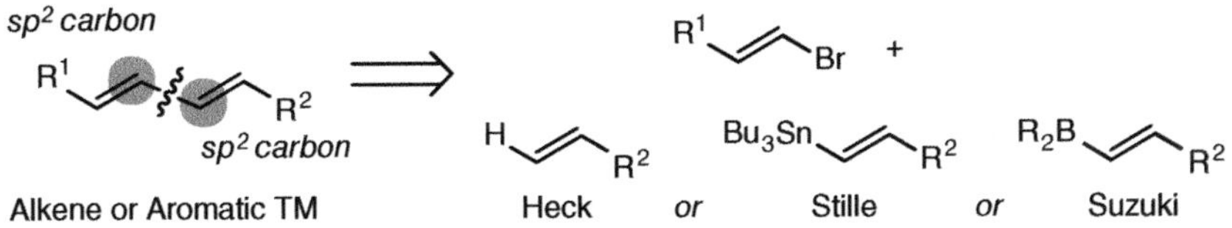

Transition-Metal Facilitated Retrosynthesis

Introduction to Strategies for Organic Synthesis, Second Edition. Laurie S. Starkey.
© 2018 John Wiley & Sons, Inc. Published 2018 by John Wiley & Sons, Inc.

TRANSITION METAL-MEDIATED SYNTHESIS

8-1. For each of the following organometallic compounds, give the electron count, the formal oxidation state of the metal, and the metal's d^n configuration.

A

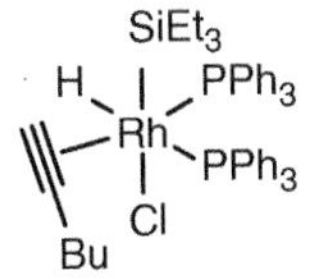

B

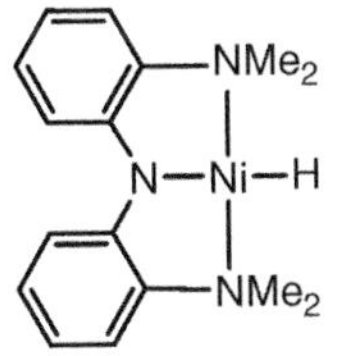

C

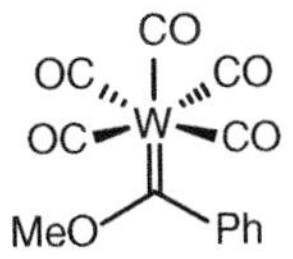

D

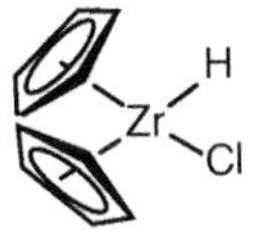

Cp₂ZrHCl
Schwartz's reagent

E

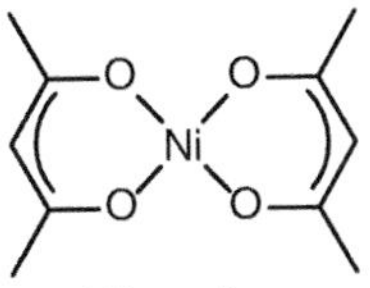

Ni(acac)₂
(acac, acetylacetonate)

F

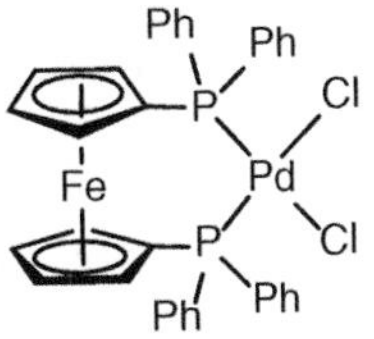

PdCl₂(dppf) (dppf,
1,1'-Bis(diphenylphosphino)
ferrocene)

G

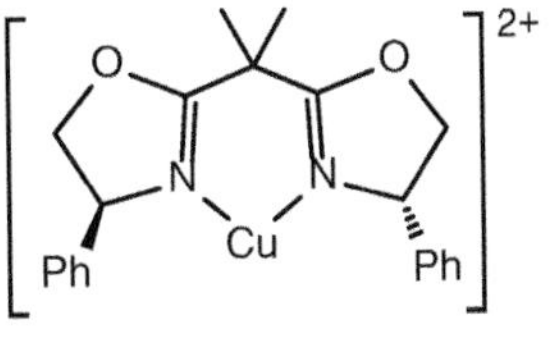

H

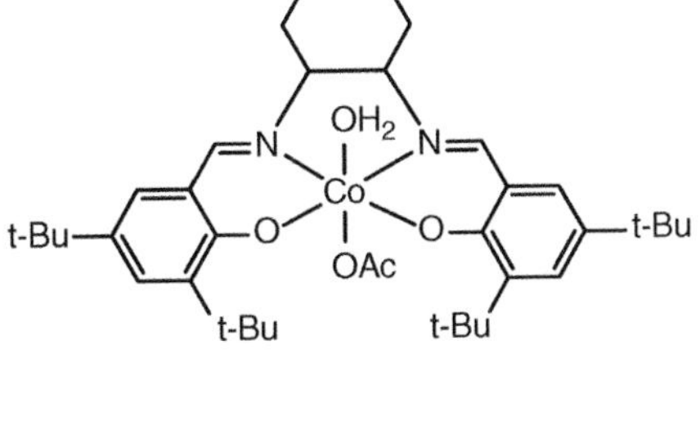

Introduction to Strategies for Organic Synthesis, Second Edition. Laurie S. Starkey.
© 2018 John Wiley & Sons, Inc. Published 2018 by John Wiley & Sons, Inc.

8-2. Chiral transition metal complexes are commonly employed in asymmetric synthesis. BINAP-derived ruthenium catalysts such as the one shown below, developed by Ryoji Noyori, are used in enantioselective hydrogenation reactions. Noyori shared the Nobel Prize in Chemistry in 2001 with William S. Knowles for this work. For the complex shown, give the electron count, the formal oxidation state of the metal, and the metal's d^n configuration.

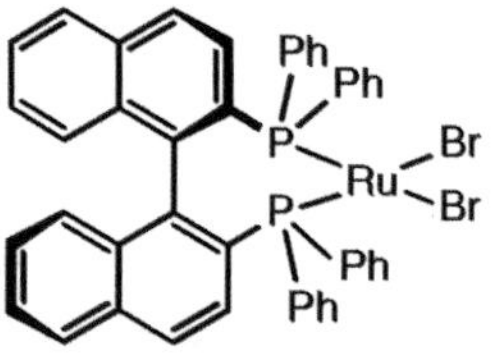

(R)-BINAP-Ru

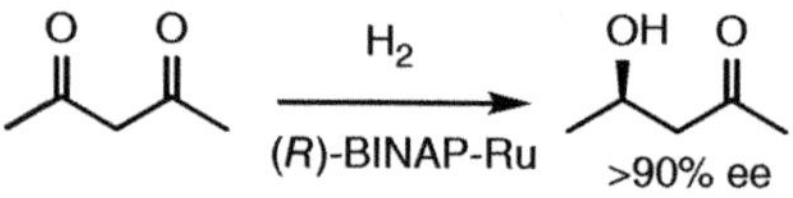

8-3. Shown below is the $[Rh(cod)Cl]_2$ complex.

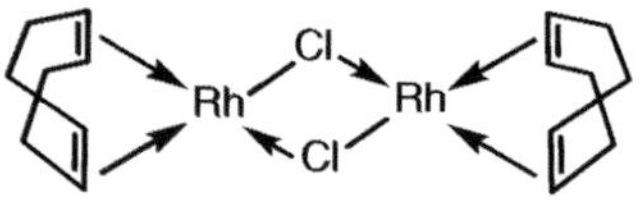

a. What is the formal oxidation state of each Rh atom?
b. Is each Rh atom coordinatively saturated or unsaturated?
c. Why do you think the name "cod" is used to describe this bidentate ligand?

8-4. Ei-ichi Negishi, the co-recipient of the Nobel Prize in Chemistry in 2010 (along with Heck and Suzuki), developed a palladium-catalyzed cross-coupling reaction that joins an organozinc compound and an organohalide. Shown below is the catalytic cycle proposed for a Negishi coupling. Describe each step of the mechanism.

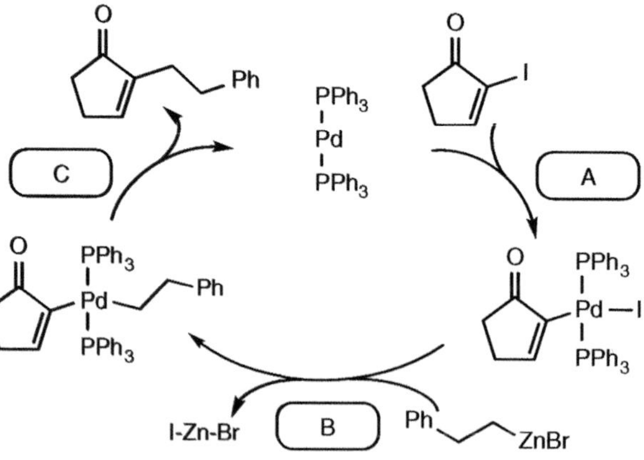

8-5. Isomerization of allylic alcohols results in aldehyde products, as shown below. Provide a reasonable mechanism for the conversion of allyl alcohol to propionaldehyde (propanal).

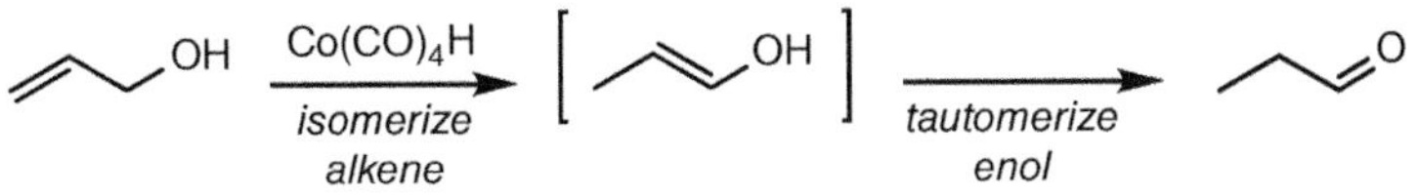

8-6. Predict the product(s) or provide the reactant and give the name of each reaction.

A

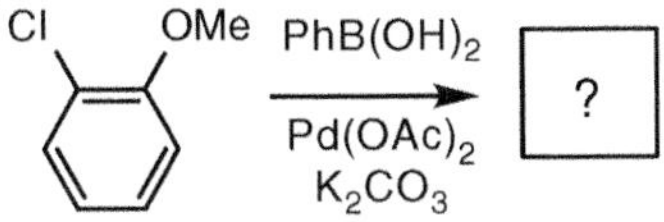

B

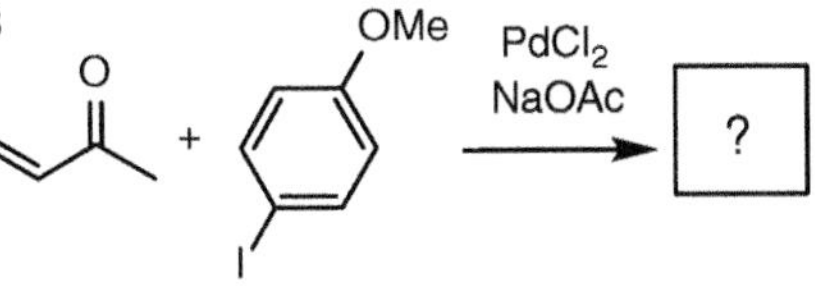

C

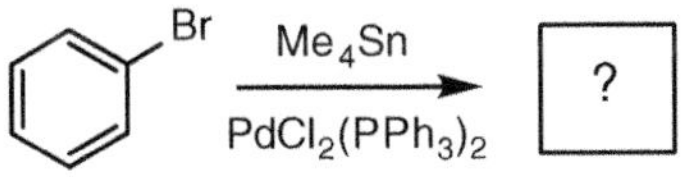

D

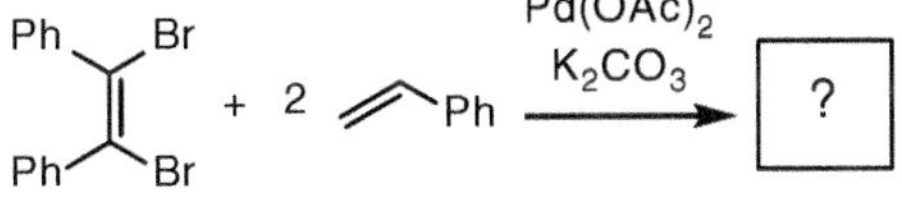

E

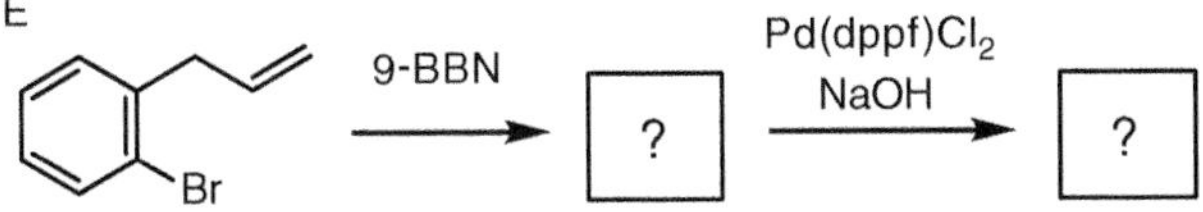

F

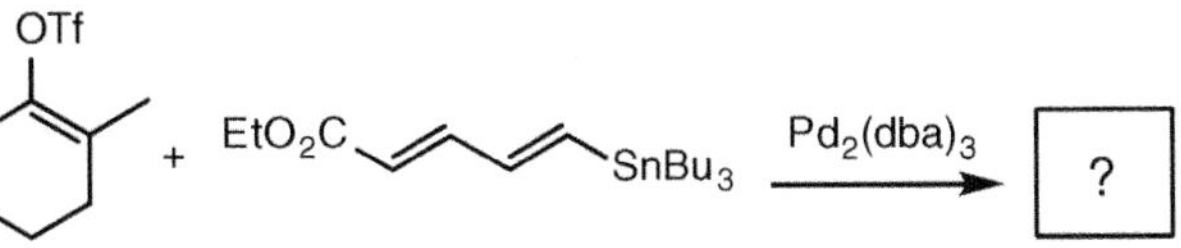

G

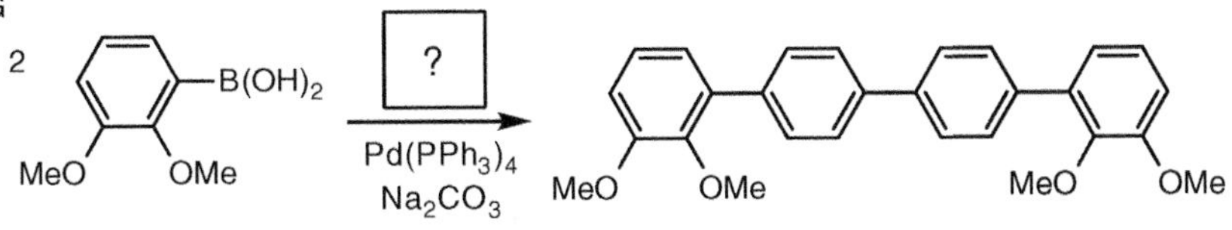

8-7. For each reaction, name the catalyst, describe the reaction, and predict the product.

A

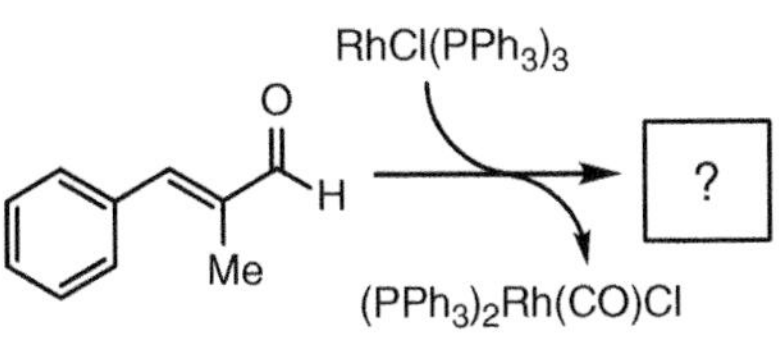

B

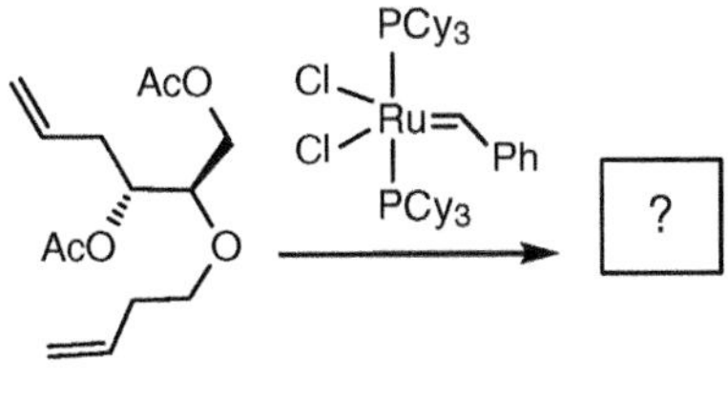

8-8. The reaction sequence shown is from a total synthesis of (−)-rhizopodin, a natural product with anticancer activity (*Angew. Chem. Int. Ed.* **2013**, *52*, 6517). Provide the missing reactant and reagent.

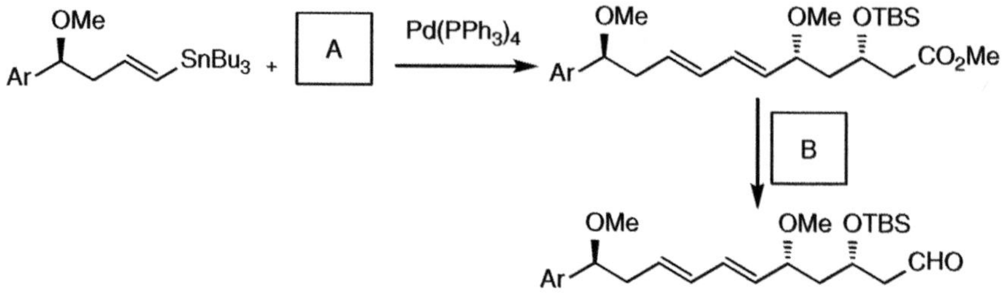

8-9. Complete the synthetic transformation given by providing the missing reagents.

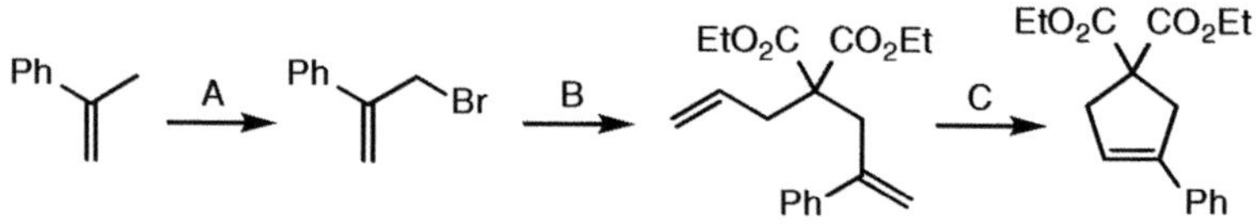

8-10. Provide a synthesis for the target molecule, using the metals and the sources of carbon provided.

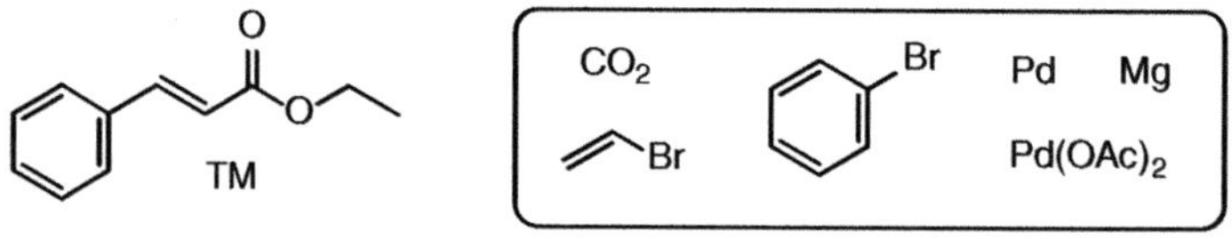

SOLUTIONS TO PROBLEMS

SOLUTIONS TO CHAPTER 1: PROTECTIVE GROUPS

1-1. Predict the major product (PG = Protective Groups).

(a) Bn group is removed by hydrogenation.

OH   OCH$_3$

CH$_3$O ⟍⟋⟍⟋OCH$_3$

ŌTBS

(b) N/R (all of these PGs are stable to Grignard and mild aqueous workup)

(c) Acetal and TBS are removed by acidic hydrolysis.

OBn   O

CH$_3$O ⟍⟋⟍⟋H

ŌH

Note: methoxy group is stable to all conditions! This is not a commonly used PG because it is not easy to remove.

(d) TBS group is removed by fluoride.

OBn OCH$_3$

CH$_3$O ⟍⟋⟍⟋OCH$_3$

ŌH

(e) N/R (all of these PGs are stable to strong base)

1-2. Select the appropriate protective groups (PG).

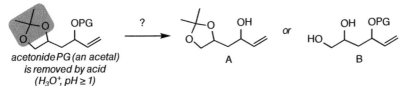

acetonide PG (an acetal) is removed by acid (H$_3$O$^+$, pH ≥ 1)

A

or

B

PG	structure	Can remove PG but keep acetonide to produce A? Conditions?	Can PG withstand H$_3$O$^+$ for acetonide removal to produce B?
TMS	–SiMe$_3$	**yes**—TBAF	**no**—unstable to acid (PG removed)
MOM	acetal	**no** (both are acetals)	**no**—unstable to acid (PG removed)
Tr	–CPh$_3$	**no** (acid needed to remove)	**no**—unstable to acid (PG removed)

Introduction to Strategies for Organic Synthesis, Second Edition. Laurie S. Starkey.
© 2018 John Wiley & Sons, Inc. Published 2018 by John Wiley & Sons, Inc.

MPM	$-CH_2C_6H_4OCH_3$	**yes**—DDQ oxidation	**yes** (benzyl ether stable if pH ≥ 1)
THP	acetal	**no** (both are acetals)	**no**—unstable to acid (PG removed)
Bz	$-COPh$	**yes**—NaOH, CH_3OH	**yes** (esters are stable if pH ≥ 1)
Bn	$-CH_2Ph$	**no** (H_2 would reduce alkene)	**yes** (benzyl ether stable if pH ≥ 1)
Ac	$-COCH_3$	**yes**—NaOH, CH_3OH	**yes** (esters are stable if pH ≥ 1)

1-3. Accomplish synthesis using protective groups.

A

TMSCI
base

add any PG

O

OTMS
hide acidic OH

1. CH_3MgBr

2. aq. wkup

TBAF

*remove
PG*

OH

OH

B

HO   OH
(1 equiv.)

TsOH
add PG

*protect aldehyde in
presence of ketone*

1. $LiAlH_4$

2. aq. wkup

$H_3O^{\oplus}$

*remove
PG*

O   OH

H

C

HO   OH

TsOH
add PG

OEt

*protect ketone
(N/R with ester)*

1. PhMgBr

2. aq. wkup

$H_3O^{\oplus}$

*remove
PG*

$\left[\begin{array}{c} O\ HO\ Ph \\ Ph \end{array}\right]$ $-H_2O$

*undergoes spontaneous
dehydration to give
conjugated product*

O   Ph

Ph

D

TBSCI
base

TBSO   OH

selectively protect 1° ROH

PCC

TBAF

*remove
PG*

HO   O

E

1. $SOCl_2$

2. CH_3OH

add any PG

MeO   O

Br
hide acidic OH

NaCN

NaOH, H_2O

(mild)

*remove ester PG, but no
hydrolysis of nitrile*

O

HO

CN

F

HO   OH   Mg

TsOH
add PG

*make
Grignard*

O   O

MgBr
hide reactive C=O

1.

O=

2. $H_3O^{\oplus}$

$H_3O^{\oplus}$

*remove
PG*

O   OH

1-4. Dipeptide synthesis.

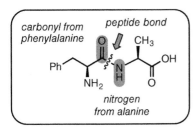

carbonyl from phenylalanine

peptide bond

nitrogen from alanine

The amide peptide bond has been formed from the carboxylic acid of phenylalanine and the amine from alanine. Protective groups must be positioned in such a way that these are the only functional groups available for the coupling reaction.

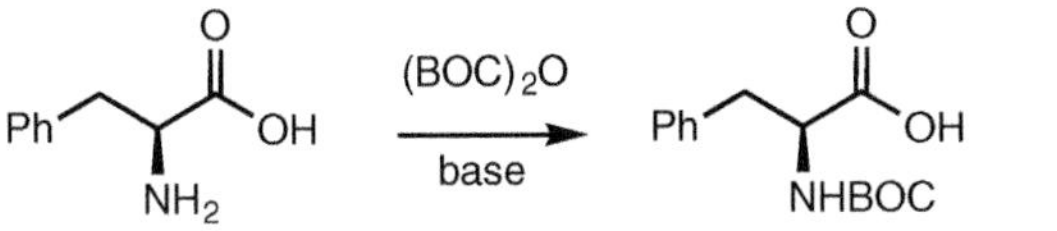

$\text{(BOC)}_2\text{O}$, base → *protect amino group*

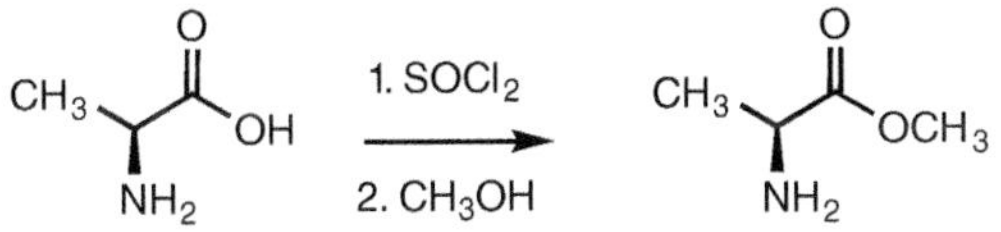

1. SOCl_2
2. CH_3OH → *protect carboxylic acid*

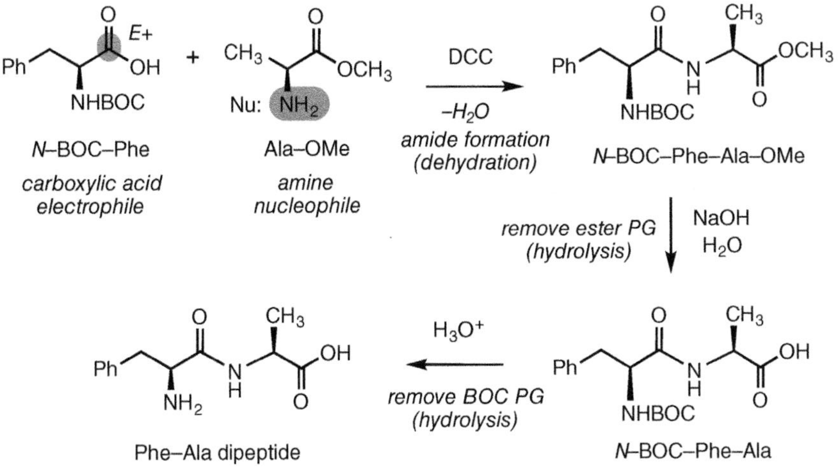

$E+$

$N\text{–}BOC\text{–}Phe$

carboxylic acid electrophile

+ Ala–OMe

Nu: NH_2

amine nucleophile

DCC

$-H_2O$

amide formation (dehydration)

$N\text{–}BOC\text{–}Phe\text{–}Ala\text{–}OMe$

remove ester PG (hydrolysis) | NaOH, H_2O

$N\text{–}BOC\text{–}Phe\text{–}Ala$

H_3O^+

remove BOC PG (hydrolysis)

Phe–Ala dipeptide

SOLUTIONS TO CHAPTER 2: NUCLEOPHILES, ELECTROPHILES, AND REDOX

2-1. Find most acidic proton and deprotonate with NaH (to make a good Nu:).

A

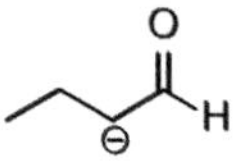

deprot. α-carbon to give resonance-stabilized enolate

B

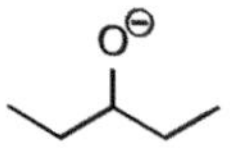

deprotonate alcohol to give alkoxide

C

deprotonate carbon alpha to cyano to give resonance-stabilized enolate

D

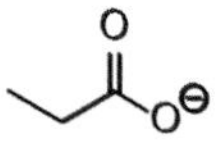

deprotonate carb.acid to give carboxylate

E

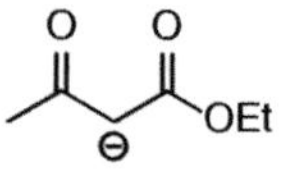

deprotonate carbon that is alpha to two EWGs to give enolate with extra resonance

F

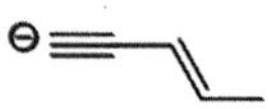

deprotonate alkyne to give an acetylide anion

G

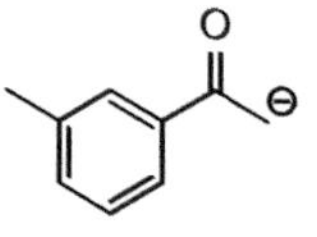

deprot. α-carbon to give resonance-stabilized enolate

H

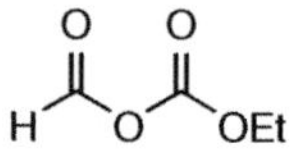

NO acidic protons! (N.R.with base)

I

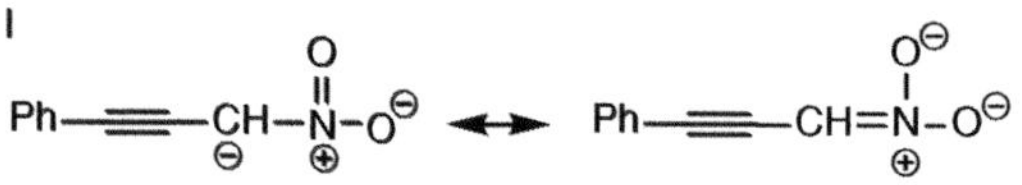

Ph—≡—CH—N–O⁻ ↔ Ph—≡—CH=N–O⁻

deprotonate carbon alpha to nitro to give resonance-stabilized enolate

J

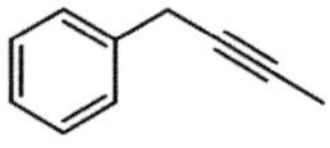

NO acidic protons! no EWGs, so conj. base would have poor resonance (N.R. with base)

K

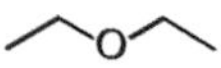

NO acidic protons! (N.R. with base)

L

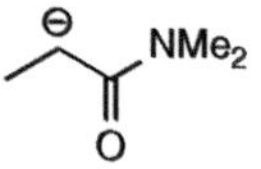

deprotonate α-carbon to give resonance-stabilized enolate

M

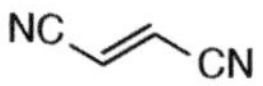

NO acidic protons! (N.R. with base)

N

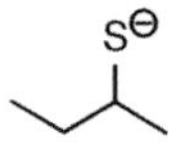

deprotonate thiol to give sulfide

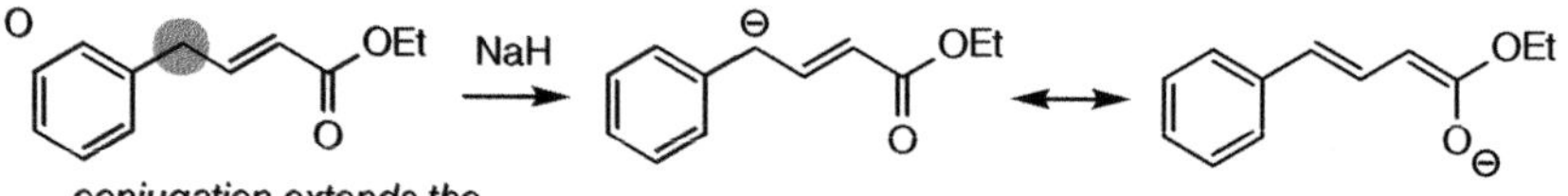

conjugation extends the effects of the EWG, making this methylene acidic

deprotonation gives a resonance-stabilized "vinylogous" enolate

2-2. Identify electrophilic (electron-deficient) sites with δ+.

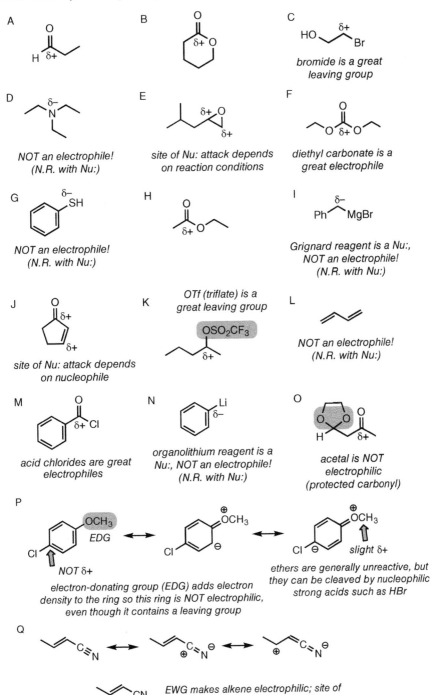

A

B

C

*bromide is a great
leaving group*

D

*NOT an electrophile!
(N.R. with Nu:)*

E

*site of Nu: attack depends
on reaction conditions*

F

*diethyl carbonate is a
great electrophile*

G

*NOT an electrophile!
(N.R. with Nu:)*

H

I

*Grignard reagent is a Nu:,
NOT an electrophile!
(N.R. with Nu:)*

J

*site of Nu: attack depends
on nucleophile*

K   *OTf (triflate) is a
great leaving group*

OSO₂CF₃

L

*NOT an electrophile!
(N.R. with Nu:)*

M

*acid chlorides are great
electrophiles*

N

*organolithium reagent is a
Nu:, NOT an electrophile!
(N.R. with Nu:)*

O

*acetal is NOT
electrophilic
(protected carbonyl)*

P

OCH₃   EDG

NOT δ+

ÖCH₃

OCH₃   *slight δ+*

*electron-donating group (EDG) adds electron
density to the ring so this ring is NOT electrophilic,
even though it contains a leaving group*

*ethers are generally unreactive, but
they can be cleaved by nucleophilic
strong acids such as HBr*

Q

*EWG makes alkene electrophilic; site of
Nu: attack depends on nucleophile*

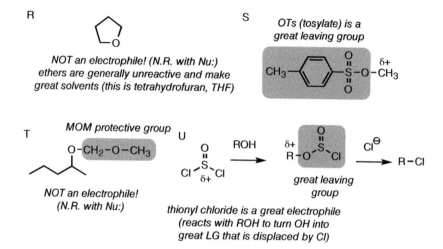

R

NOT an electrophile! (N.R. with Nu:)
ethers are generally unreactive and make
great solvents (this is tetrahydrofuran, THF)

S

OTs (tosylate) is a
great leaving group

T

MOM protective group

O-CH₂-O-CH₃

NOT an electrophile!
(N.R. with Nu:)

U

ROH

great leaving group

thionyl chloride is a great electrophile
(reacts with ROH to turn OH into
great LG that is displaced by Cl)

2-3. Identify the reductions and oxidations involved in each transformation
and provide the necessary reagents.

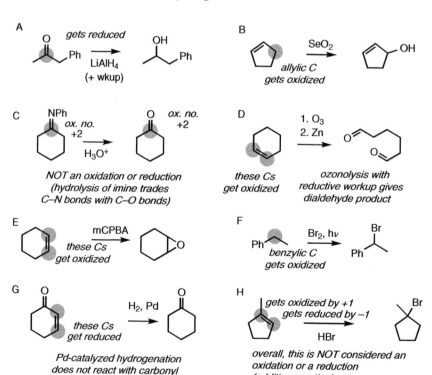

A

gets reduced

LiAlH₄
(+ wkup)

B

SeO₂

allylic C
gets oxidized

C

ox. no.
+2

H₃O⁺

ox. no.
+2

NOT an oxidation or reduction
(hydrolysis of imine trades
C–N bonds with C–O bonds)

D

1. O₃
2. Zn

these Cs
get oxidized

ozonolysis with
reductive workup gives
dialdehyde product

E

mCPBA

these Cs
get oxidized

F

Br₂, hν

benzylic C
gets oxidized

G

H₂, Pd

these Cs
get reduced

Pd-catalyzed hydrogenation
does not react with carbonyl

H

gets oxidized by +1
gets reduced by –1

HBr

overall, this is NOT considered an
oxidation or a reduction
(addition reaction)

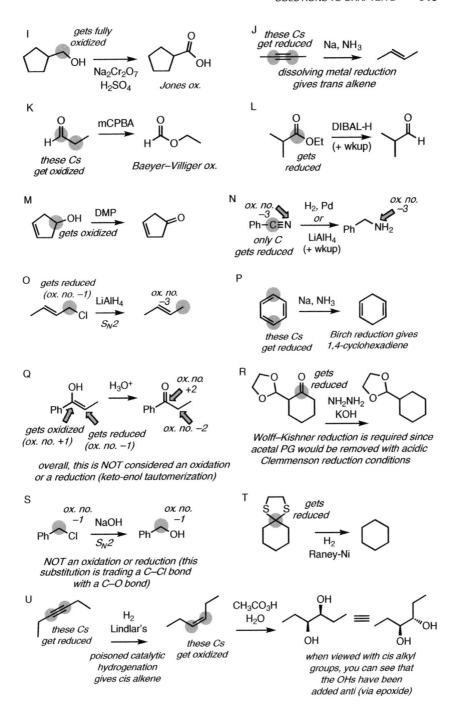

2-4. Provide the missing reagents for each of the following synthetic steps.

A excess $LiAlH_4$ (workup)
reduce carb. acids to alcohols

B TsOH *protect 1,2-diol as acetonide*

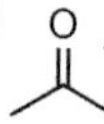

C Swern (or DMP, etc.)
oxidize alcohol to aldehyde

D
1. $LiAlH_4$ (workup); 2. TBSCl, base
reduce ester (excess DIBAL used in journal article); protect alcohol

E $NaBH_4$, MeOH (or LAH)
reduce aldehyde to alcohol

Kazuyuki Miyashita *et al.*, *Tet. Lett.* **2007**, *48*, 3829.

SOLUTIONS TO CHAPTER 3: 1-FG TMs

3.1a Provide the reagents needed.

A 1. $Hg(OAc)_2$, H_2O; 2. $NaBH_4$

B 1. $LiAlH_4$; 2. H_3O^+

C

**NaOH would give E2 as major product*

D 1. BH_3-THF 2. H_2O_2, NaOH

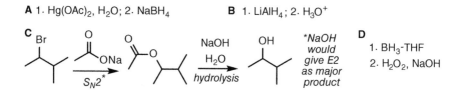

3.1b Provide the starting materials needed to create the indicated bond.

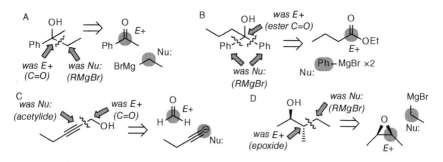

3.2 Provide the missing starting material or reagent.

A 1. $Hg(OAc)_2$, H_2O; 2. $NaBH_4$; 3. $SOCl_2$
(note: HCl would give rearranged product)

B PBr_3
(inversion)

C 1. HBr, ROOR *(anti-Mark.)*
2. Mg

D

(rotate product to reflect anti-addn of Br atoms)

cis alkene gives correct product stereochem.

E

F

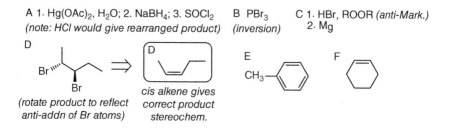

3.3 Provide the reagents needed.

A 1. NaH; 2. PhCH₂Br
(Williamson ether synthesis: deprotonate;S_N2)

B H₂
Lindlar's

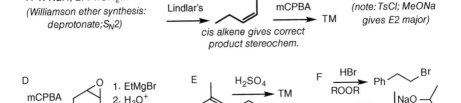

mCPBA
TM

cis alkene gives correct product stereochem.

C 1. NaH; 2. MeI
(note: TsCl; MeONa gives E2 major)

D mCPBA

1. EtMgBr
2. H₃O⁺
TM

E H₂SO₄
TM

F HBr
ROOR
Ph
Br

TM ← NaO

3.5 Provide the reagents needed.

A CH₃I(excess)

B 1. PhCH₂NH₂; 2. LiAlH₄; 3. H₂O(wkup)

C 1. NaCN; 2. H₂, Pd

D NH₃,

E PhNH₂, NaBH₃CN

F 1. NaN₃
2. H₂, Pd

G 1.
2. NaOH
H₂O
S_N2
hydrolysis

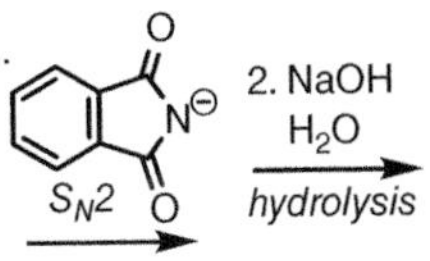

3.6a Provide the missing starting material and name reaction.

A 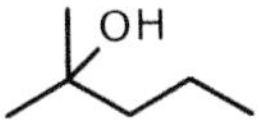
OH
alcohol dehydration

B 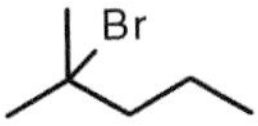
Br
need halide LG for E2

C
dissolving metal reduction (trans)

D 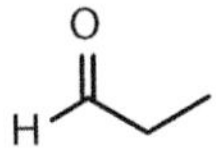
O
H
need carbonyl for Wittig reaction

3.6b Provide the reagents needed.

A 1. TsCl, pyridine; 2. *t*-BuOK, heat
(make good LG, Hofmann w/ bulky base)

B 1. PhMgBr; 2. H₃O⁺; 3. H₂SO₄, heat
(Grignard, then dehydration of alcohol)

C 1. Br₂ *hv* ; 2. EtONa, heat
(add LG for E2, Zaitsev w/ strong base)

D 1. HBr; 2. PPh₃; 3. BuLi
(Mark. addn of HBr, S_N2, deprotonate)

3.7 Provide the reagents needed.

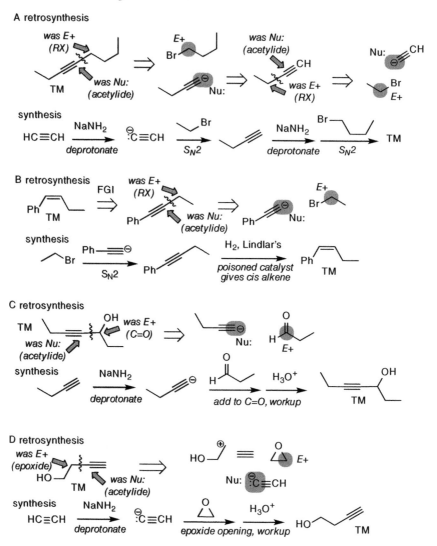

A retrosynthesis

B retrosynthesis

C retrosynthesis

D retrosynthesis

3.8A Provide the missing reagents.

A DMP *(oxidation)*
(or Swern, PCC, etc.)

B   Zn/Hg, HCl *(Clemmenson)*
   or
NH₂NH₂, KOH *(Wolff–Kishner)*

or

2. Ni-H₂

1. SH   SH, BF₃
*thioacetal formation,
followed by Raney Ni*

C H₂SO₄, heat *(dehydration)*   D H₂, Pd *(cat. hydrogenation)*

E PBr₃ *(substitution)*   F LiAlH₄ *(S_N2)*   G Mg *(make Grignard)*   H H₂O *(protonate)*

3.8B Provide the reagents needed.

A retrosynthesis

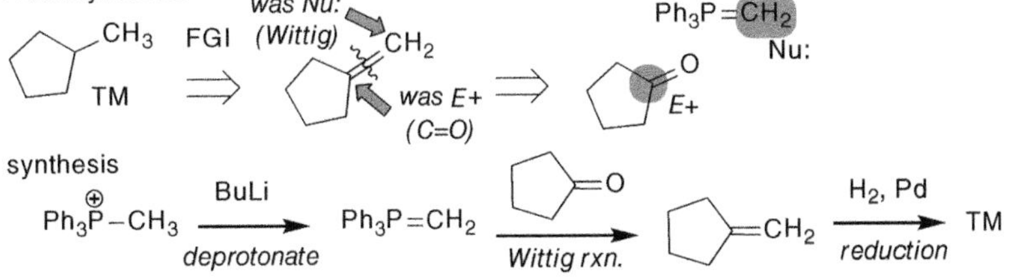

synthesis

B retrosynthesis

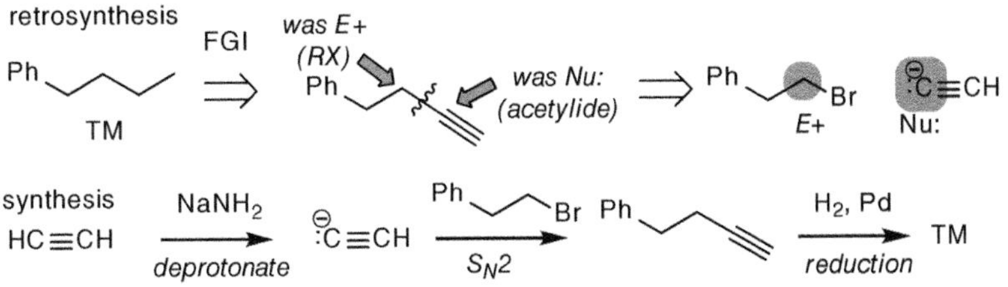

synthesis

C

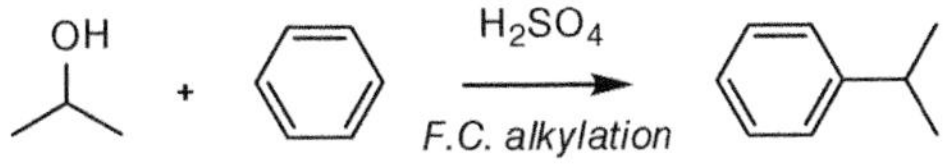

D retrosynthesis

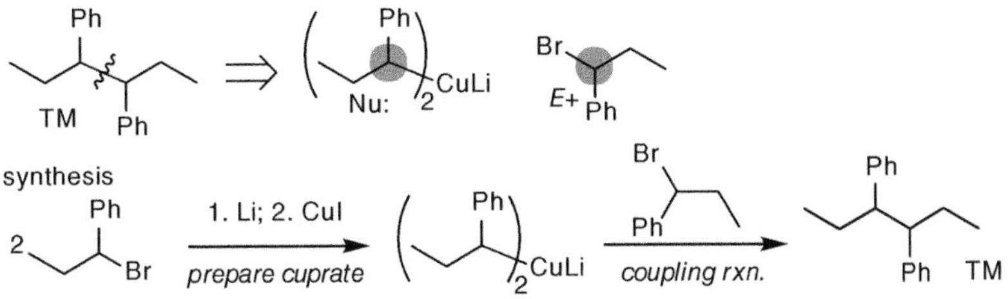

synthesis

3.9A Provide the missing reagents.

A NaCN (S_N2) B 1. MeMgBr; 2. H_3O^+ *(Grignard, workup/hydrolysis)*

C 1. NaOH (S_N2) D 1. *t*-BuOK *(E2)*; 2. O_3; 3. Zn, H_2O E HC≡CNa
   2. DMP *(oxidation)*   *(ozonolysis with reductive workup)* (S_N2)

F 1. $Hg(OAc)_2$, H_2O; 2. $NaBH_4$ G 1. BH_3-THF; 2. H_2O_2, NaOH
   (Mark. hydration, tautomerize)   *(anti-Mark. hydration, tautomerize)*

3.9B Provide the reagents needed.

A retrosynthesis

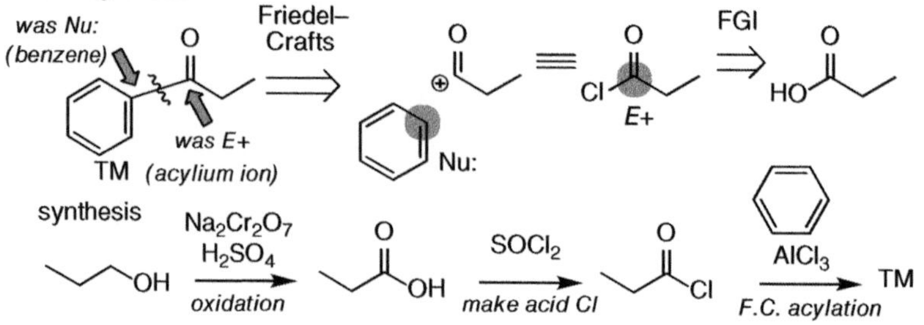

B retrosynthesis

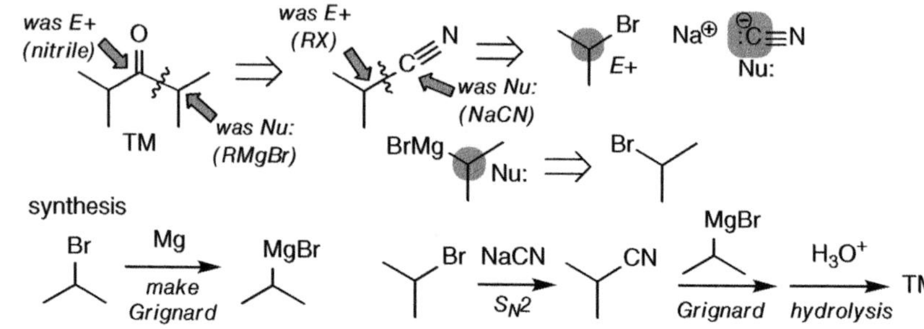

3.9C Provide the reagents needed.

A 1. LDA; 2. MeI; *(deprotonate; S_N2)*
3. LDA; 4. MeI *(kinetic enolate)*

B 1. LDA; 2. MeI; 3. TMSCl, Et$_3$N, heat;
4. TBAF; 5. MeI *(thermodynamic enolate)*

C 1. NaOEt; 2. PhCH$_2$Br
   (deprotonate; S_N2)

D 1. 2 equiv. LDA;
   2. PhCH$_2$Br *(dienolate)*

E NaOEt, Br⌃⌃⌃⌃Br
  (2 equiv)

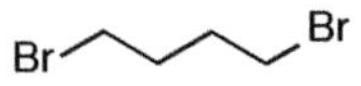

F retrosynthesis

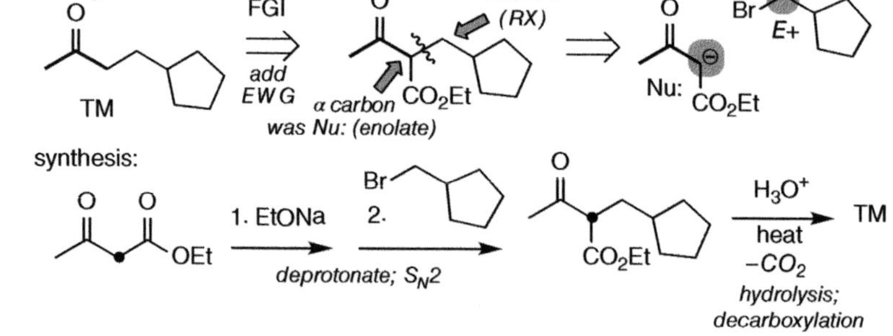

3.10 Provide a synthesis of the TM using the given retrosynthesis.

A retrosynthesis

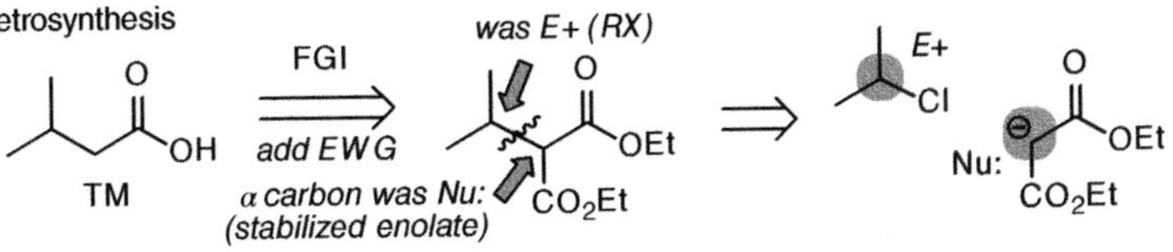

synthesis:

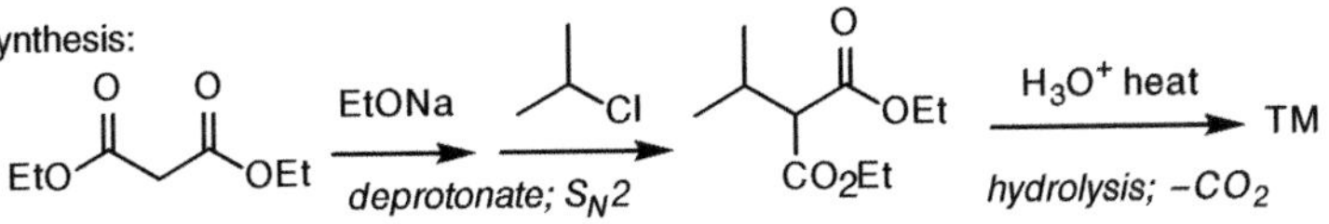

B retrosynthesis

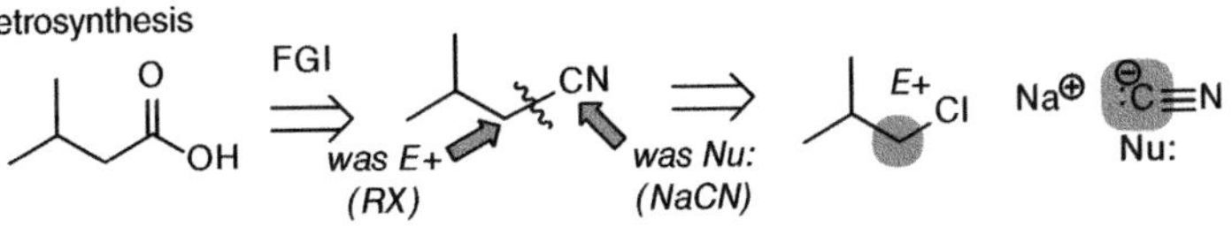

synthesis:

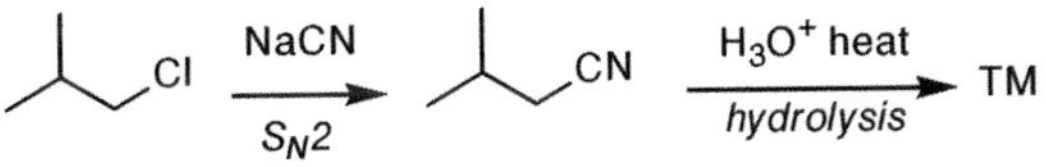

C retrosynthesis

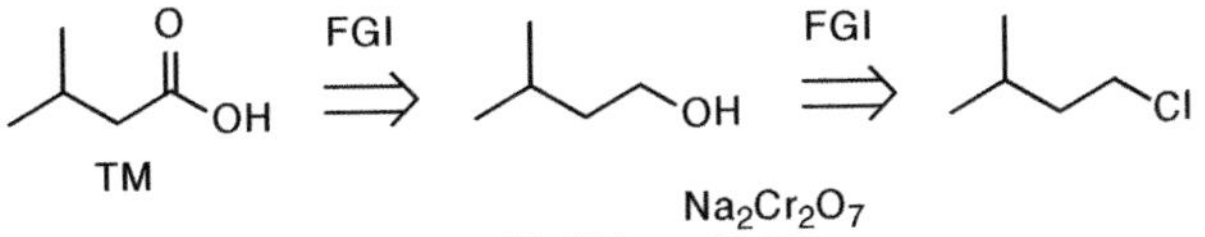

synthesis:

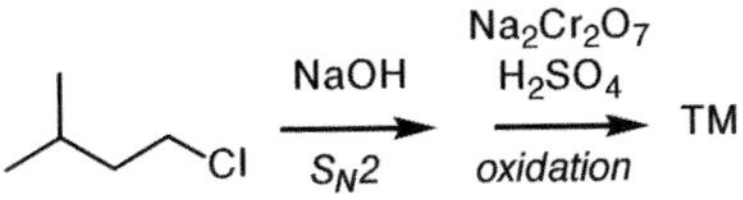

D retrosynthesis

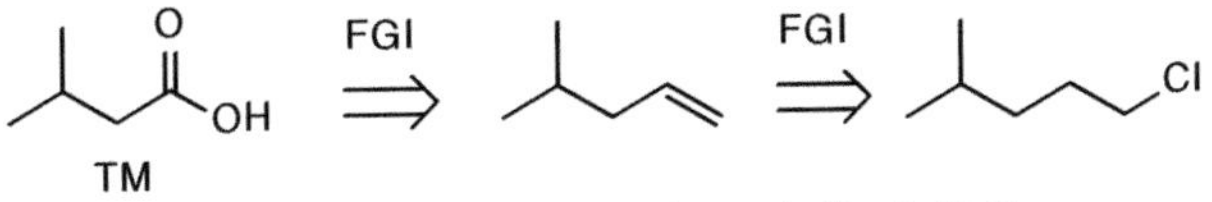

synthesis:

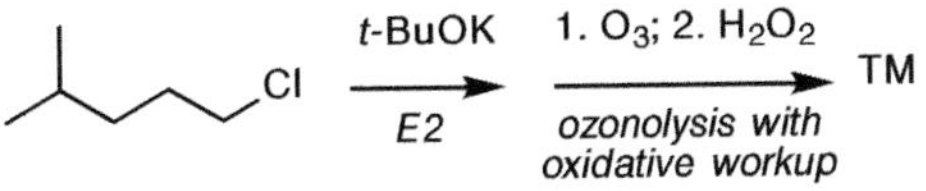

3.11A Provide four possible syntheses of the given TM using the disconnections shown.

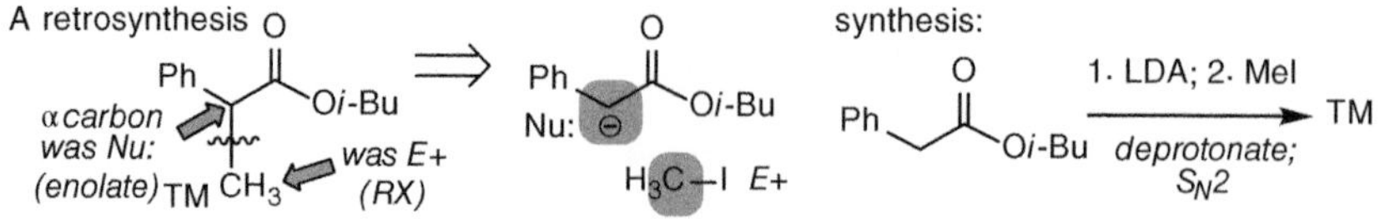

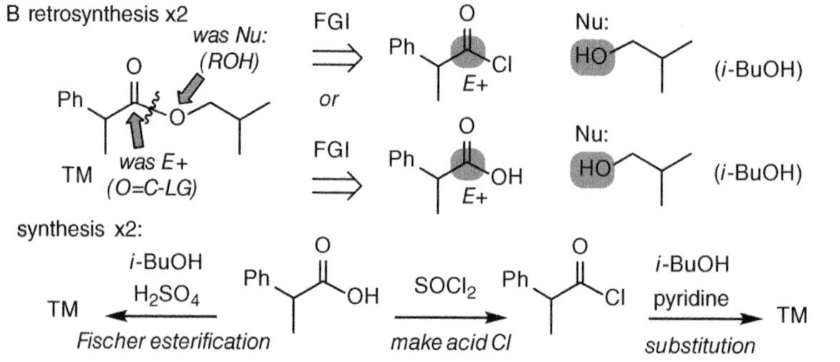

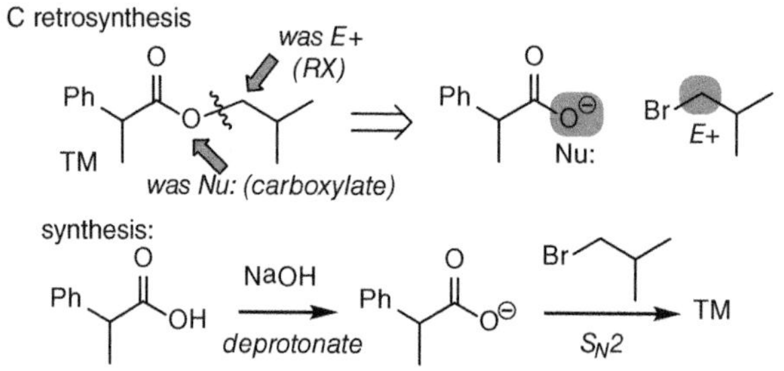

A retrosynthesis

B retrosynthesis x2

synthesis x2:

C retrosynthesis

synthesis:

3.11B Provide the reagents needed.

A 1. $KMnO_4$, heat *(oxidize to carb.acid)*; 2. $SOCl_2$, pyridine; 3. NH_3 (excess) *(make amide)*

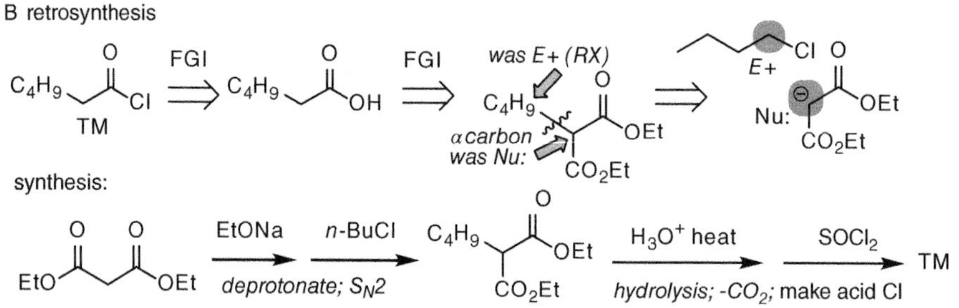

B retrosynthesis

synthesis:

C retrosynthesis

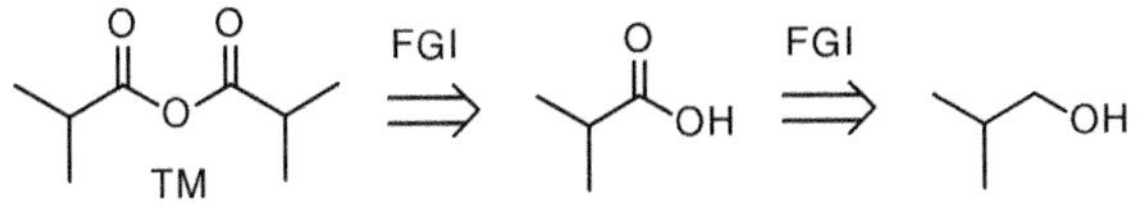

synthesis:

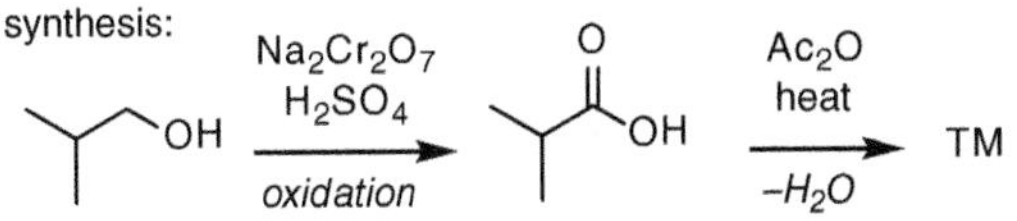

D retrosynthesis

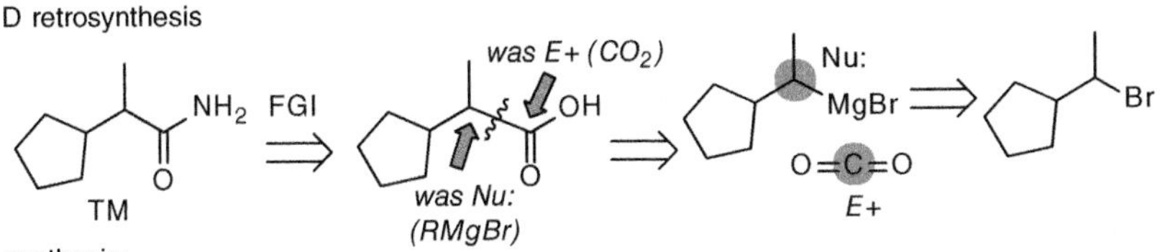

synthesis:

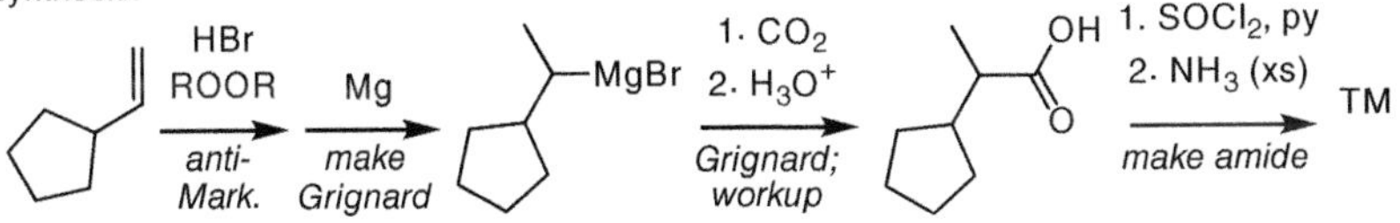

E retrosynthesis

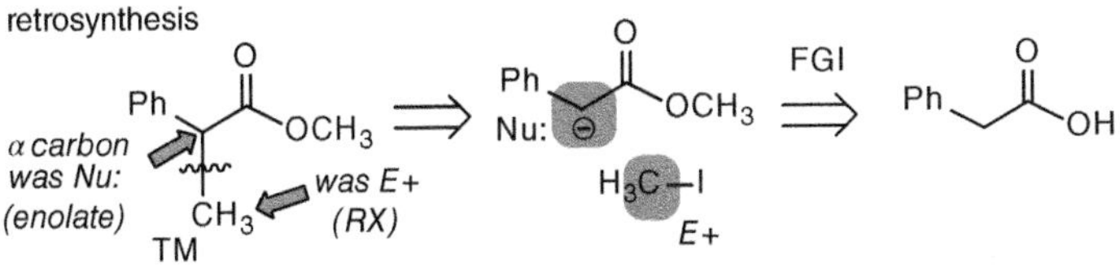

synthesis:

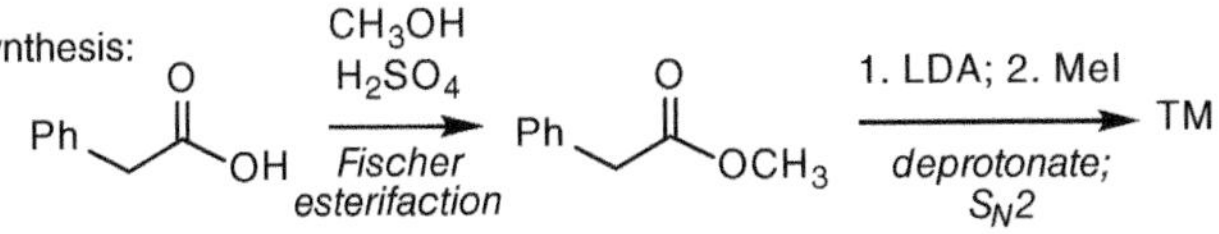

F retrosynthesis

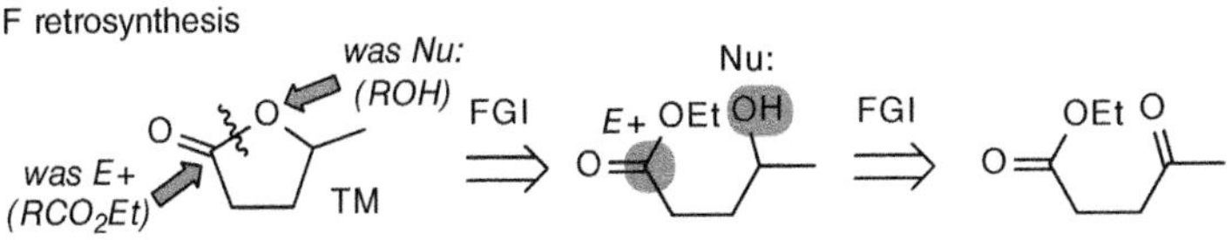

synthesis:

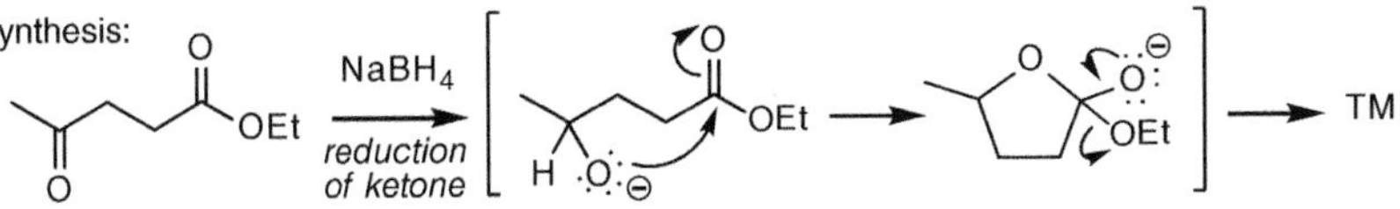

End-of-Chapter Problems

3-1. Provide the synthetic equivalent for each synthon.

A

add a metal (e.g., Grignard) to
make a carbon group a Nu:

B

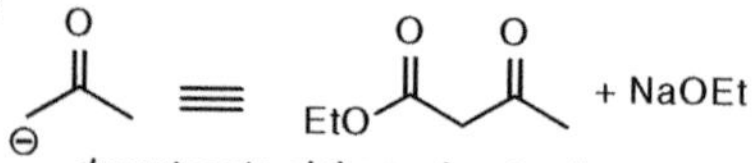

deprotonate alpha carbon to give
enolate (second EWG can be removed)

C

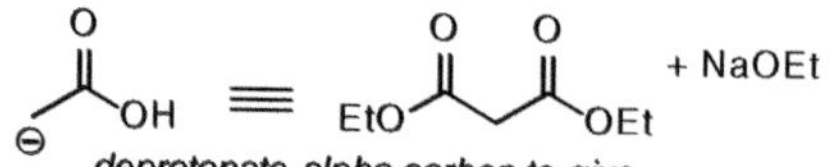

deprotonate alpha carbon to give
enolate (second EWG can be removed)

D

add a leaving group to
make an alkyl group an E+

E

after Nu: attacks, OH remains
on same carbon

F

after Nu: attacks, OH remains
on adjacent carbon

G

add LG (Cl or OEt) so
after Nu: attacks, C=O remains

H

after Nu: attacks,
C=O and OH remain

I

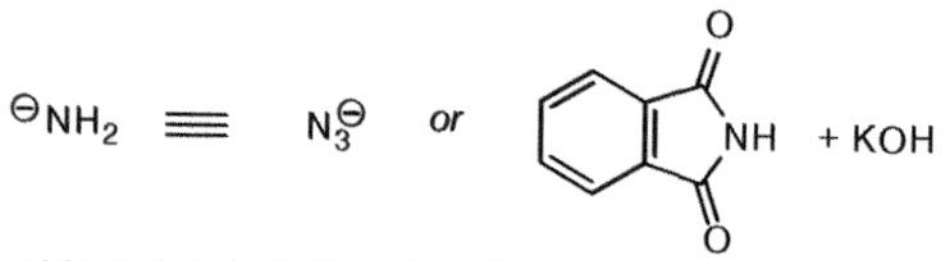

$^\ominus NH_2$ ≡ $N_3^\ominus$ or [image] NH + KOH

azide and phthalimide anion (Gabriel synthesis) are both
useful synthetic equivalents for the amide ion (or ammonia)

3-2. Provide reasonable disconnections/retrosyntheses.

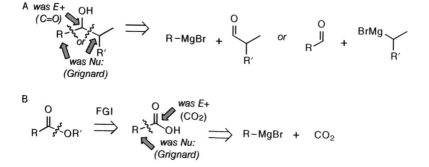

A was E+
(C=O)
R— or R'
was Nu:
(Grignard)

⟹ R−MgBr + [ketone with R'] or [aldehyde R] + BrMg [R']

B

FGI ⟹ was E+ (CO₂) ... was Nu: (Grignard) ⟹ R−MgBr + CO₂

C

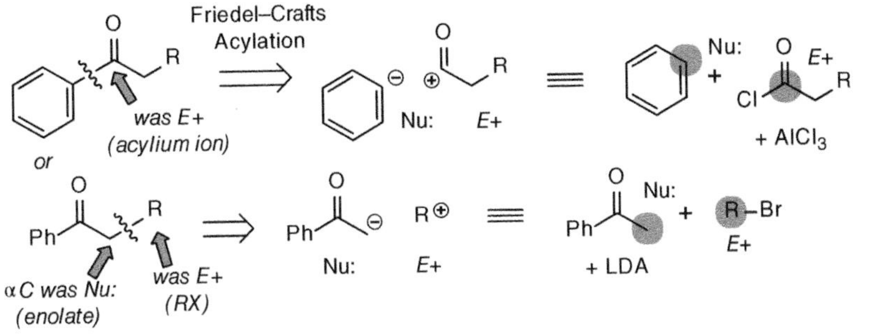

Friedel–Crafts
Acylation

was E+
(acylium ion)

Nu: E+

Nu:

+ AlCl₃

or

α C *was Nu:*   *was E+*
(enolate)   *(RX)*

Nu: E+

Ph

Nu:

R–Br
E+

+ LDA

D

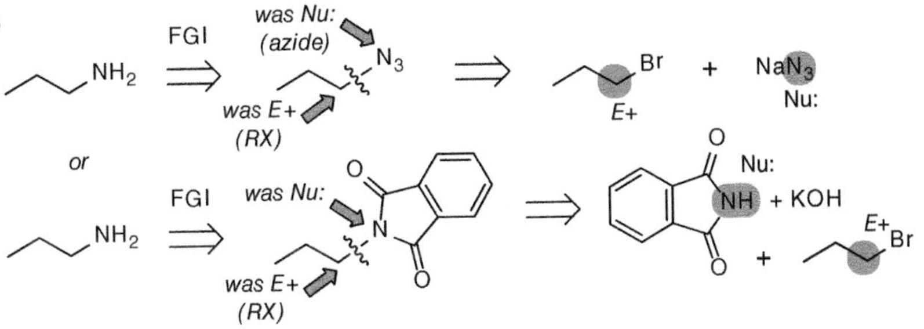

FGI

was Nu:
(azide)

was E+
(RX)

Br   +   NaN₃
E+   Nu:

or

FGI

was Nu:

was E+
(RX)

Nu:

NH + KOH

+

E+
Br

E

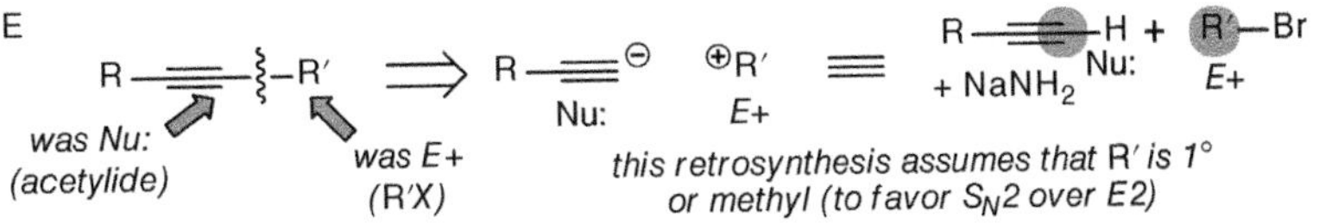

R ≡ R′

was Nu:
(acetylide)

was E+
(R′X)

Nu: E+

R≡H + R′–Br
Nu: E+
+ NaNH₂

this retrosynthesis assumes that R′ is 1°
or methyl (to favor S_N2 over E2)

F

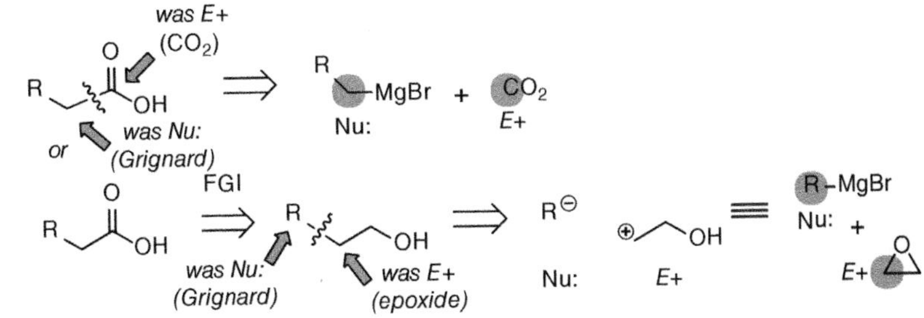

was E+
(CO₂)

was Nu:
(Grignard)

MgBr   +   CO₂
Nu: E+

or

FGI

R

was Nu:
(Grignard)

OH

was E+
(epoxide)

R⁻

Nu:

OH
E+

R–MgBr
Nu: +

E+

G

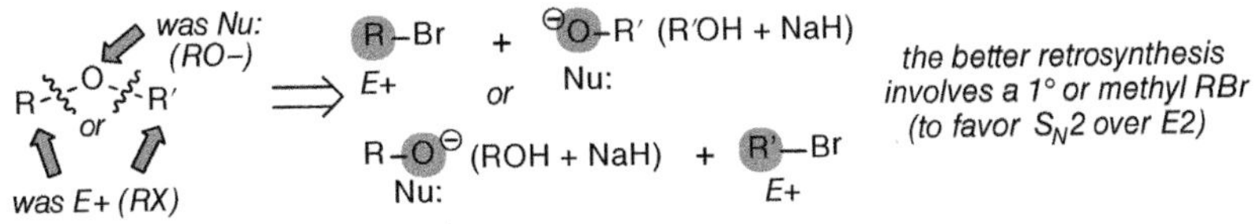

was Nu:
(RO–)

or

was E+ (RX)

R–Br   +   ⁻O–R′ (R′OH + NaH)
E+   Nu:
or

R–O⁻ (ROH + NaH)   +   R′–Br
Nu: E+

the better retrosynthesis
involves a 1° or methyl RBr
(to favor S_N2 over E2)

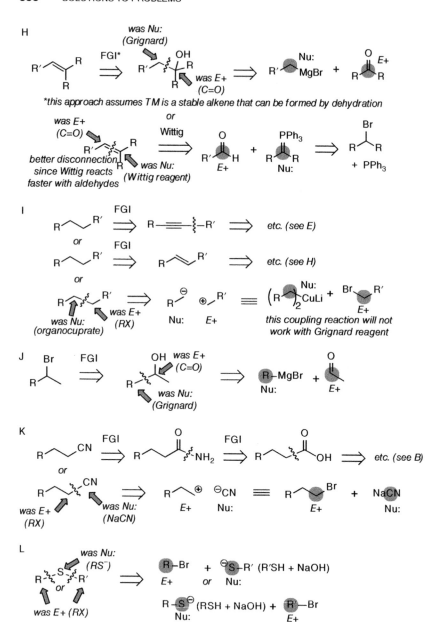

3-3. Provide reagents to transform the given starting material into the desired product.

A

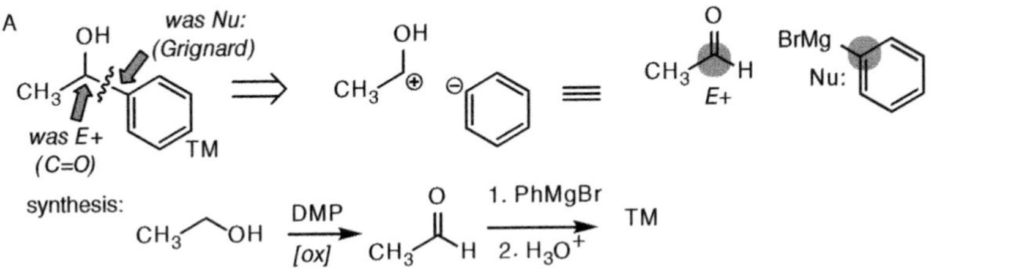

was Nu:
(Grignard)

OH

CH₃   TM

was E+
(C=O)

OH

CH₃ ⊕   ⊖

≡

CH₃ ─ H (O)   E+

BrMg

Nu:

synthesis:

CH₃ OH   DMP [ox]   CH₃ ─ H (O)   1. PhMgBr   TM
2. H₃O⁺

B

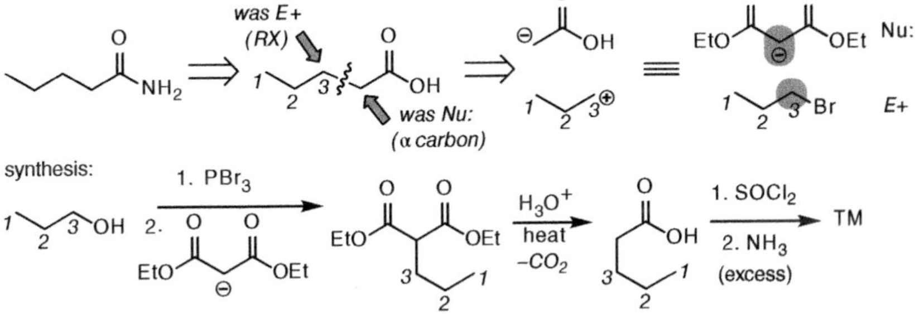

O
╱╲╱╲ NH₂

was E+
(RX)

⟹

O
1 ╱ 2 ╲ 3 ╱ OH

was Nu:
(α carbon)

⟹

⊖ O
OH

1 ╱ 2 ╲ 3 ⊕

≡

O   O
EtO   OEt
⊖

Nu:

1 ╱ 2 ╲ 3 Br   E+

synthesis:

1 ╱ 2 ╲ 3 OH   1. PBr₃
2.
O   O
EtO   OEt
⊖

O   O
EtO   OEt
3 ╲ 1
2

H₃O⁺
heat
─CO₂

O
OH
3 ╲ 1
2

1. SOCl₂
2. NH₃
(excess)

TM

C *this carbon was the Nu: (Wittig)*

this carbon was the E+ (C=O)

Nu: CH₂=PPh₃
+
E+ O
Ph

was E+
(RX)

⟹

O
Ph

was Nu:
(α carbon)

O
⊖

⊕ Ph

≡

O
Nu:
+ LDA

Br ╱ Ph
E+

synthesis:

OH   DMP [ox]   O   1. LDA, −78°C
2.
Br ╱ Ph
deprotonate; SN2

O
Ph

CH₂=PPh₃
Wittig

TM

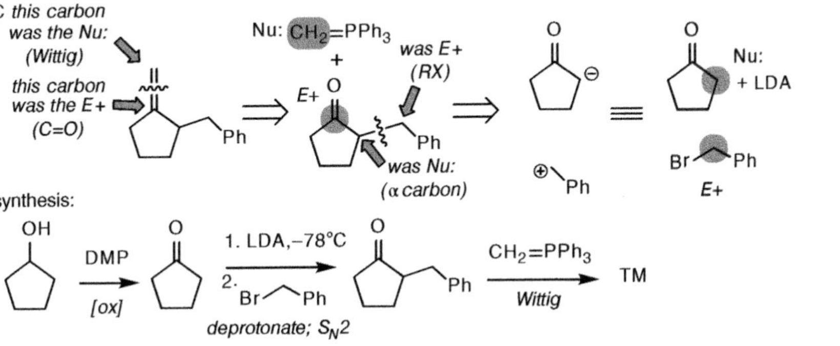

D

OH *this carbon was the E+ (C=O)*
O=

this carbon was the Nu: (RMgBr)

⟹

O=   O   ⊖

carbonyl must be protected from Grignard reagent

OH
⊕

≡

O
E+

O   O
─MgBr
Nu:

synthesis:

O=   ─Br   HO   OH
acid

form acetal to protect carbonyl
(tosic acid, TsOH, is often used)

O   O
─Br   Mg   O   O
─MgBr

acidic aqueous workup will also hydrolyze PG   TM

=O
H₃O⊕

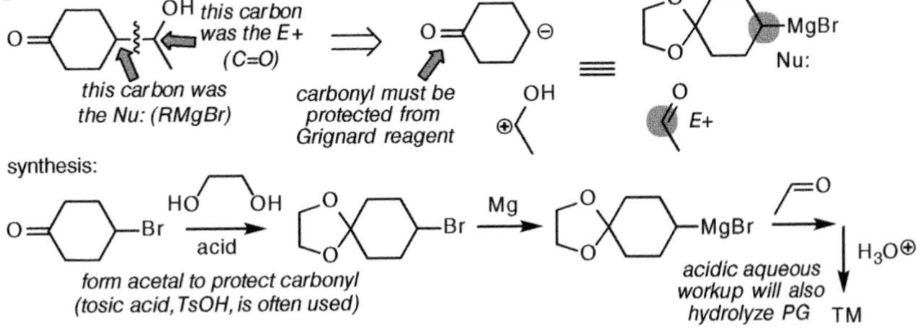

E

*this carbon was
the Nu: (RMgBr)*

OH   OH

⟹   ⊕   ⊖CH₃   ≡   E+

CH₃   CH₃MgBr   Nu:

*this carbon
was the E+*   *trans stereochemistry must have
come from epoxide ring opening*

synthesis:

   mCPBA   →   ⁗O   CH₃MgBr   H₃O⊕ →   TM
[ox]   (racemic)

F

$\overset{1}{}\overset{2}{}\overset{3}{}\overset{4}{}\overset{5}{}$ FGI ⟹   OH   FGI ⟹ $\overset{1}{}\overset{2}{}\overset{3}{}\overset{4}{}\overset{5}{}$

no change in carbon chain, so just FGI

synthesis:

  1. BH₃-THF →   OH   DMP → TM
  2. H₂O₂, NaOH   [ox]

*anti-Markovnikov
addition of water*

G

O   *was E+*
     (RX)

$\overset{1}{}\overset{2}{}\overset{3}{}$ FGI ⟹ EtO $\overset{1}{}\overset{2}{}\overset{3}{}$ ⟹ EtO $\overset{1}{}\overset{2}{}\overset{3}{}$

     was Nu: *was Nu:*   *(α carbon)*
     (α carbon)   *was E+*
               (RX)

synthesis:

  1. EtONa   1. 2 equiv.   H₃O⊕
           LDA*   OEt → TM

O O   2.   O O   2. ─Br   heat

$\overset{3}{}\overset{2}{}\overset{1}{}$OEt   Br $\overset{3}{}\overset{2}{}\overset{1}{}$OEt   −CO₂

                    **make dienolate to alkylate
less acidic alpha carbon (C-3)*

H

O─$\overset{}{}$─CH₃

⟹   O⊖   OH   Nu:

was Nu: *was E+*   ⊕CH₃   ≡
(RO−)   *(RX)*   CH₃I

                   + NaH   E+

*must retain stereochemistry so
disconnect on methyl side of ether*

synthesis:

OH   NaH →   O⊖   CH₃I → TM   *no change in
configuration of chiral
center*

    deprotonate   SN2

I *not a logical*
disconnection

this carbon was the
E+ (acylium ion)

FGI

Ph
2
1
3
4

this carbon was
the Nu: (PhH)

Friedel–Crafts
acylation

+ PhH

FGI

FGI

HO
1
2
3
4

synthesis:

OH

1. Na₂Cr₂O₇,
H₂SO₄

2. SOCl₂,
base

[ox]; make acid
chloride

AlCl₃
Friedel–Crafts
acylation

Ph

NH₂NH₂, KOH

(or Zn/Hg, HCl
or H₂, Pd)
reduction

TM

J

this carbon was
the E+ (C=O)

Ph
O

FGI

HO

this carbon was
the Nu: (RMgBr)

+
Ph
Cl

HO

E+

MgBr
Nu:

Br

synthesis:

1. Br₂,
FeBr₃

2. Mg

MgBr

O

H₃O⊕

attack C=O; wkup

HO

Ph
O
Cl

+ base
make ester

TM

K

O

FGI

OH
OH

FGI

FGI

OH

acetal
(acetonide)

+

O

synthesis:

OH

H₂SO₄,
heat

dehydration

OsO₄

syn
dihydroxylation

OH
OH

+ acid

O

add PG; tosic acid (TsOH) is
often used to make acetals

TM

L

FGI

need alkyne to
form trans
alkene

this was the
E+ (RX)

⊕

≡

Br
E+

this was the Nu:
(acetylide)

C≡CH

Nu:

FGI

Br

Br

+ NaNH₂

synthesis:

Br₂

Br
Br

OH
heat

double E2
elimination
(−HBr x2)

1. NaNH₂

2.
Br

deprotonate; SN2

Na, NH₃

dissolving
metal
reduction

TM

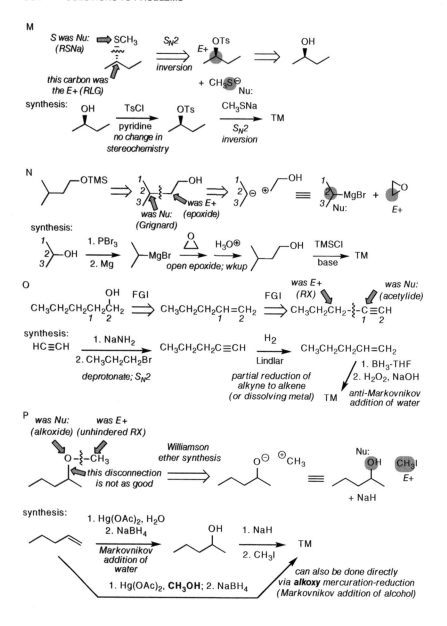

Q

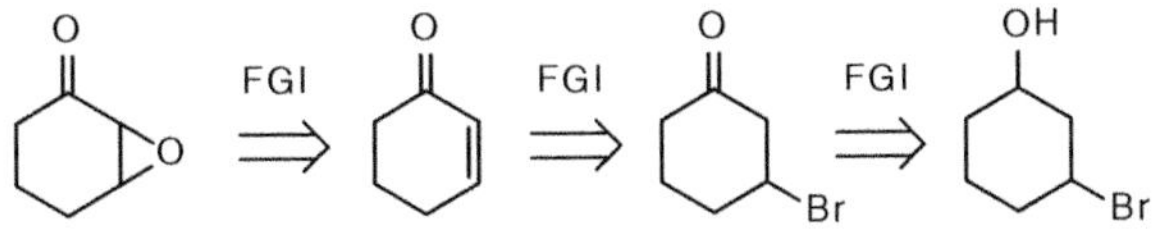

FGI
FGI
FGI
O
O
O
OH
Br
Br

synthesis:

OH
Br
DMP
O
Br
NaOEt
E2 or
β-elim.
(E1cb)
O
mCPBA
TM
oxidize first, so carbonyl can direct
elimination to give conjugated product

R

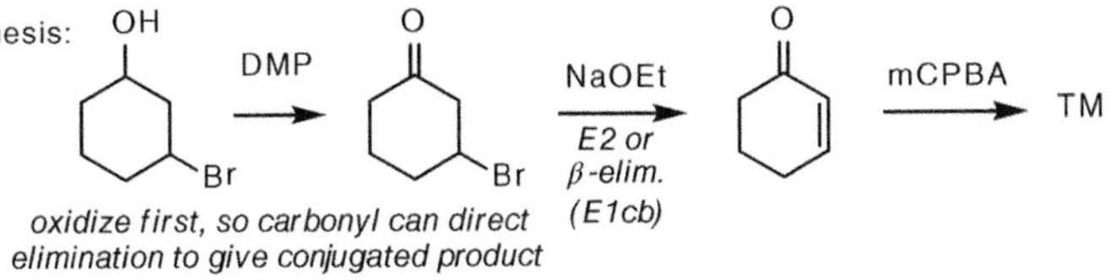

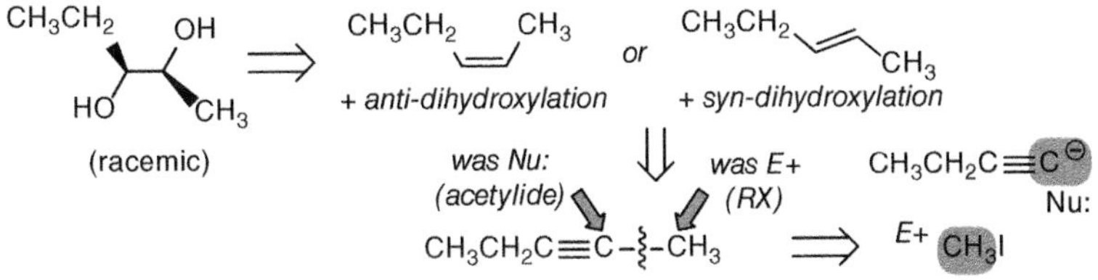

CH3CH2
OH
HO
CH3
(racemic)
CH3CH2
CH3
+ anti-dihydroxylation
or
CH3CH2
CH3
+ syn-dihydroxylation
was Nu:
(acetylide)
was E+
(RX)
CH3CH2C≡C-CH3
CH3CH2C≡C⊖
Nu:
E+ CH3I

synthesis:

CH3CH2C≡CH
1. NaNH2
2. CH3I
deprotonate; SN2
CH3CH2C≡CCH3
H2, Lindlar
add Hs syn
CH3CO3H, H2O
add OHs anti
TM
or
Na, NH3
add Hs anti
OsO4
add OHs syn
TM

S

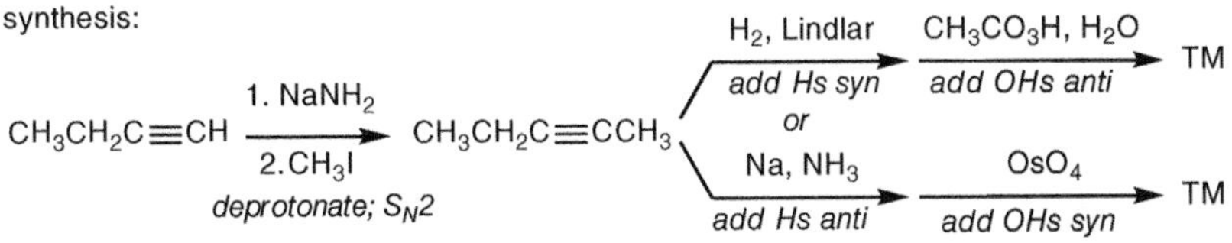

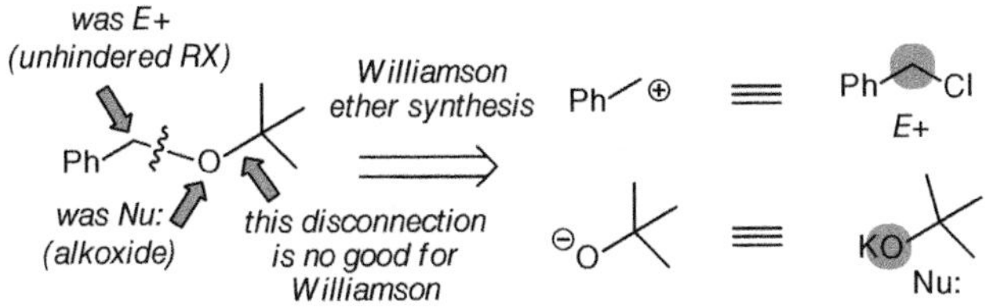

was E+
(unhindered RX)
Ph
O
was Nu:
(alkoxide)
this disconnection
is no good for
Williamson
Williamson
ether synthesis
Ph
Ph
Cl
E+
KO
Nu:

synthesis:

Ph
OH
SOCl2
base
Ph
Cl
t-BuOK
SN2
TM
+ acid
can also be done directly via acid-
catalyzed addition of the alcohol to an
alkene (involves stable 3° carbocation)

T

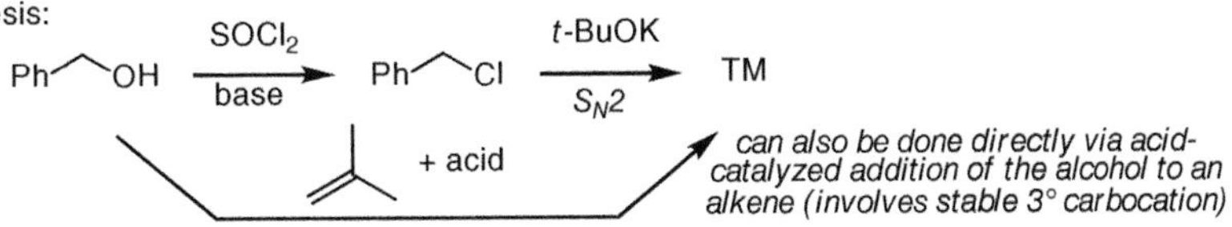

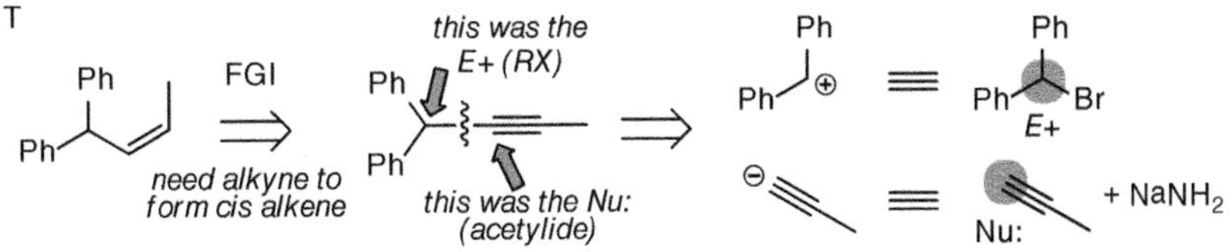

Ph
Ph
FGI
need alkyne to
form cis alkene
Ph
Ph
this was the
E+ (RX)
this was the Nu:
(acetylide)
Ph
Ph
Ph
Ph
Br
E+
Nu:
+ NaNH2

synthesis:

Ph
Ph
Br2
hν
free-radical
halogenation
Ph
Ph
Br
Ph
Ph
H2, Lindlar
add Hs syn
TM

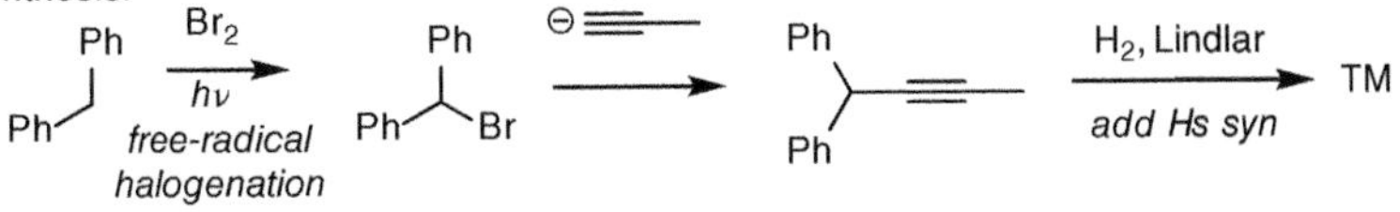

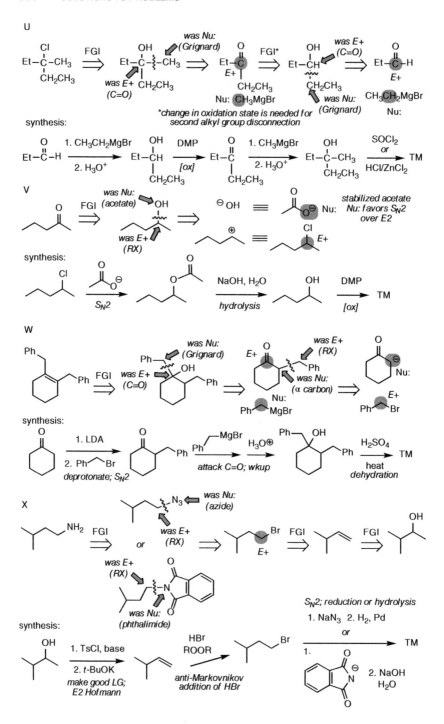

U

synthesis:

V

synthesis:

W

synthesis:

X

synthesis:

Y

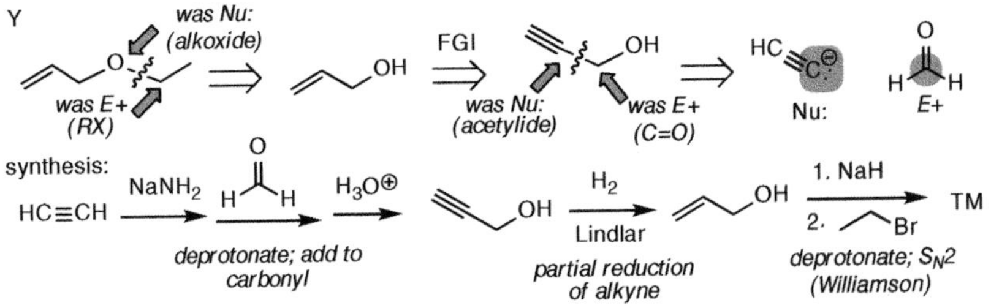

synthesis:

$HC\equiv CH$ $\xrightarrow{NaNH_2}$ $\xrightarrow{H_3O^\oplus}$

deprotonate; add to carbonyl

$\xrightarrow[\text{Lindlar}]{H_2}$ $\xrightarrow[\text{2.}\ \diagup\text{Br}]{\text{1. NaH}}$ TM

partial reduction of alkyne   *deprotonate; S_N2 (Williamson)*

Z

synthesis:

$\xrightarrow[\text{2. H}_3\text{O}^+]{\text{1. PhMgBr}}$ Ph$\diagdown\diagup$OH $\xrightarrow{PBr_3}$ Ph$\diagdown\diagup$Br $\xrightarrow[\text{2. BuLi}]{\text{1. PPh}_3}$ TM

S_N2; deprotonate

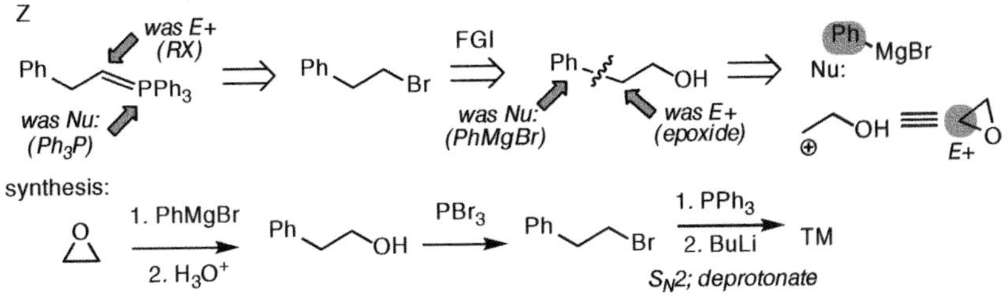

3-4. Provide a synthesis, using NaCN and any alkyl halide(s).

A retrosynthesis

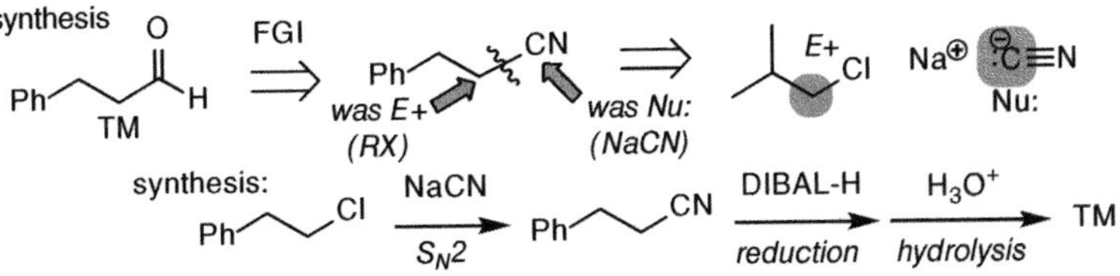

synthesis:

Ph$\diagdown\diagup$Cl $\xrightarrow[\text{S}_N2]{\text{NaCN}}$ Ph$\diagdown\diagup$CN $\xrightarrow[\text{reduction}]{\text{DIBAL-H}}$ $\xrightarrow[\text{hydrolysis}]{\text{H}_3\text{O}^+}$ TM

B retrosynthesis

synthesis:

$\xrightarrow[\text{S}_N2]{\text{NaCN}}$ CN $\xrightarrow[\text{reduction}]{\text{H}_2,\ \text{Pd}}$ TM

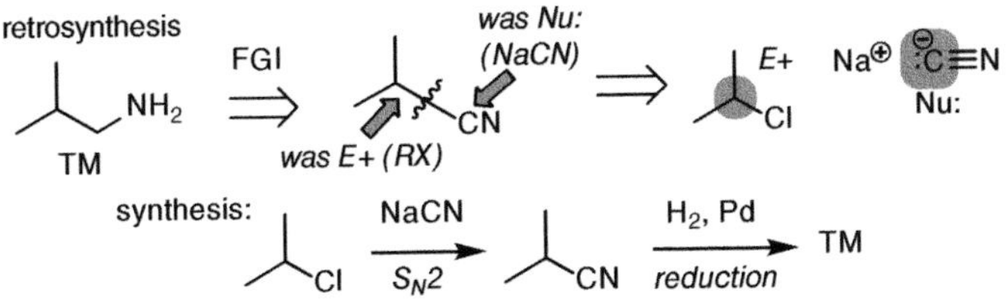

C retrosynthesis

synthesis:

Ph$\diagdown\diagup$Cl $\xrightarrow[\text{S}_N2]{\text{NaCN}}$ Ph$\diagdown\diagup$CN $\xrightarrow[\text{hydrolysis}]{\text{H}_3\text{O}^+\ \text{heat}}$ TM

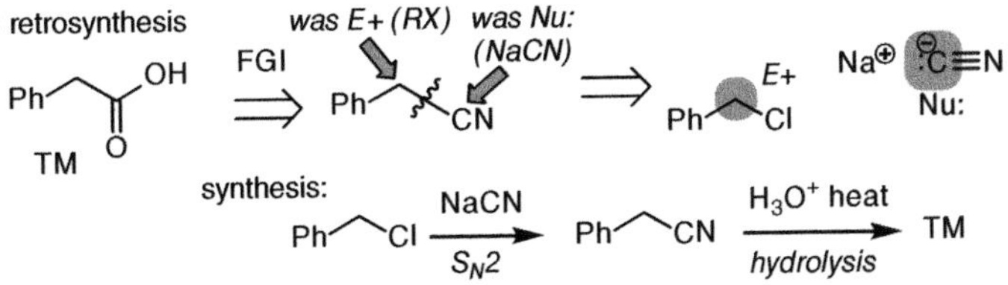

D retrosynthesis

TM

synthesis

3-5. C-14 Synthesis Game.

A $HC\equiv CH$ $\xrightarrow{NaNH_2,\ NH_3}$ $HC\equiv CNa$ $\xrightarrow{*CH_3Br}$ $\equiv\overset{*}{-}CH_3$

B $*CH_3Br$ $\xrightarrow{Mg}$ $*CH_3MgBr$ $\xrightarrow[\text{2. workup}]{1.\ \overset{O}{\triangle}}$ $*\!\!\diagup\!\!\diagup\!\!OH$

C $*CH_3Br$ $\xrightarrow{NaOH}$ $*CH_3OH$ $\xrightarrow{DMP}$ $O{=}\overset{*}{C}H_2$ $\xrightarrow[\text{2. workup}]{1.\ HC\equiv CNa\ (from\ A)}$ $\equiv\!\!\diagup\!\!\overset{*}{\diagup}\!\!OH$

D $\equiv\overset{*}{-}CH_3$ *(from A)* $\xrightarrow[\substack{2.\ \overset{O}{\triangle} \\ 3.\ \text{workup}}]{1.\ NaNH_2}$ $HO\diagdown\!\!\diagdown\!\!\equiv\!\!\diagdown\!\!\overset{*}{C}H_3$ $\xrightarrow{H_2,\ Pd}$ $HO\diagdown\!\!\diagdown\!\!\diagdown\!\!\diagdown^*$

E $HC\equiv CH$ $\xrightarrow[\substack{2.\ \overset{O}{\triangle} \\ 3.\ \text{workup}}]{1.\ NaNH_2}$ $\equiv\!\!\diagdown\!\!\diagdown OH$ $\xrightarrow{DMP}$ $\overset{H}{\underset{O}{\diagdown\!\!\diagdown}}\!\!\equiv$ *reduce first to remove acidic proton before Grignard* $\xrightarrow[\substack{2.\ *CH_3MgBr\ (from\ B) \\ 3.\ \text{workup}}]{1.\ H_2,\ Pd}$ $*\!\!\diagdown\!\!\overset{OH}{\diagdown}\!\!\diagdown\!\!\diagdown$

F $*\!\!\diagdown\!\!\diagdown OH$ *(from B)* $\xrightarrow{PBr_3}$ $*\!\!\diagdown\!\!\diagdown Br$ $\xrightarrow[\substack{\text{(from A)}}]{HC\equiv CNa}$ $*\!\!\diagdown\!\!\diagdown\!\!\equiv$ $\xrightarrow[\substack{2.\ mCPBA}]{1.\ H_2\ \text{Lindlar's}}$ $*\!\!\diagdown\!\!\diagdown\!\!\overset{O}{\triangle}$

G $\equiv\!\!\diagdown\!\!\diagdown OH$ *(from E)* *reduce first to remove acidic proton before NaH* $\xrightarrow[\substack{2.\ NaH \\ 3.\ BnBr}]{1.\ H_2\ \text{Lindlar's}}$ $\diagdown\!\!\diagdown\!\!\diagdown OBn$ $\xrightarrow[\substack{2.\ *CH_3MgBr\ (from\ B) \\ 3.\ \text{workup}}]{1.\ mCPBA}$ $*\!\!\diagdown\!\!\overset{OH}{\diagdown}\!\!\diagdown\!\!\diagdown OBn$

H (from C) $\overset{*}{\underset{\text{OH}}{\text{propargyl}}}$ $\xrightarrow[\substack{\text{reduce first to remove acidic}\\\text{proton before NaH}}]{\text{H}_2, \text{Pd}}$ $\overset{*}{\underset{}{\text{OH}}}$ $\xrightarrow[\substack{2. \triangle\text{O}}]{1. \text{NaH}}$ $\overset{*}{\underset{}{\text{O}\text{OH}}}$

I $\underset{CH_3}{\overset{*}{}}\text{OH}$ (from D) $\xrightarrow[\substack{2. \text{DMP}}]{1. \text{Na, NH}_3}$ $\overset{*}{\underset{\substack{\textit{trans alkene}\\\textit{needed}}}{}}\overset{O}{\underset{H}{}}$ $\xrightarrow[\substack{2. \text{Br}_2\\\textit{add acetal PG;}\\\textit{anti bromination}}]{\substack{1. \text{HO}\quad\text{OH}\\\text{TsOH}}}$ $\underset{\substack{Br}}{\overset{Br}{}}\overset{*}{\underset{}{\text{O}}}$

J $\overset{*}{\underset{\text{OH}}{}}$ (from B) $\xrightarrow{\text{DMP}}$ $\overset{*}{\underset{H}{\overset{O}{}}}$ $\xrightarrow[\substack{2. \text{workup}}]{\substack{1. \text{HC}\equiv\text{CNa}\\\text{(from A)}}}$ $\overset{*}{\underset{\text{OH}}{}}$

K $\overset{*}{\underset{\text{OH}}{}}$ (from J) $\xrightarrow[\substack{2. \text{DMP}}]{\substack{1. \text{H}_2\\\text{Lindlar's}}}$ $\overset{*}{\underset{O}{}}$ $\xrightarrow{\text{OsO}_4}$ $\overset{*}{\underset{O}{}}\overset{\text{OH}}{\underset{\text{OH}}{}}$ $\xrightarrow[\substack{\text{TsOH}\\\textit{acetonide PG}}]{\overset{O}{}}$ $\overset{*}{\underset{O}{\overset{O}{}}}$

L $\overset{}{\underset{\text{OH}}{}}$ (from E) $\xrightarrow[\substack{\text{base}}]{\text{TBSCl}}$ $\overset{}{\underset{\substack{\text{OTBS}\\\textit{protect first to remove}\\\textit{acidic proton before NaNH}_2}}{}}$ $\xrightarrow[\substack{2. \text{O}=\text{CH}_2\\\text{(from C)}\\3. \text{workup}}]{1. \text{NaNH}_2}$ $\underset{\text{TBSO}}{\overset{\text{HO}}{}}\overset{*}{}$

M $\overset{\text{HO}}{\underset{\substack{\text{(from L)}\\\text{TBSO}\\\textit{leave PG on to}\\\textit{distinguish between OHs}}}{}}\overset{*}{}$ $\xrightarrow[\substack{2. \text{TsCl}\\\text{base}}]{\substack{1. \text{H}_2\\\text{Lindlar's}}}$ $\underset{\substack{\text{OTBS}\\\textit{need cis alkene}}}{\overset{\text{OTs}}{}}\overset{*}{}$ $\xrightarrow[\substack{2. \text{NaOH}\\\textit{deprotect;}\\\textit{intramolecular}\\\textit{Williamson}}]{1. \text{TBAF}}$ $\overset{O}{\underset{}{}}\overset{*}{}$

N $\overset{\text{HO}}{\underset{\substack{\text{(from L)}\\\text{TBSO}\\\textit{leave PG on to}\\\textit{differentiate OHs}}}{}}\overset{*}{}$ $\xrightarrow{\substack{\text{H}_2\\\text{Pd}}}$ $\underset{\substack{\text{TBSO}}}{\overset{\text{OH}}{}}\overset{*}{}$ $\xrightarrow[\substack{2. \text{TBAF}\\\text{Jones [ox];}\\\textit{deprotect}}]{\substack{1. \text{KMnO}_4\\\text{heat}}}$ $\underset{\text{HO}}{\overset{\text{HO}}{}}\overset{O}{\underset{}{}}\overset{*}{}$ $\xrightarrow[\substack{\textit{Fischer}\\\textit{esterification}}]{\substack{\text{HA}\\\text{heat}}}$ $\overset{O}{\underset{O}{}}$

since TBS is unlikely to survive acidic Jones oxidation conditions, permanganate oxidation is used instead

3-6. Provide necessary reagents.

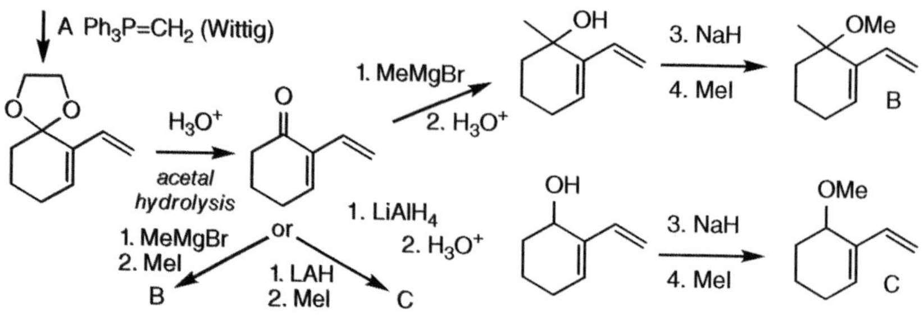

3-7. Provide missing reagents/products. (Possible reagents are given in quotes, and the actual reagents used in the literature are provided in parentheses.)

A MOMCl
B $LiAlH_4$; workup
C "DMP" [ox] (TEMPO/NaOCl)

D

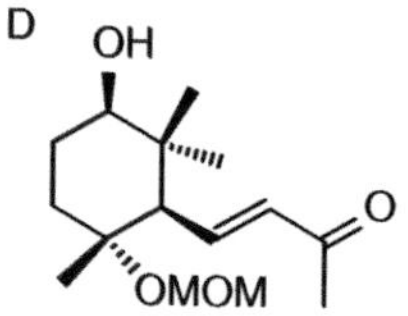

E H_2, Pd
F "TBSCl" (TBSOTf)
G 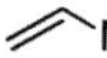 MgBr; workup

SOLUTIONS TO CHAPTER 4: 2-FG TMs

4.1A Provide the missing starting material(s) or products.

A

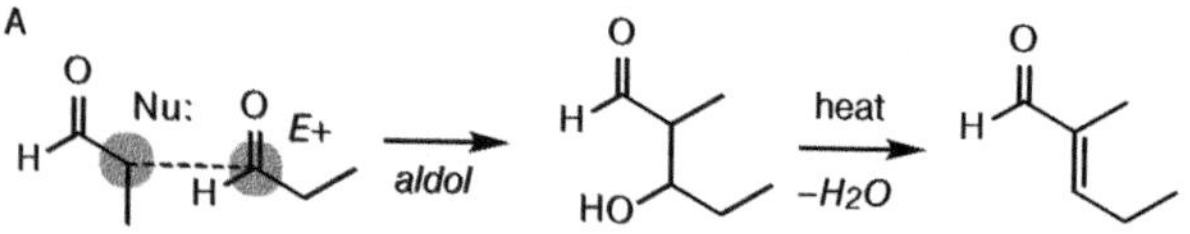

B

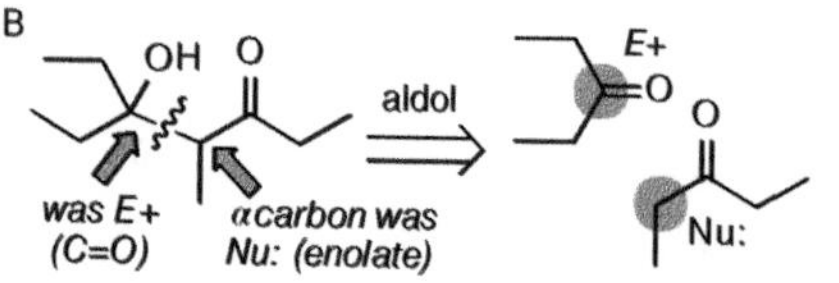

C

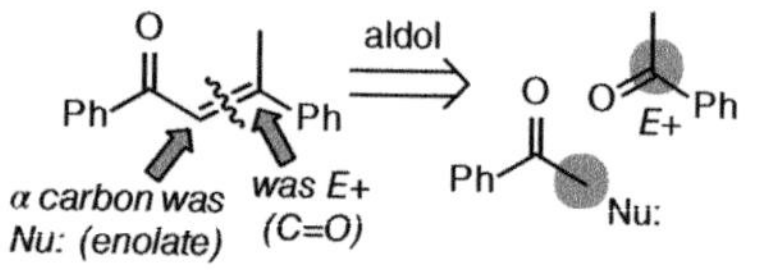

D

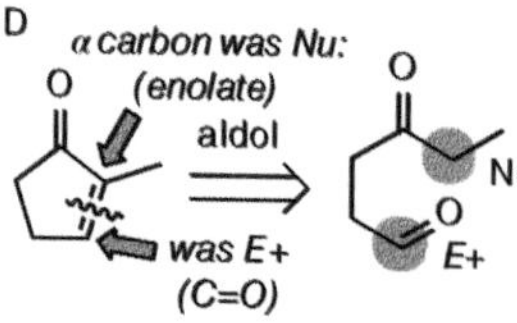

E

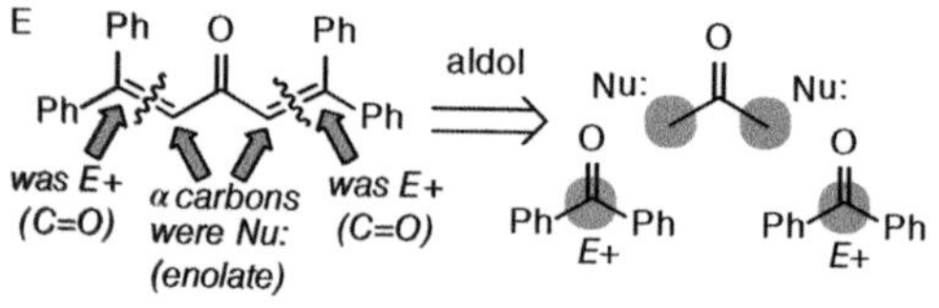

4.1B Provide the reagents needed for transformations.

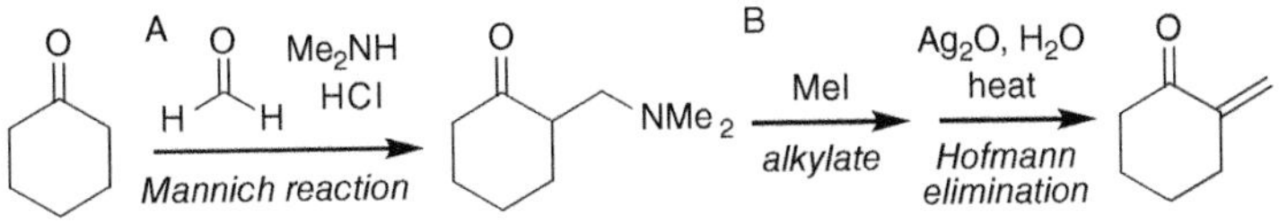
A
Me2NH
HCl
Mannich reaction
NMe2
B
MeI
alkylate
Ag2O, H2O
heat
Hofmann
elimination

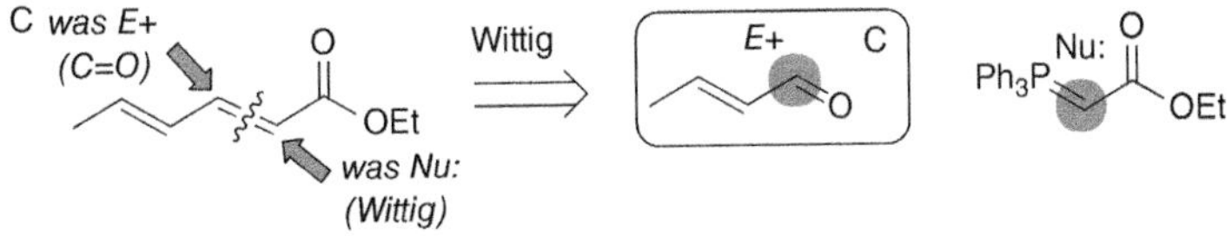
C was E+
(C=O)
OEt
was Nu:
(Wittig)
Wittig
E+
C
Ph3P
Nu:
OEt

4.2 Provide the missing starting material(s) or products.

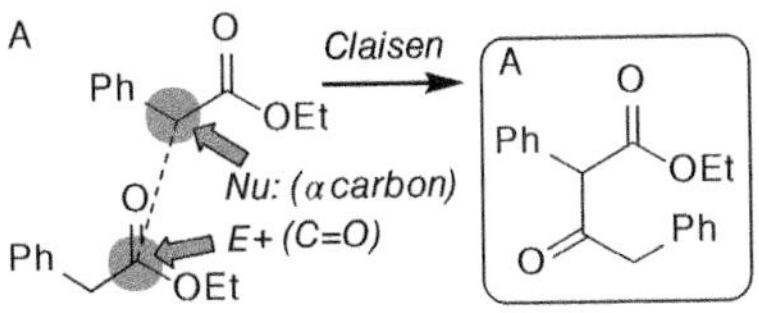
A
Ph
OEt
Claisen
Nu: (α carbon)
E+ (C=O)
Ph
OEt
A
Ph
OEt
Ph

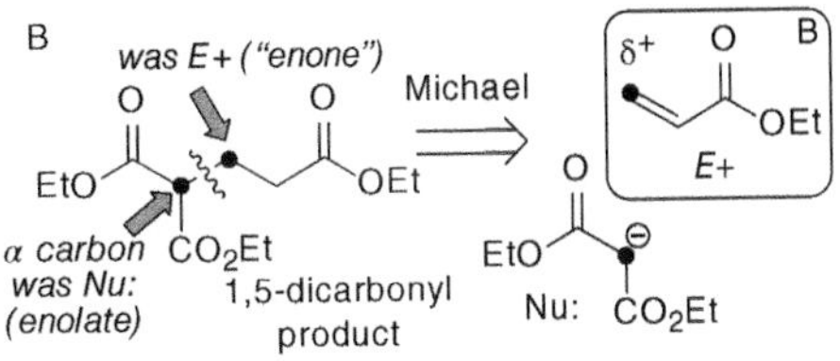
B
was E+ ("enone")
Michael
EtO
OEt
α carbon
was Nu:
(enolate)
CO2Et
1,5-dicarbonyl
product
δ+
B
OEt
E+
EtO
Nu:
CO2Et

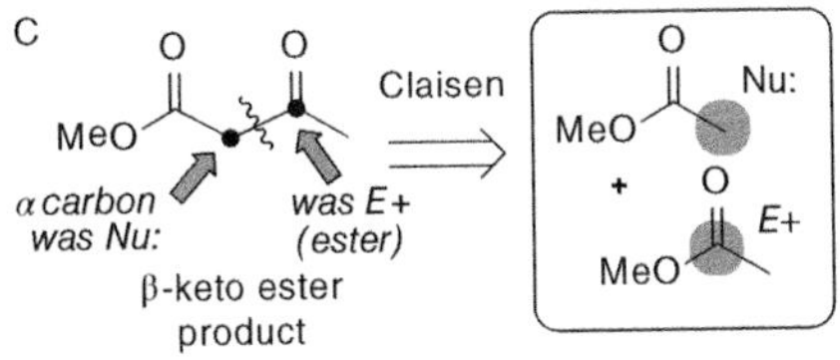
C
MeO
α carbon
was Nu:
was E+
(ester)
β-keto ester
product
Claisen
MeO
Nu:
+
E+
MeO

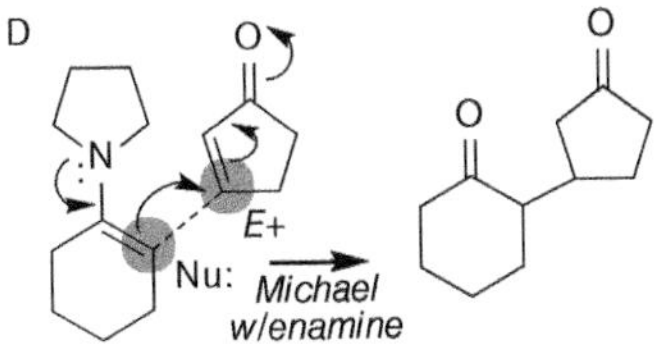
D
N
E+
Nu:
Michael
w/enamine

4.3 Provide the reagents needed for transformations.

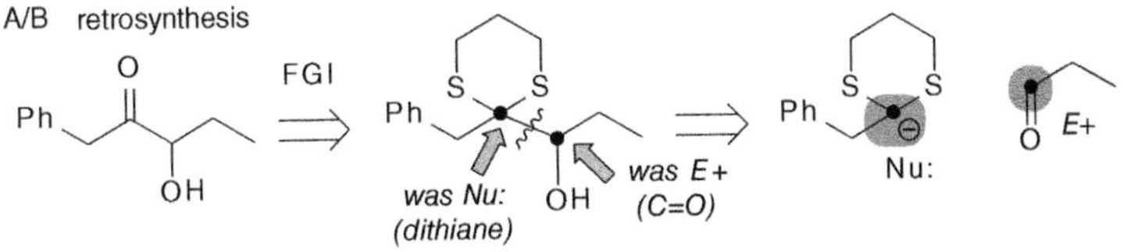
A/B retrosynthesis
Ph
OH
FGI
Ph
S
S
was Nu:
(dithiane)
OH
was E+
(C=O)
Ph
S
S
Nu:
E+

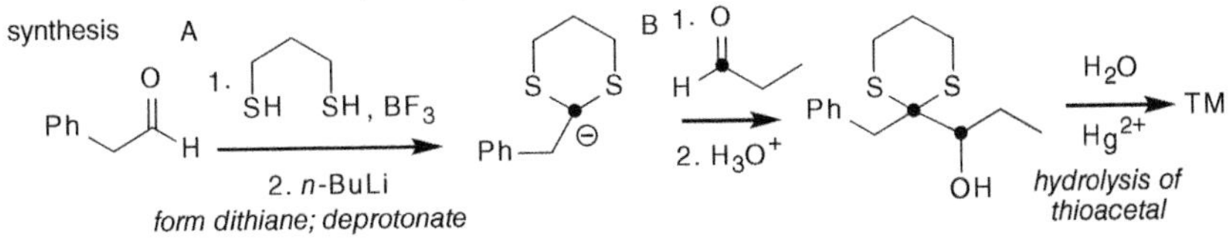
synthesis A
Ph
H
1. SH SH, BF3
2. n-BuLi
form dithiane; deprotonate
S
S
Ph
B 1.
H
2. H3O+
Ph
S
S
OH
H2O
Hg2+
TM
hydrolysis of
thioacetal

C retrosynthesis:

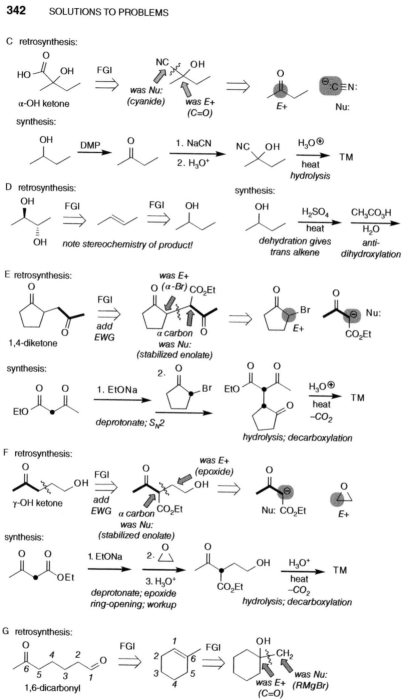

End-of-Chapter Problems

4-1. Provide the corresponding synthetic equivalent for each of the following synthons.

A

dithiane anion is the equivalent of
the umpolung acyl anion

B

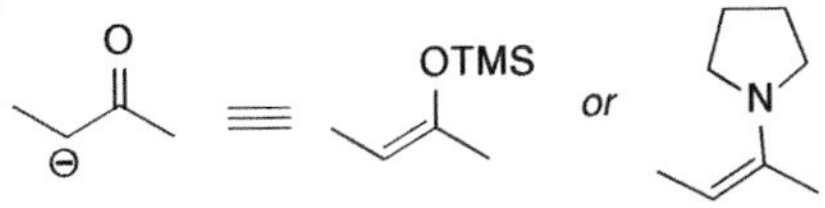

or

silyl enol ethers and enamines are
thermodynamic enolate equivalents

C

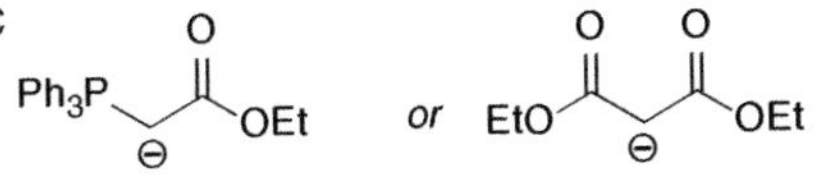

or

stabilized Wittig or
Horner–Wadsworth–
Emmons (HWE)

decarboxylation removes
second EWG, followed by
re-esterification

D

add a conjugated pi bond to
make the beta carbon an E+

E

after Nu: attacks, OH remains
on same carbon

F

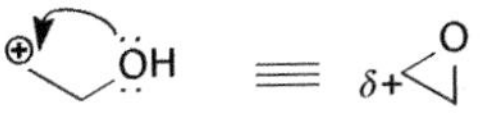

after Nu: attacks, OH remains
on adjacent carbon

G

add LG to make alpha
carbon an E+ (umpolung)

H

umpolung: cyanide is a good Nu: that
can be hydrolyzed to a carboxylic acid

I

add LG so after Nu: attacks, C=O
remains (as in Claisen condensation)

J

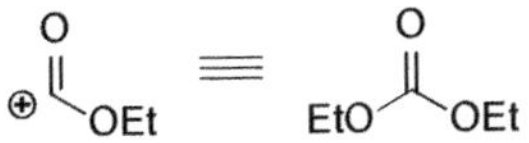

diethylcarbonate has two LGs, so
after Nu: attacks, ester FG remains

K

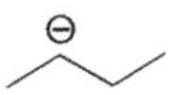

cuprates add 1,4- to
α,β-unsaturated
systems (Michael
acceptors)

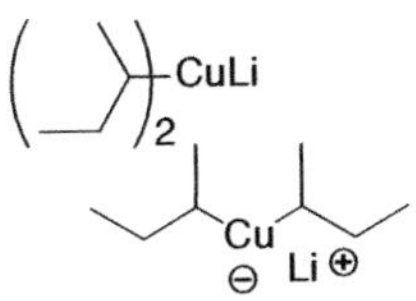

L

Mannich reagent gives better
aldol yields than formaldehyde

4-2. What disconnections/retrosyntheses are possible for each TM?

A

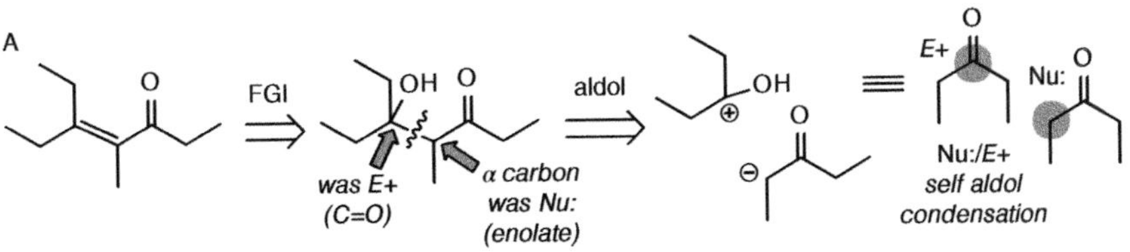

B *was E+*
   (epoxide)

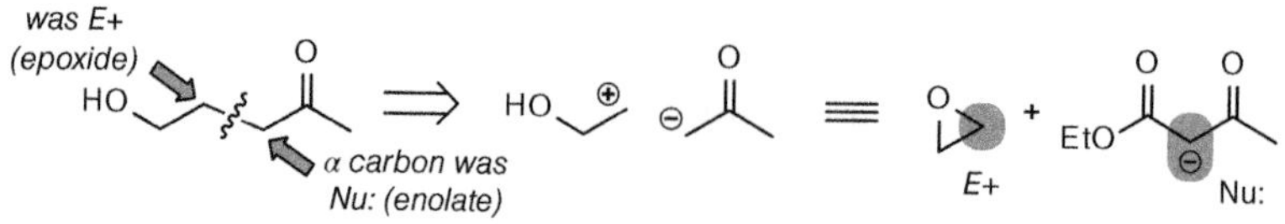

α carbon was
Nu: (enolate)

E+

Nu:

C

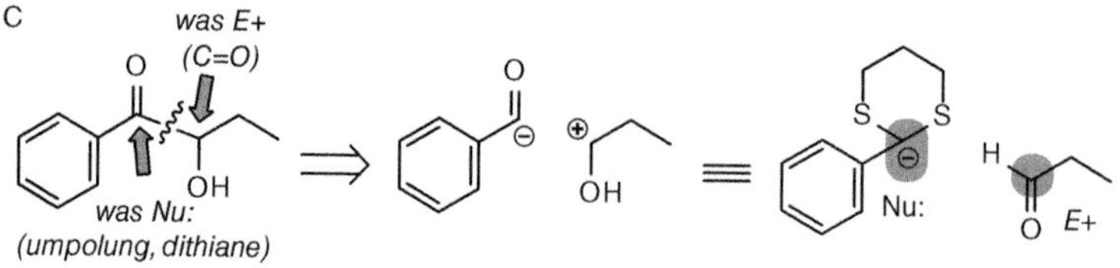

was E+
(C=O)

was Nu:
(umpolung, dithiane)

Nu:

E+

D α carbon was
Nu: (enolate)   *was E+*
(C=O)   Robinson annulation

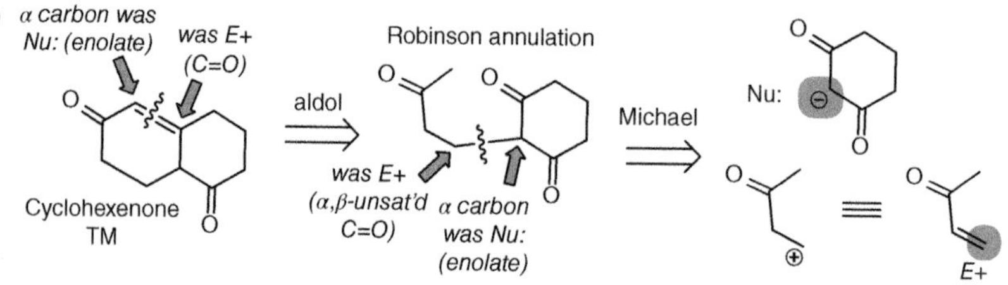

aldol

was E+
*(α,β-unsat'd   α carbon
C=O)   was Nu:
(enolate)*

Michael

Nu:

Cyclohexenone
TM

E+

E

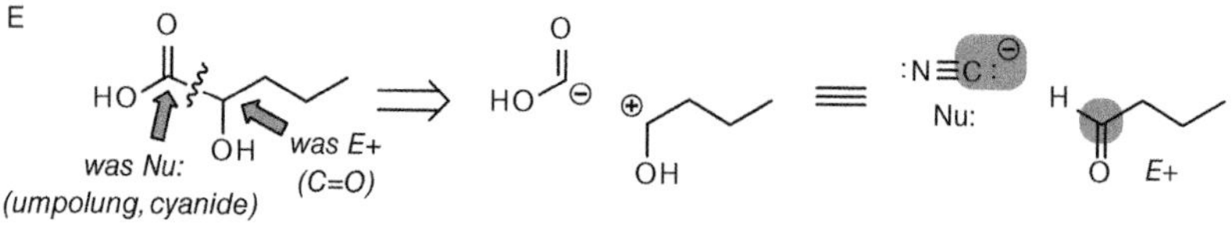

HO

was Nu: was E+
OH   *(C=O)*
(umpolung, cyanide)

:N≡C:

Nu:

E+

F *was E+*
(ester)

Ph   Ph

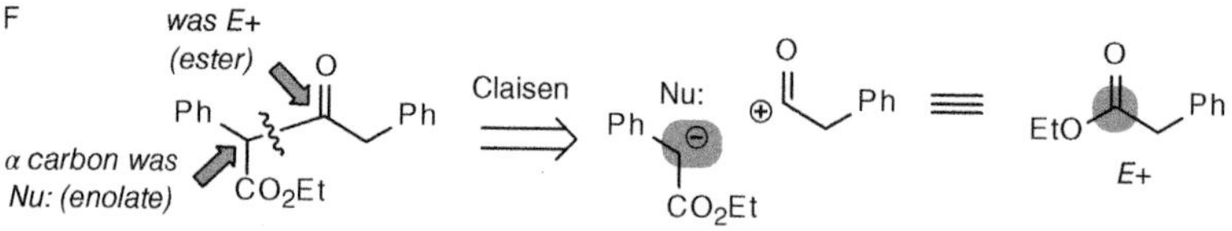

α carbon was
Nu: (enolate)   CO₂Et

Claisen

Nu:
Ph   Ph

CO₂Et

EtO   Ph

E+

G   *was E+*
*(umpolung;
α-bromo)*

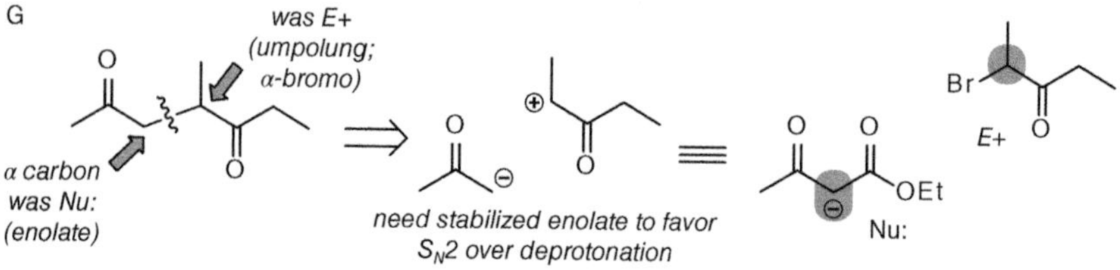

α carbon
was Nu:
(enolate)

*need stabilized enolate to favor
S_N2 over deprotonation*

Br

E+

OEt

Nu:

H   α carbon was
Nu: (enolate)

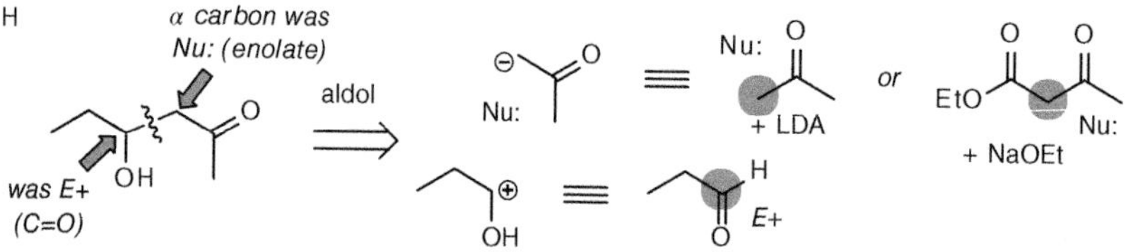

aldol

was E+
OH
(C=O)

Nu:

Nu:

Nu:
+ LDA   or   EtO

Nu:
+ NaOEt

E+

I   α carbon with two
EWGs was Nu:
(enolate)   CO₂Et

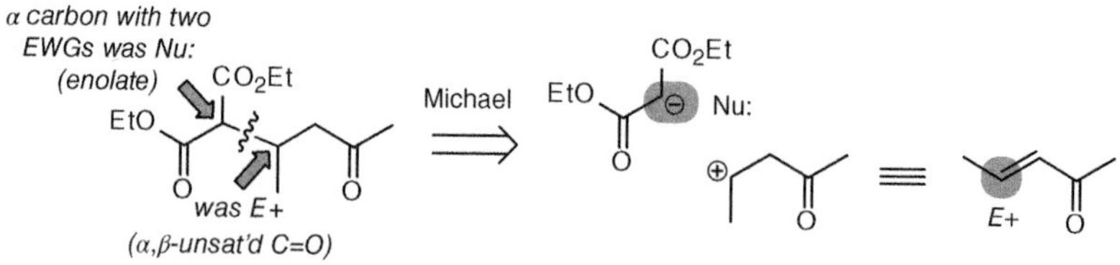

EtO

CO₂Et

Michael   EtO   Nu:

was E+
(α,β-unsat'd C=O)

E+

4-3. Starting with cyclohexanone, provide a synthesis for each of the following target molecules.

A

FGI

+ H_2O

α, β-unsat'd ketone came from β-hydroxy ketone

this carbon was the E+ (C=O)

aldol

this α carbon was the Nu: (enolate)

OH

Nu:

E+

synthesis:

only Nu: best E+

base
heat

TM

stepwise approach (LDA) not needed for this mixed aldol since benzaldehyde has no alpha protons; cat. KOH/MeOH is a common base; addition of heat promotes dehydration

B

was E+ (RX)

FGI

α carbon was Nu: (enolate)

E+
CH_3I

was E+ (C=O)

Claisen

α carbon was Nu: (enolate)

Nu:

EtO

E+

synthesis:

+ EtO Ph

1. EtONa, EtOH
2. mild $H_3O^{\oplus}$

mixed Claisen (stepwise/LDA not needed)

Ph

1. NaOH
2. CH_3I

TM

α-alkylation; LDA not needed since proton is alpha to two EWGs

C

FGI

was E+ (C=O)

aldol

α carbon was Nu: (enolate)

Nu:

+ H H

E+

$^{\oplus}NMe_2$

H H

E+

Mannich reagent works better than formaldehyde

synthesis:

H H
Me_2NH
base

NMe₂

1. CH_3I
2. NaOH

mCPBA

TM

D

was E+ (C=O)

CH_3 OH

was Nu: (dithiane)

CH_3 $\ominus$

OH

S S

CH_3 $\ominus$

Nu:

E+

synthesis:

S S

CH_3 $\ominus$

S S

CH_3 OH

H_2O
Hg^{2+}

TM

dithiane prep.

1.

O

CH_3 H

SH SH

S S

2. BuLi CH_3 $\ominus$

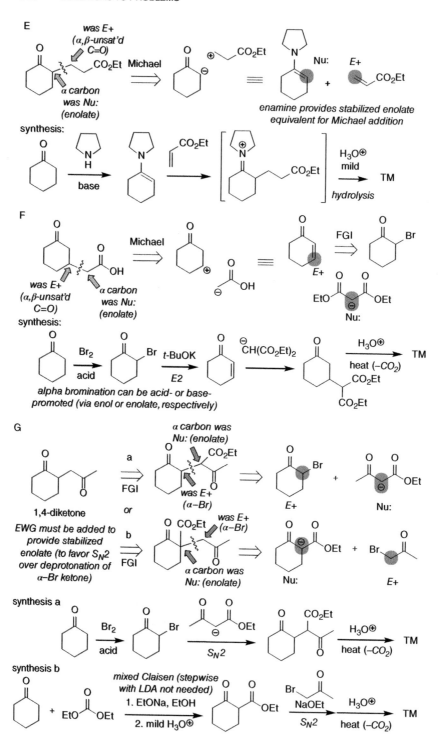

E

was E+
(α,β-unsat'd C=O)

CO₂Et **Michael**

α carbon
was Nu:
(enolate)

enamine provides stabilized enolate equivalent for Michael addition

synthesis:

base

H₃O⊕
mild
TM

hydrolysis

F

Michael

FGI

was E+
(α,β-unsat'd C=O)

α carbon
was Nu:
(enolate)

Nu:

synthesis:

Br₂
acid

Br *t*-BuOK
E2

⊖CH(CO₂Et)₂

H₃O⊕
heat (−CO₂)
TM

alpha bromination can be acid- or base-promoted (via enol or enolate, respectively)

G

α carbon was
Nu: (enolate)

CO₂Et

a
FGI

or

b
FGI

was E+
(α−Br)

1,4-diketone

EWG must be added to
provide stabilized
enolate (to favor S_N2
over deprotonation of
α−Br ketone)

was E+
(α−Br)

α carbon was
Nu: (enolate)

E+

Nu:

Nu:

E+

synthesis a

Br₂
acid

Br

SN2

H₃O⊕
heat (−CO₂)
TM

synthesis b

+ EtO OEt

mixed Claisen (stepwise
with LDA not needed)
1. EtONa, EtOH
2. mild H₃O⊕

OEt

NaOEt
SN2

H₃O⊕
heat (−CO₂)
TM

H

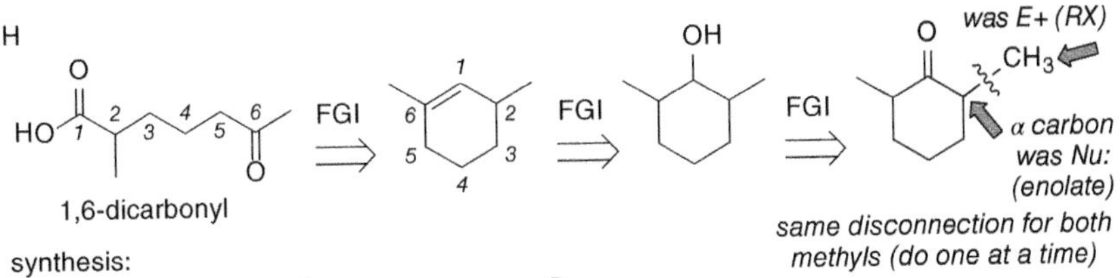

HO (with numbers 1 2 3 4 5 6) — 1,6-dicarbonyl

FGI (numbers 1 6 2 5 3 4)

FGI

FGI

was E+ (RX)
CH₃
*α carbon
was Nu:
(enolate)*

*same disconnection for both
methyls (do one at a time)*

synthesis:

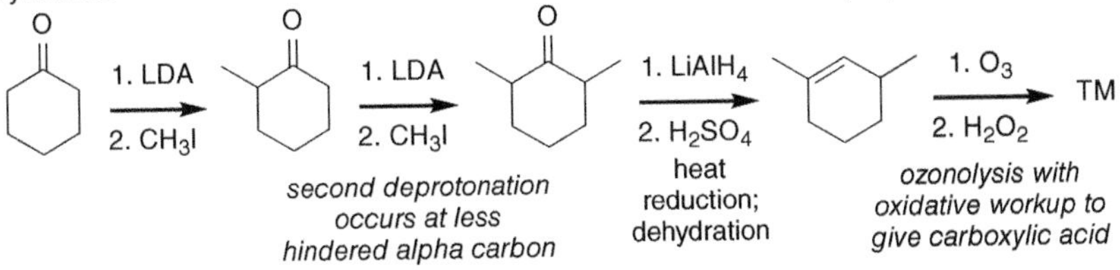

1. LDA
2. CH₃I

1. LDA
2. CH₃I

*second deprotonation
occurs at less
hindered alpha carbon*

1. LiAlH₄
2. H₂SO₄
heat
*reduction;
dehydration*

1. O₃
2. H₂O₂
*ozonolysis with
oxidative workup to
give carboxylic acid*

TM

I

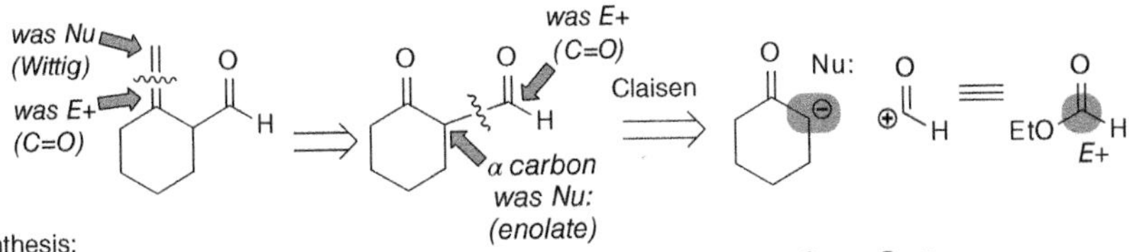

*was Nu
(Wittig)*

*was E+
(C=O)*

*was E+
(C=O)*

*α carbon
was Nu:
(enolate)*

Claisen

Nu:
EtO   H
E+

synthesis:

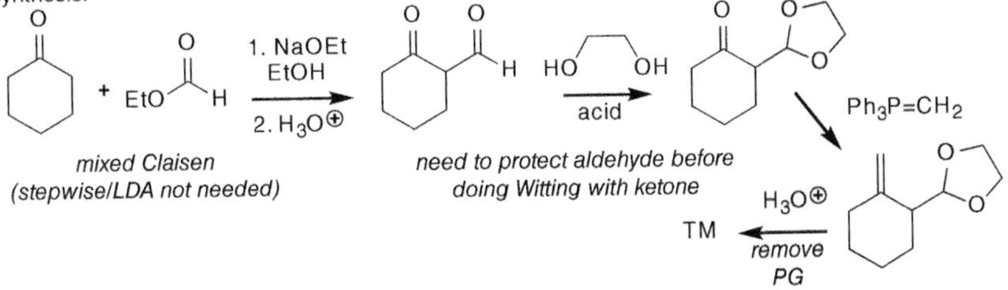

+ EtO   H

1. NaOEt
EtOH
2. H₃O⊕

*mixed Claisen
(stepwise/LDA not needed)*

HO   OH
acid

*need to protect aldehyde before
doing Witting with ketone*

Ph₃P=CH₂

H₃O⊕

TM ← *remove
PG*

J *α carbon
was Nu:
(enolate)*

*was E+
(C=O)*

aldol

*was E+
(enone)*

Michael

*α carbon was
Nu: (enolate)*

CO₂Et

CO₂Et

-CO₂Et
Nu:

E+

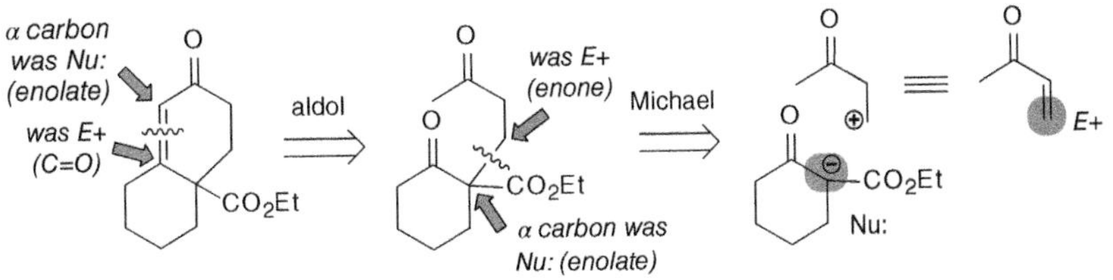

synthesis:

+ EtO   OEt

1. NaOEt
EtOH
2. H₃O⊕

*mixed Claisen
(stepwise/LDA not needed)*

OEt

*base
heat
Robinson*

TM

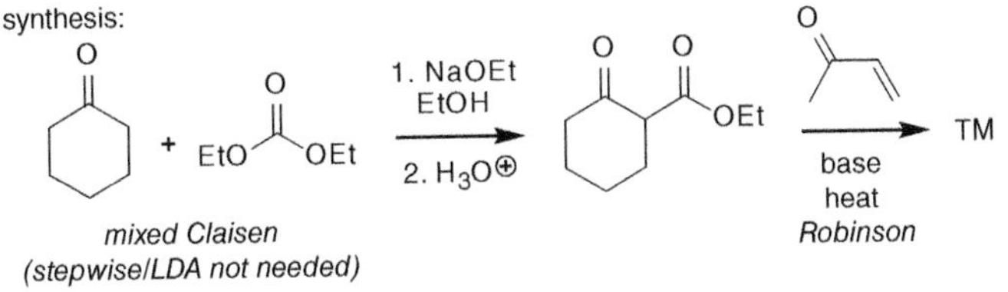

K

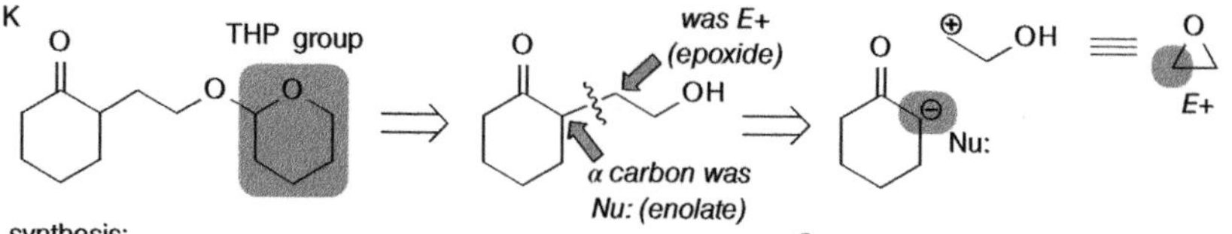

synthesis:

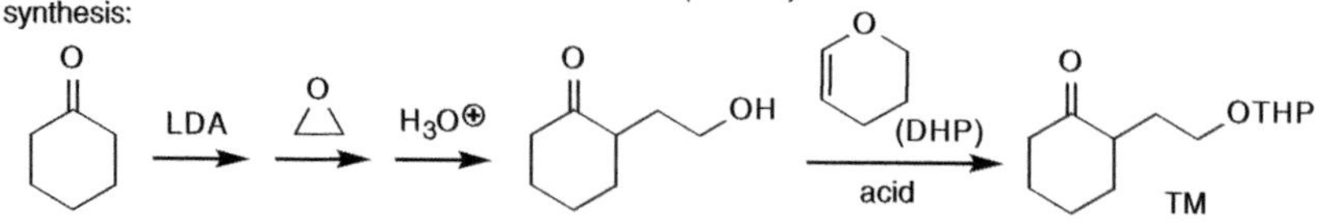

L

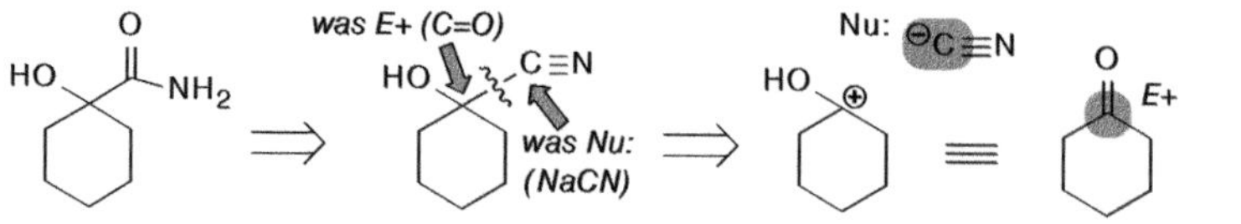

synthesis:

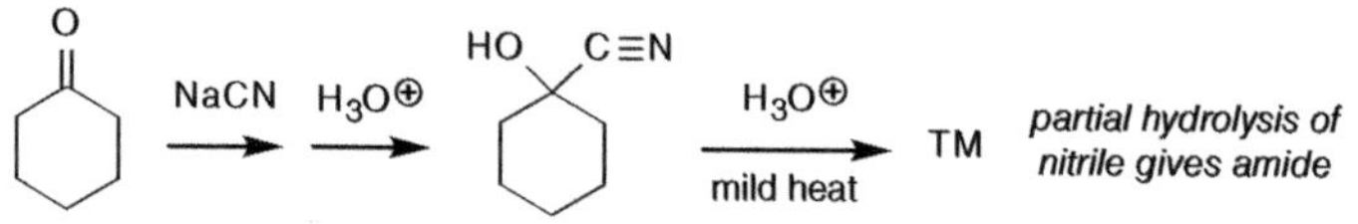

4-4. Provide the reagents necessary to transform the given starting material into the desired product.

A

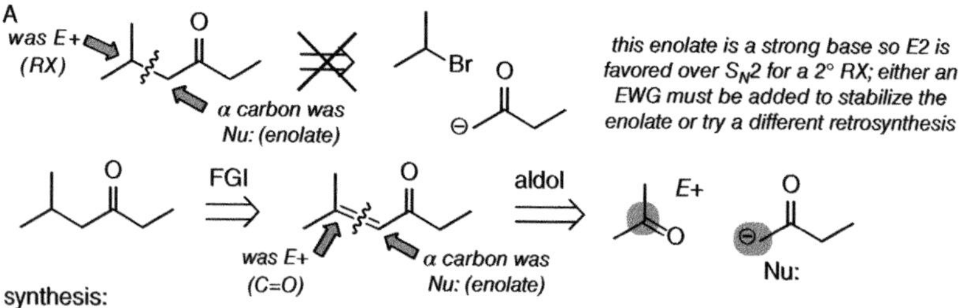

this enolate is a strong base so E2 is favored over S_N2 for a 2° RX; either an EWG must be added to stabilize the enolate or try a different retrosynthesis

synthesis:

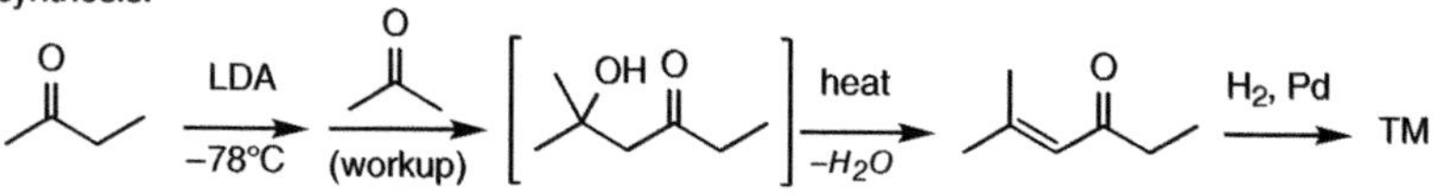

B

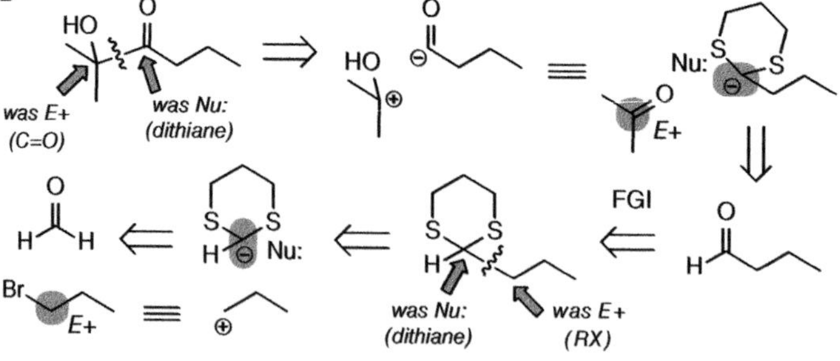

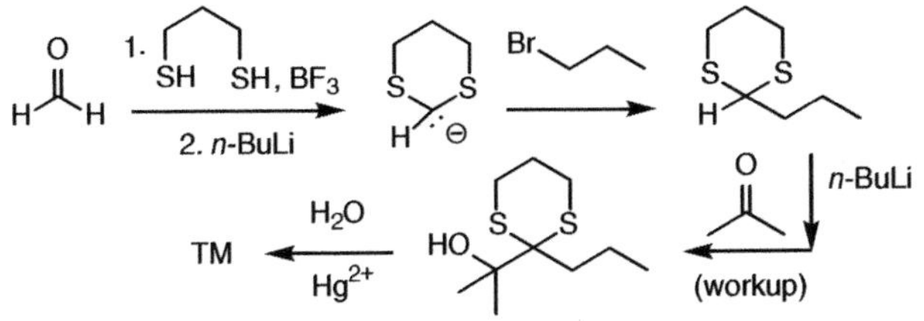

C

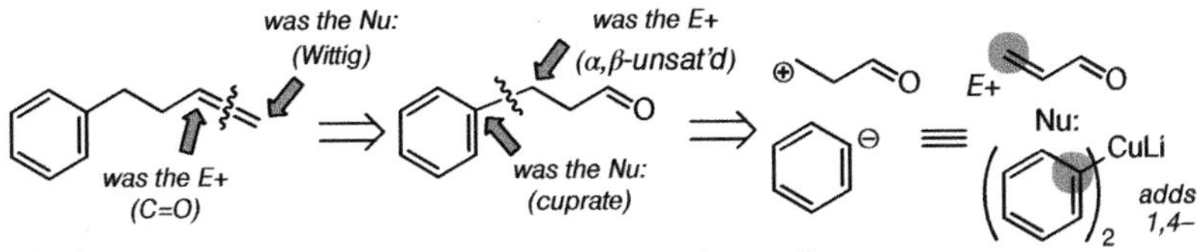

was the Nu:
(Wittig)

was the E+
(α,β-unsat'd)

was the E+
(C=O)

was the Nu:
(cuprate)

Nu:

adds
1,4–

synthesis:

$$\xrightarrow[\text{FeBr}_3]{\text{Br}_2}$$

1. Li
2. CuI

1.
2. $H_3O^{\oplus}$

$Ph_3P=CH_2$ → TM

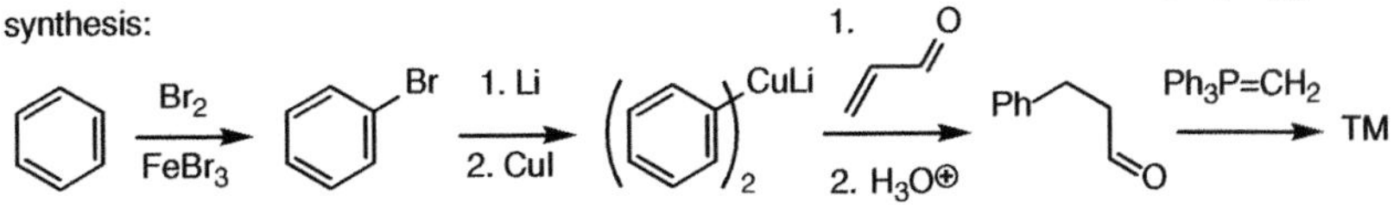

D

FGI ⟹ FGI ⟹

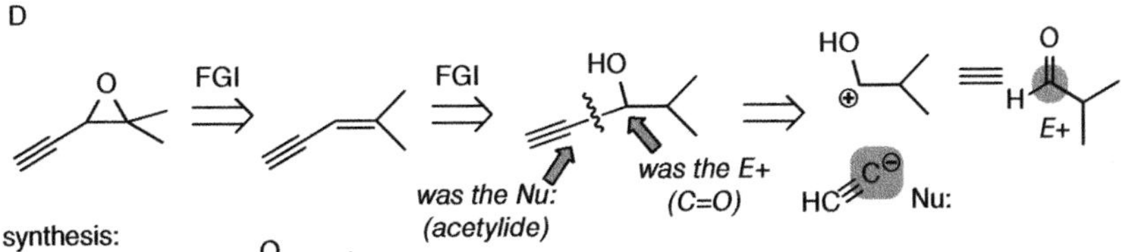

was the Nu:
(acetylide)

was the E+
(C=O)

E+

Nu:

synthesis:

$HC\equiv CH$ $\xrightarrow{NaNH_2}$ $\xrightarrow{H_3O^{\oplus}}$ $\xrightarrow[\text{heat}]{H_2SO_4}$ $\xrightarrow{mCPBA}$ TM

E

α carbon
was Nu:
(enolate)

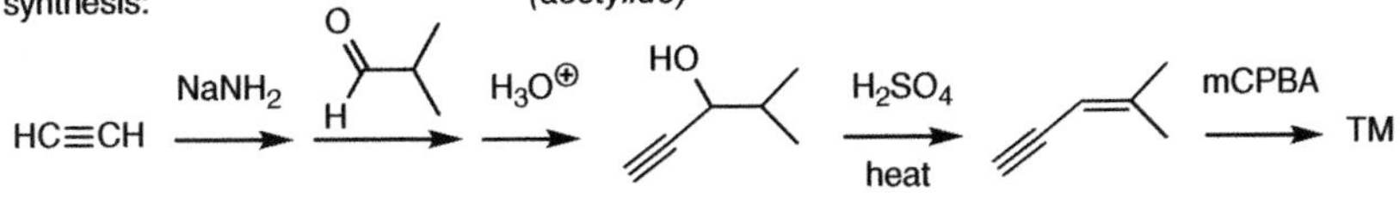

was E+
(C=O)

aldol ⟹

α carbon was
Nu: (enolate)

was the E+
(-unsat'd),αβ

Michael ⟹

E+

Nu:

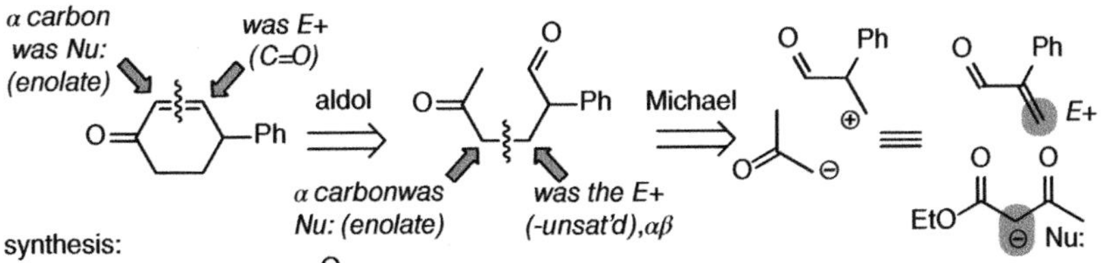

synthesis:

$\xrightarrow[\substack{\text{EtONa} \\ \text{EtOH} \\ \text{Michael}}]{}$ $\xrightarrow[\text{aldol}]{}$ $\xrightarrow[\substack{\text{heat} \\ -CO_2}]{H_3O^{\oplus}}$ TM

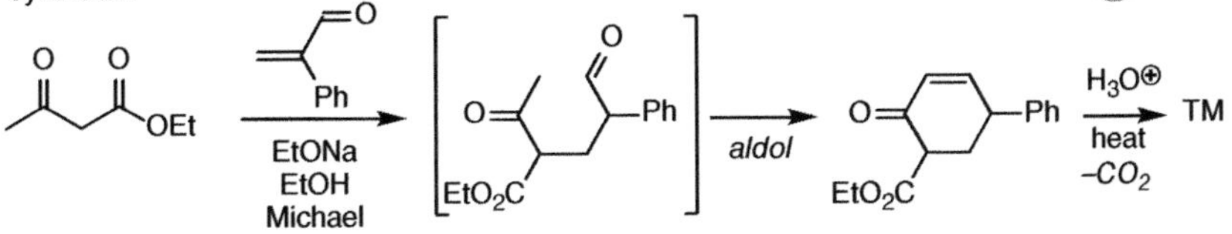

F

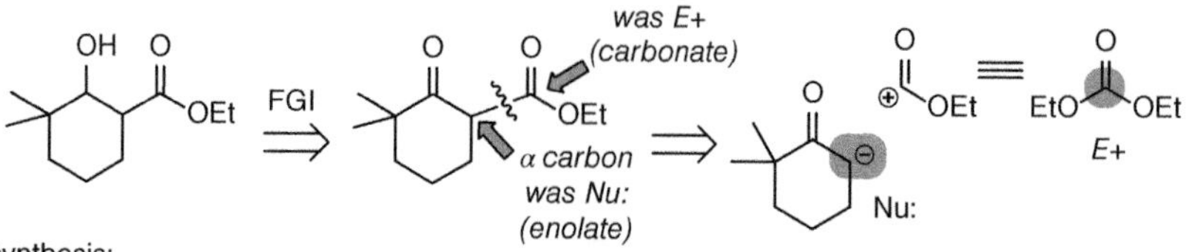

was E+
(carbonate)

α carbon
was Nu:
(enolate)

E+

Nu:

synthesis:

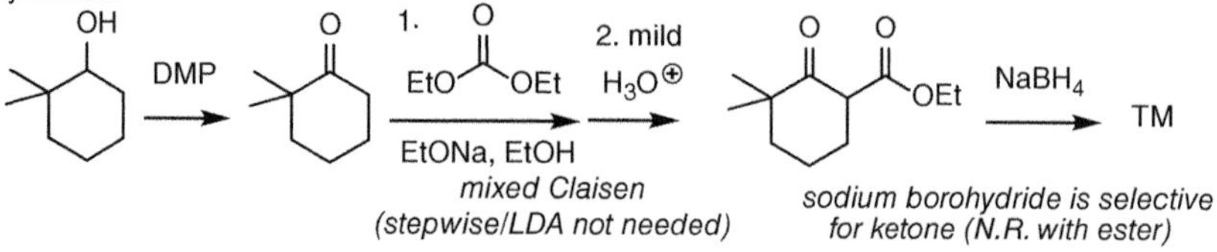

DMP

1.

EtONa, EtOH
mixed Claisen
(stepwise/LDA not needed)

2. mild
H₃O⊕

NaBH₄
TM

sodium borohydride is selective
for ketone (N.R. with ester)

G

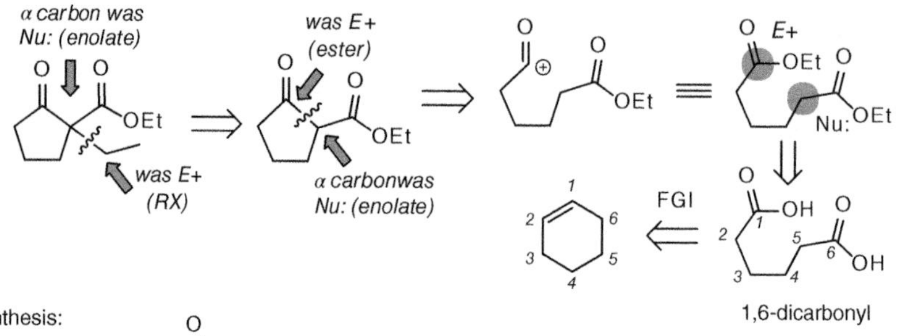

α carbon was
Nu: (enolate)

was E+
(ester)

O E+

was E+
(RX)

α carbon was
Nu: (enolate)

Nu:

FGI

1,6-dicarbonyl

synthesis:

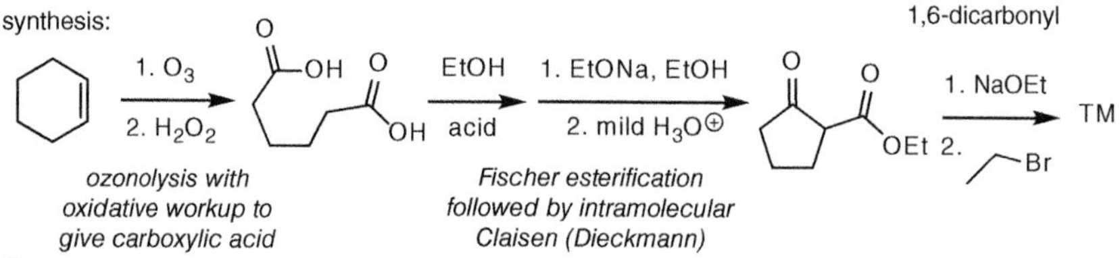

1. O₃
2. H₂O₂

ozonolysis with
oxidative workup to
give carboxylic acid

EtOH
acid

Fischer esterification
followed by intramolecular
Claisen (Dieckmann)

1. EtONa, EtOH
2. mild H₃O⊕

1. NaOEt
2. ⌐Br
TM

H

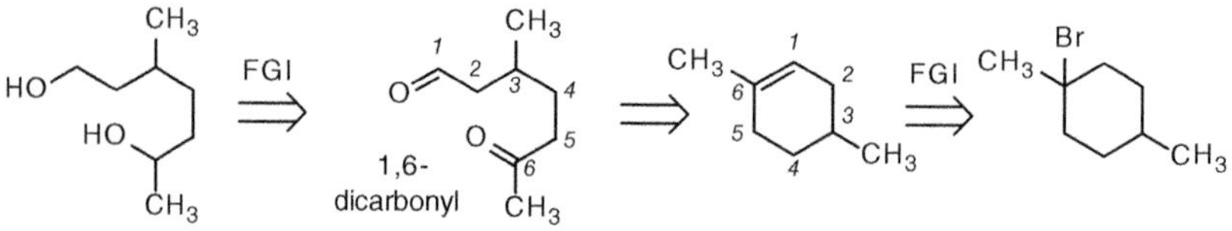

CH₃
HO
HO
CH₃

FGI

CH₃

1,6-
dicarbonyl

CH₃

FGI

CH₃
Br
CH₃

synthesis:

CH₃
CH₃

Br₂
hv

Br CH₃
CH₃

KOH
E2

CH₃
CH₃

1. O₃
2. NaBH₄
TM

ozonolysis with NaBH₄
workup gives alcohols

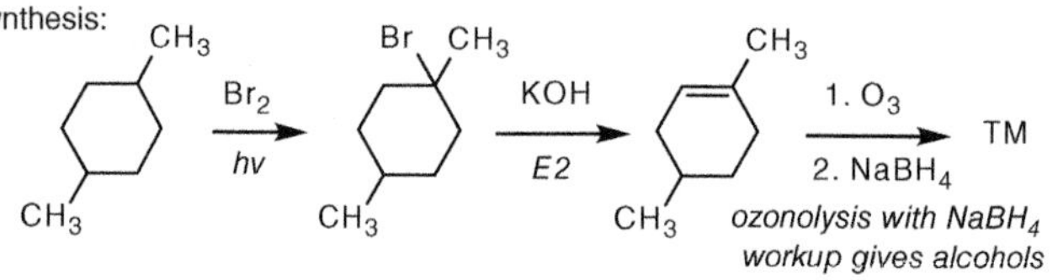

I

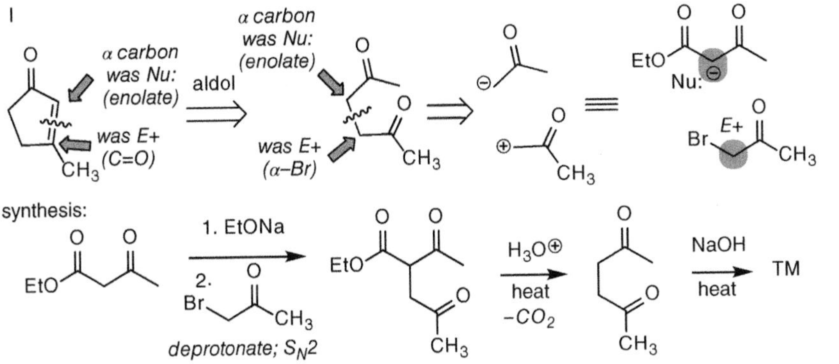

α carbon
was Nu:
(enolate)

α carbon
was Nu:
(enolate)   aldol

was E+
(C=O)   CH₃

was E+
(α-Br)   CH₃

CH₃

EtO
Nu: ⊖

E+
Br   CH₃

synthesis:

EtO

1. EtONa

2.
Br   CH₃

deprotonate; S_N2

EtO

CH₃

H₃O⊕
heat
−CO₂

CH₃

NaOH
heat   TM

CH₃

J

NC
NC   was E+
(C=O)   aldol

α carbon was Nu:
(CN=EWG)

NC
NC   ⊖ Nu:

O   E+

FGI

synthesis:

1. O₃

2. (CH₃)₂S   O

can use either reductive or oxidative
ozonolysis workup to give ketone

NC   CN

base
−H₂O*

TM

*in this type of aldol
involving two electron-
withdrawing groups,
known as a Knoevenagel
condensation, dehydration is
favored to give the doubly
conjugated alkene product

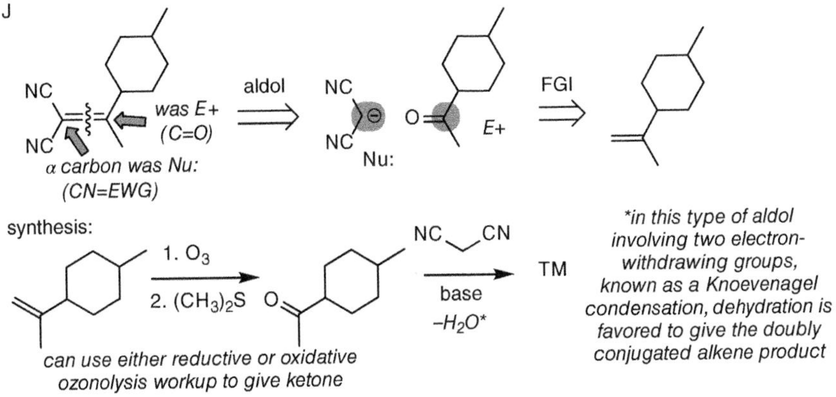

4-5. Provide a synthesis for each of the following target molecules.

A

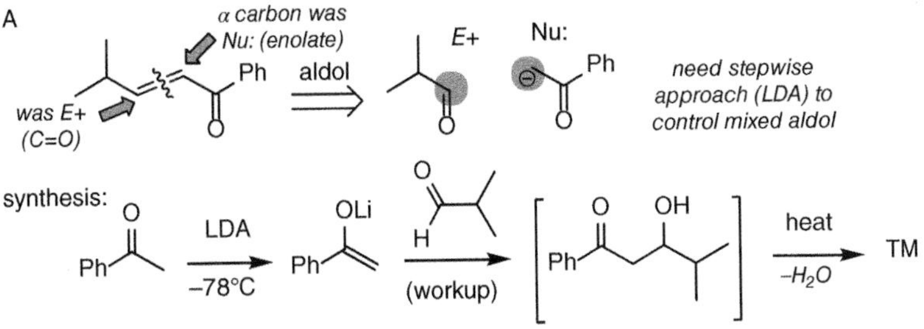

α carbon was
Nu: (enolate)

Ph   aldol

was E+
(C=O)

E+   Nu:

Ph

need stepwise
approach (LDA) to
control mixed aldol

synthesis: O

Ph   LDA
−78°C

OLi

Ph   O
H

(workup)

O   OH
Ph

heat
−H₂O   TM

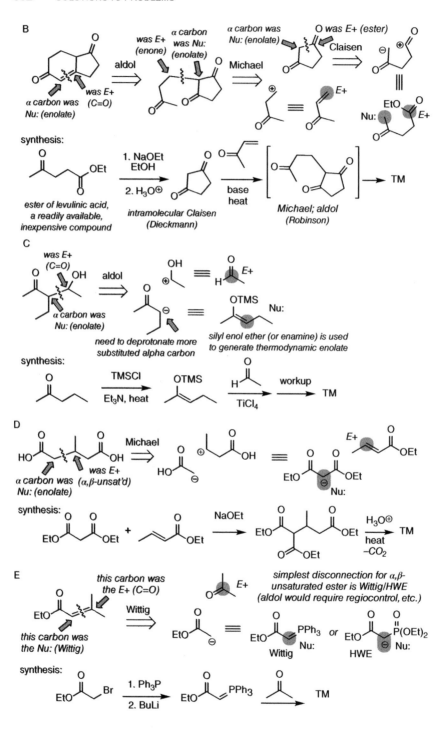

F

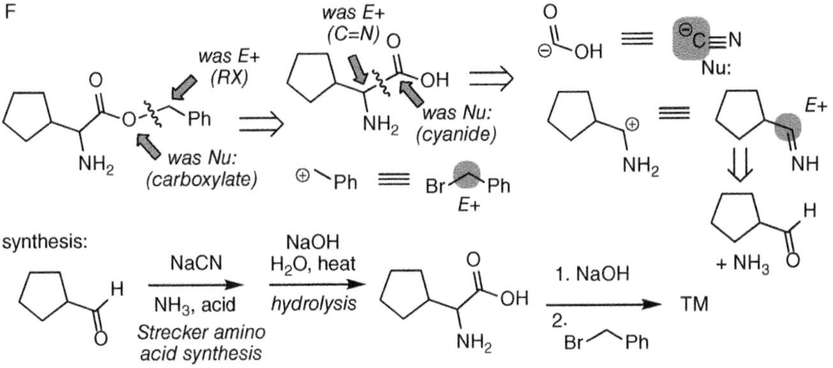

was E+ (RX)

was E+ (C≡N)

was Nu: (carboxylate)

was Nu: (cyanide)

Nu:

E+

E+

⊕ Ph ≡ Br Ph

E+

synthesis:

NaCN
NH₃, acid
Strecker amino acid synthesis

NaOH
H_2O, heat
hydrolysis

1. NaOH
2. Br Ph

TM

+ NH_3

G

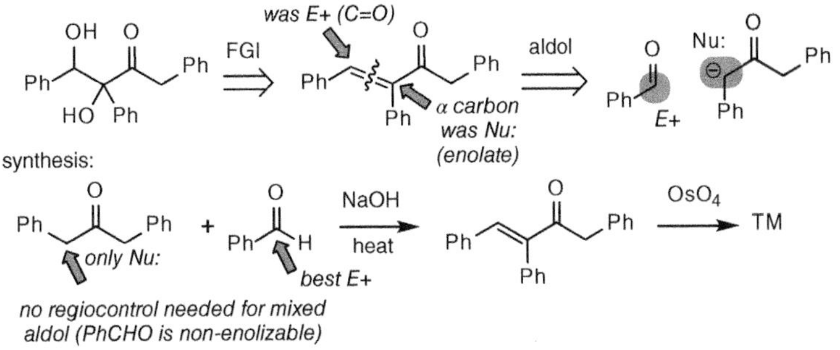

was E+ (C=O)

FGI

aldol

α carbon was Nu: (enolate)

Nu:

E+

synthesis:

only Nu:

best E+

NaOH
heat

OsO_4

TM

no regiocontrol needed for mixed aldol (PhCHO is non-enolizable)

H was E+ (acid)

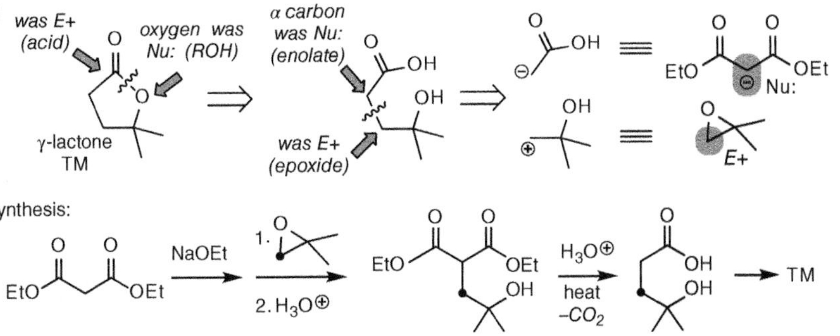

oxygen was Nu: (ROH)

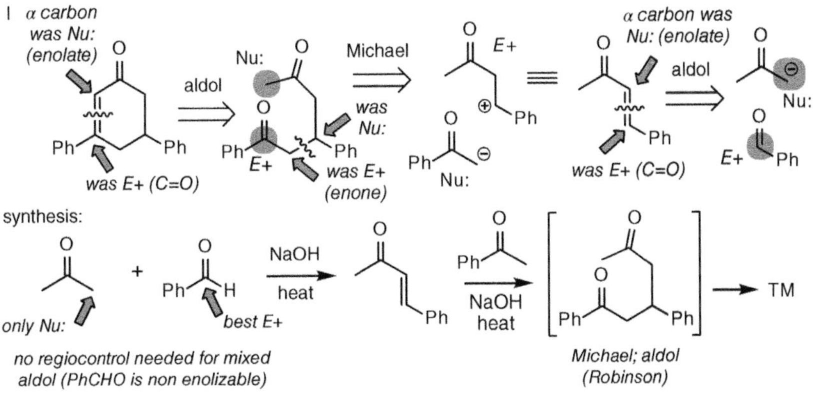

α carbon was Nu: (enolate)

was E+ (epoxide)

Nu:

E+

γ-lactone TM

synthesis:

NaOEt
1. O
2. $H_3O^⊕$

$H_3O^⊕$
heat
$-CO_2$

TM

I α carbon was Nu: (enolate)

aldol

Michael

α carbon was Nu: (enolate)

aldol

was E+ (C=O)

was Nu:

E+

was Nu:

was E+ (enone)

Nu:

was E+ (C=O)

E+

Nu:

synthesis:

only Nu:

best E+

NaOH
heat

Ph
NaOH
heat

Michael; aldol (Robinson)

TM

no regiocontrol needed for mixed aldol (PhCHO is non enolizable)

4-6. In which order should acetophenone, 2-butanone, and LDA be added to the reaction flask to generate the desired product? An aldol disconnection helps to determine the required nucleophile and electrophile.

A stepwise approach is needed to control the regiochemistry of this mixed aldol reaction. First, we need to make the kinetic enolate shown by combining 1-butanone and LDA at low temperature (typically done at −78°C with a dry ice/acetone bath). To decide which reagent to add first, keep in mind that enolates are very reactive nucleophiles that will attack any electrophiles that are present. If the LDA is added to a solution of the 2-butanone, then as the enolate is formed it will immediately react with the 2-butanone that is present and a self-aldol product is expected.

Instead, if the 2-butanone is added to a solution of LDA, then the resulting enolate will be formed in the absence of any electrophiles. A small excess of LDA ensures that the kinetically favored enolate is the major one formed, by deprotonation at the less hindered alpha carbon, and no equilibration can occur to give the more stable, thermodynamic enolate. The target molecule is prepared by starting with an LDA solution, adding 2-butanone to make the enolate, and then adding acetophenone.

4-7. A retrosynthesis of the product will direct us to the appropriate mechanism.

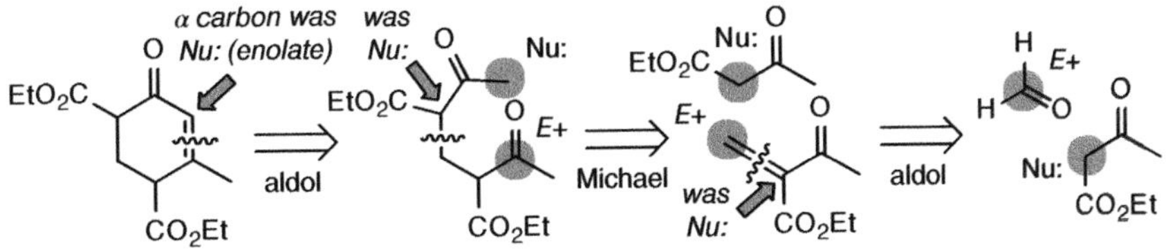

The mechanism involves an aldol reaction with dehydration, followed by a Michael addition, and then an intramolecular aldol with dehydration (all base-catalyzed). Loss of the two ester groups begins with an acid-catalyzed hydrolysis, followed by decarboxylation of each of the resulting carboxylic acids.

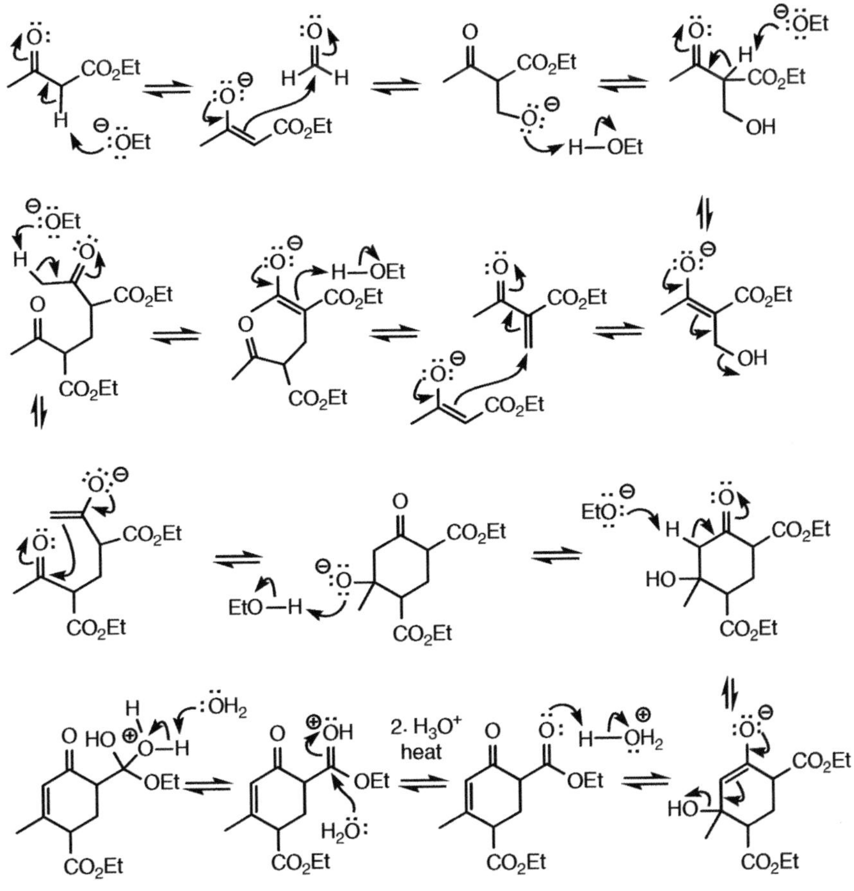

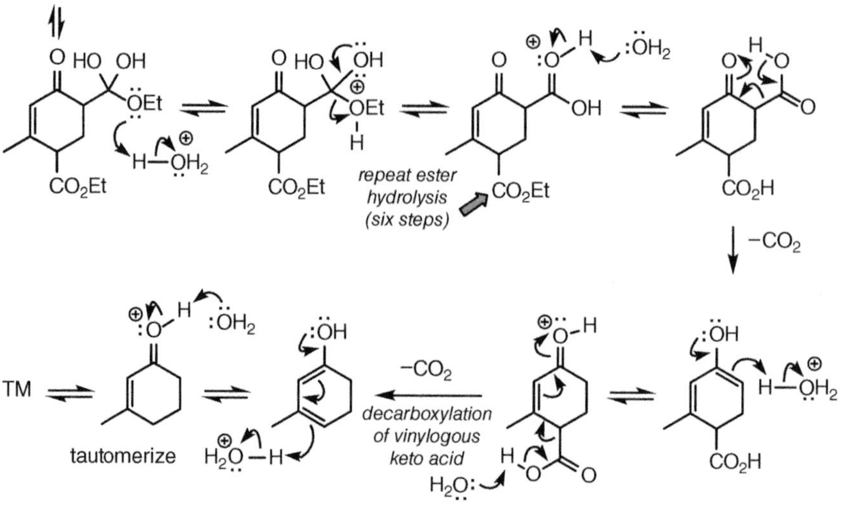

4-8. A retrosynthesis of the product will direct us to the appropriate mechanism.

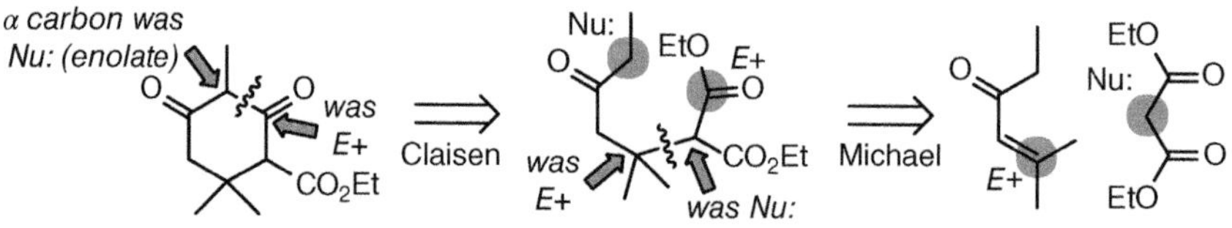

The mechanism involves a Michael reaction followed by a Dieckmann reaction (intramolecular Claisen), all base-promoted. Loss of the ester group begins with an acid-catalyzed hydrolysis, followed by decarboxylation of the resulting carboxylic acid. Finally, a Robinson annulation with methyl vinyl ketone affords the final product.

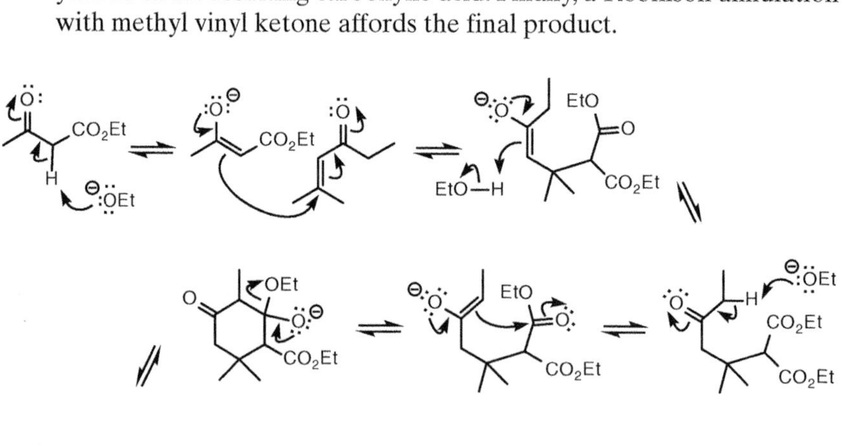

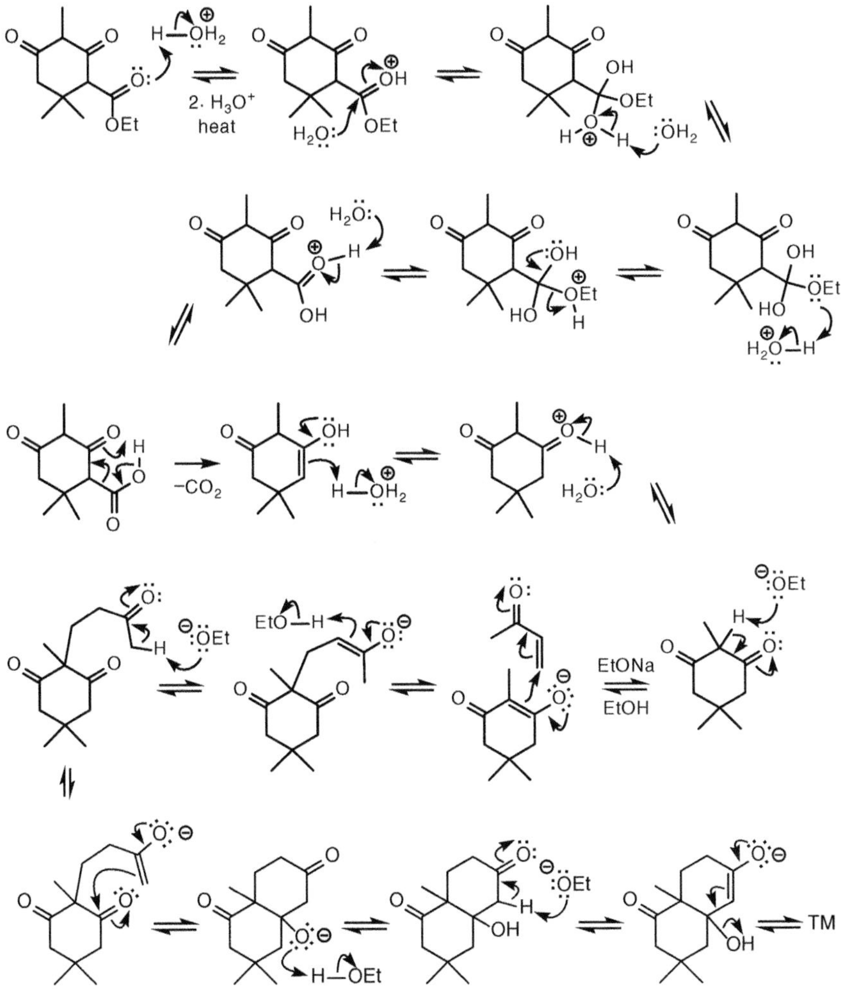
2. H_3O^+
heat
$-CO_2$
EtONa
EtOH
TM

SOLUTIONS TO CHAPTER 5: AROMATIC TMs

5.1 Predict the major product(s) of the following reactions.

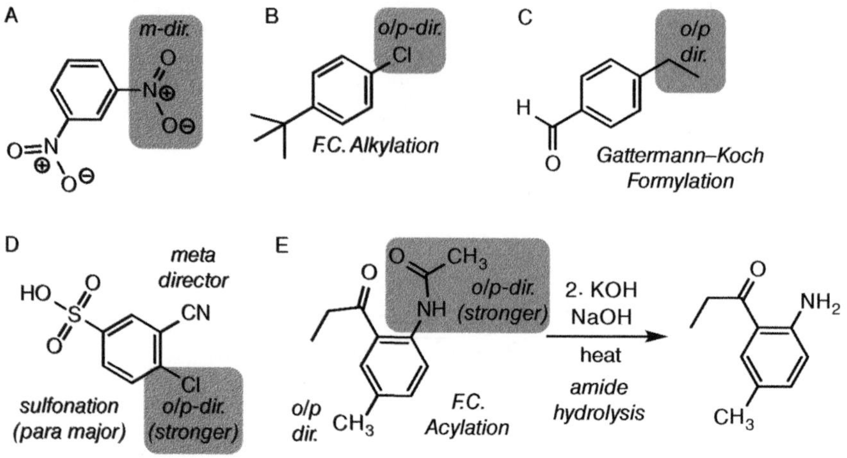

A

B

F.C. Alkylation

C

Gattermann–Koch
Formylation

D

sulfonation
(para major)

E

5.2 Provide the reagents needed for each transformation.

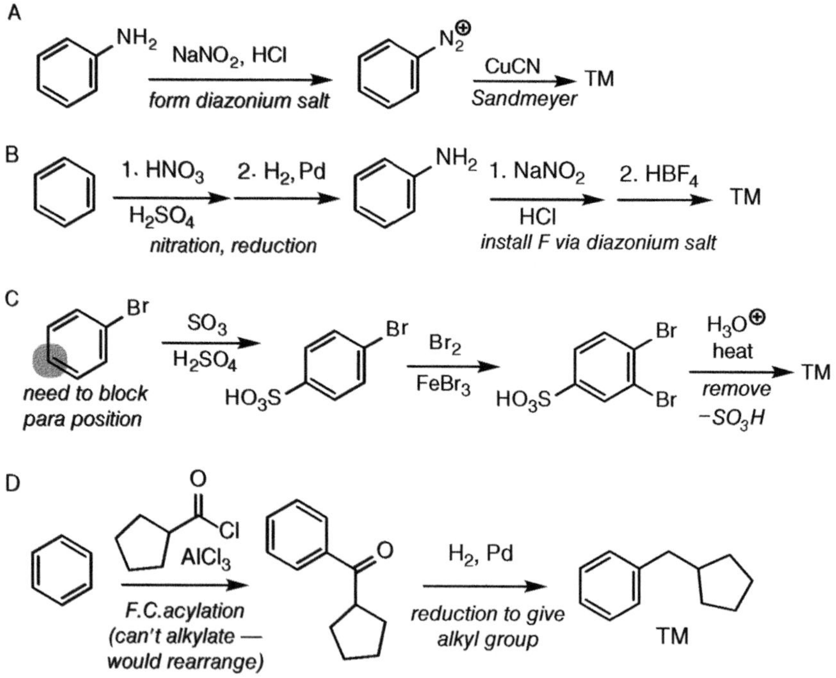

A

B

C

D

End-of-Chapter Problems

5-1. Synthesize from benzene.

A

$$\text{benzene} \xrightarrow[\substack{\text{2. } H_2, \text{Pd} \\ \textit{nitration,} \\ \textit{reduction}}]{\substack{\text{1. } HNO_3 \\ H_2SO_4}} \text{NH}_2-\text{C}_6H_5 \xrightarrow[\substack{\textit{multiple} \\ \textit{bromination with} \\ \textit{strong activating} \\ \textit{group}}]{Br_2, \text{FeBr}_3} \text{NH}_2-(\text{Br}_3) \xrightarrow[\substack{\text{2. } H_3PO_2 \\ \textit{remove -NH}_2}]{\substack{\text{1. NaNO}_2 \\ HCl}} \text{TM}$$

o/p-director

B

$$\text{benzene} \xrightarrow[\substack{\textit{F.C. Alkylation}}]{\substack{CH_3Br \\ AlCl_3}} (\text{CH}_3) \xrightarrow[\substack{\text{2. } H_2, \text{Pd} \\ \textit{nitration,} \\ \textit{reduction}}]{\substack{\text{1. } HNO_3 \\ H_2SO_4}} (\text{NH}_2, \text{CH}_3) \xrightarrow[\substack{\text{2. HBF}_4 \\ \textit{add -F via} \\ \textit{diazonium salt}}]{\substack{\text{1. NaNO}_2 \\ HCl}} (\text{F, CH}_3) \xrightarrow[\substack{\textit{free-radical} \\ \textit{halogenation}}]{Br_2, h\nu} \text{TM}$$

o/p-director

C

$$\text{benzene} \xrightarrow[\substack{\textit{F.C. Alkylation} \\ \textit{need to block} \\ \textit{para position}}]{\substack{CH_3Br \\ AlCl_3}} (\text{CH}_3) \xrightarrow[\substack{\text{2. } CH_3Br \\ AlCl_3 \\ \textit{F.C. Alkylation} \\ \textit{slow with EWG}}]{\substack{\text{1. } SO_3 \\ H_2SO_4}} (\text{CH}_3, CH_3, SO_3H) \xrightarrow[\substack{\textit{heat} \\ \textit{remove} \\ -SO_3H}]{H_3O^{\oplus}} (\text{CH}_3, CH_3) \xrightarrow[\substack{\textit{oxidation}}]{KMnO_4} \text{TM}$$

or

$$\text{benzene} \xrightarrow[\substack{\textit{F.C. Acylation}}]{AlCl_3} \text{(phenyl ketone acid)} \xrightarrow[\substack{\textit{Wolff–Kishner} \\ \textit{removes EWG} \\ \textit{(allows 2nd F.C.)}}]{\substack{NH_2NH_2 \\ KOH}} \text{(phenyl acid)} \xrightarrow[\substack{\text{2. } AlCl_3 \\ \textit{intra-} \\ \textit{molecular} \\ \textit{F.C. Acylation}}]{\text{1. } SOCl_2} \text{(tetralone)} \xrightarrow[\substack{\textit{oxidation}}]{KMnO_4} \text{TM}$$

D

$$\text{(Cl-phenyl-N-propyl amine)} \xRightarrow{FGI} \text{(Cl-phenyl amide)} \xRightarrow{} \text{(phenyl amide)}$$

amine TM   *use protected*
              o/p-director

$$\text{benzene} \xrightarrow[\substack{\text{2. } H_2, \text{Pd} \\ \textit{nitration,} \\ \textit{reduction}}]{\substack{\text{1. } HNO_3 \\ H_2SO_4}} \text{(NH}_2) \xrightarrow[\substack{\textit{pyridine} \\ \textit{form amide,} \\ \textit{monochlorination}}]{\substack{\text{acid chloride} \\ Cl_2 \\ FeCl_3}} \text{(HN amide Cl)} \xrightarrow[\substack{\text{2. } H_2O}]{\text{1. LiAlH}_4} \text{TM}$$

E

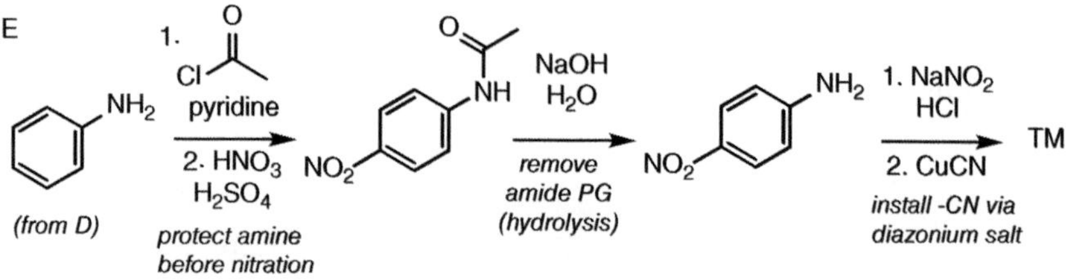

1.
$$\overset{O}{\underset{Cl}{\parallel}}\diagdown$$
pyridine

2. HNO_3
H_2SO_4

(from D)

protect amine
before nitration

NaOH
H_2O

remove
amide PG
(hydrolysis)

1. $NaNO_2$
HCl

2. CuCN

install -CN via
diazonium salt

TM

F

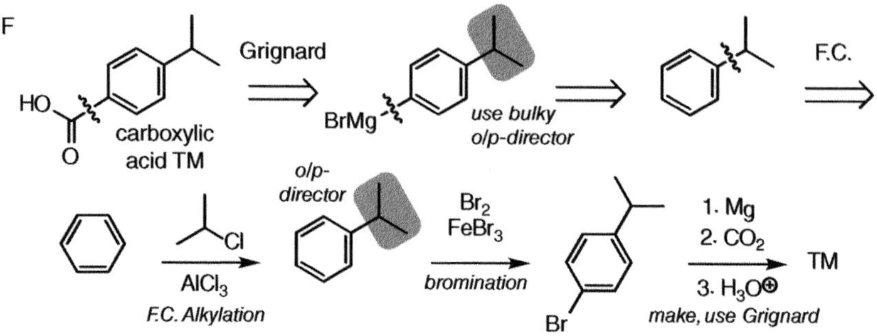

Grignard

carboxylic
acid TM

use bulky
o/p-director

F.C.

o/p-
director

$\overset{iPr}{\diagup}Cl$

$AlCl_3$

F.C. Alkylation

Br_2
$FeBr_3$

bromination

1. Mg
2. CO_2
3. $H_3O^{\oplus}$

TM

make, use Grignard

G

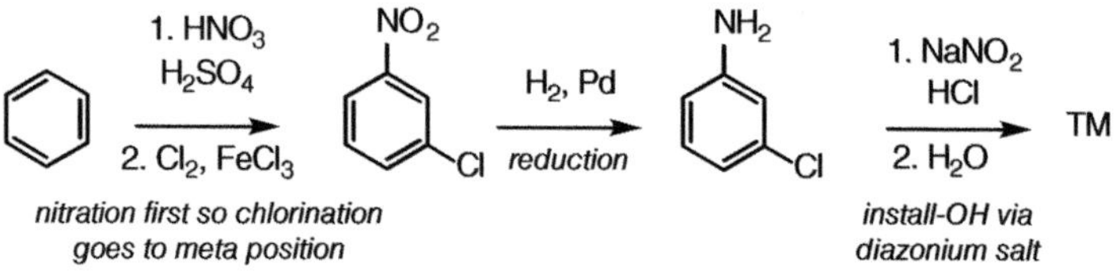

1. HNO_3
H_2SO_4

2. Cl_2, $FeCl_3$

nitration first so chlorination
goes to meta position

H_2, Pd

reduction

1. $NaNO_2$
HCl

2. H_2O

install-OH via
diazonium salt

TM

H

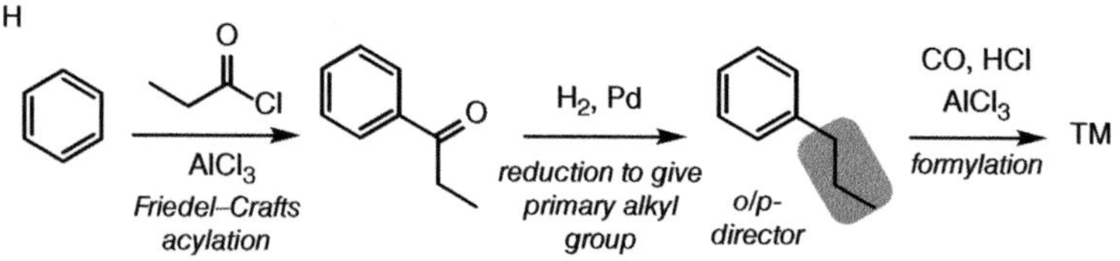

$AlCl_3$

Friedel–Crafts
acylation

H_2, Pd

reduction to give
primary alkyl
group

o/p-
director

CO, HCl
$AlCl_3$

formylation

TM

I

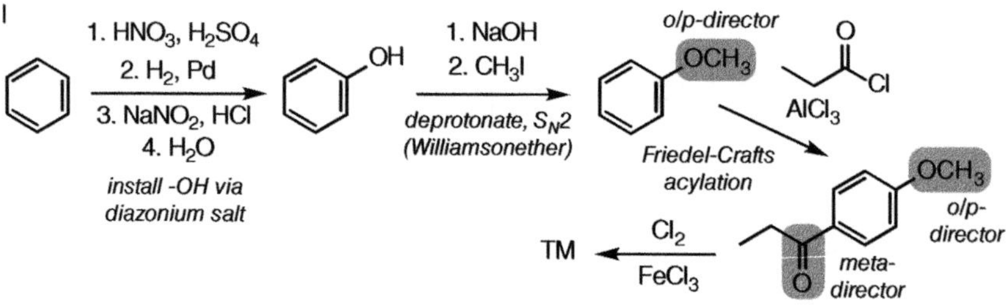

1. HNO_3, H_2SO_4
2. H_2, Pd
3. $NaNO_2$, HCl
4. H_2O

install -OH via
diazonium salt

1. NaOH
2. CH_3I

deprotonate, S_N2
(Williamson ether)

o/p-director

OCH_3

Friedel-Crafts
acylation

$AlCl_3$

OCH_3

o/p-
director

meta-
director

TM

Cl_2
$FeCl_3$

J

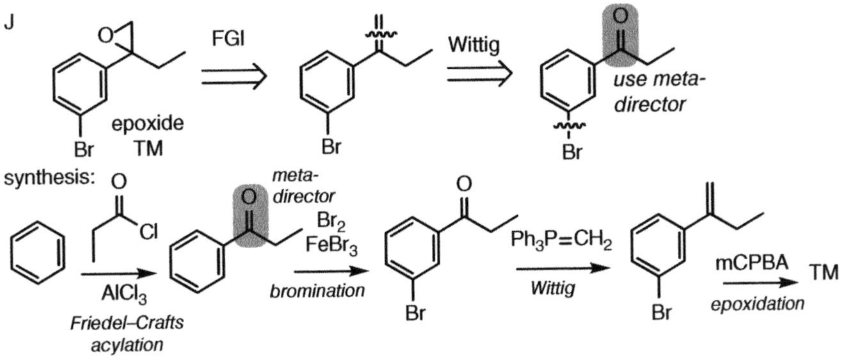

epoxide TM

FGI

Wittig

use meta-director

Br

synthesis:

AlCl$_3$
Friedel–Crafts acylation

meta-director
Br$_2$
FeBr$_3$
bromination

Ph$_3$P=CH$_2$
Wittig

mCPBA   TM
epoxidation

K

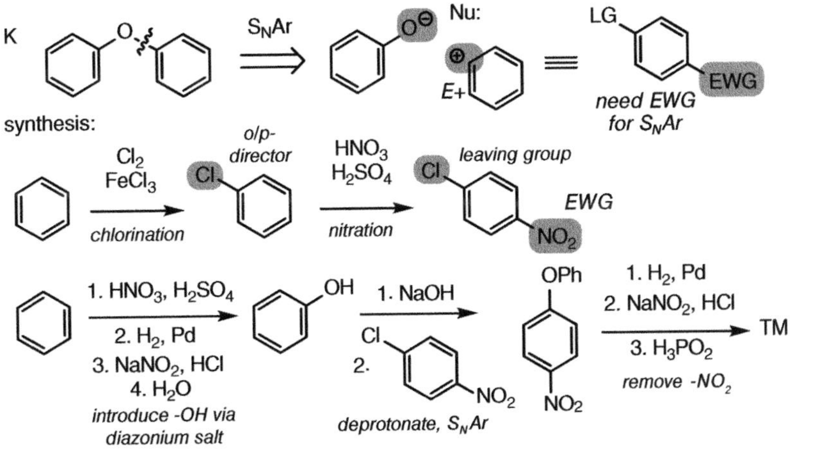

S$_N$Ar

Nu:

E+

need EWG for S$_N$Ar

synthesis:

Cl$_2$
FeCl$_3$
chlorination

o/p-director

HNO$_3$
H$_2$SO$_4$
nitration

leaving group

EWG
NO$_2$

1. HNO$_3$, H$_2$SO$_4$
2. H$_2$, Pd
3. NaNO$_2$, HCl
4. H$_2$O
introduce -OH via diazonium salt

1. NaOH
2.
deprotonate, S$_N$Ar

1. H$_2$, Pd
2. NaNO$_2$, HCl
3. H$_3$PO$_2$
remove -NO$_2$

TM

L

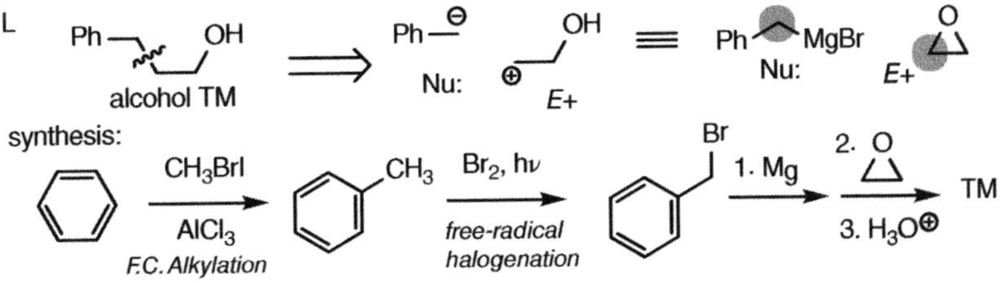

alcohol TM

Ph—
Nu:

OH
E+

Ph   MgBr
Nu:
E+

synthesis:

CH$_3$Brl
AlCl$_3$
F.C. Alkylation

CH$_3$

Br$_2$, hv
free-radical halogenation

Br

1. Mg
2.
3. H$_3$O$^\oplus$
TM

M *was E+ (C=O)*

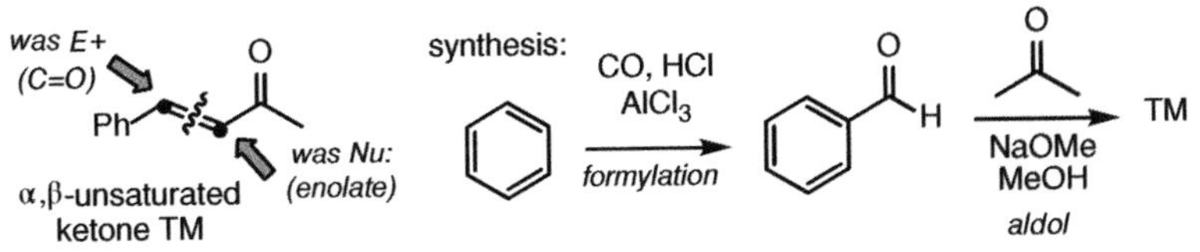

Ph

was Nu: (enolate)

α,β-unsaturated ketone TM

synthesis:

CO, HCl
AlCl$_3$
formylation

NaOMe
MeOH
aldol

TM

N

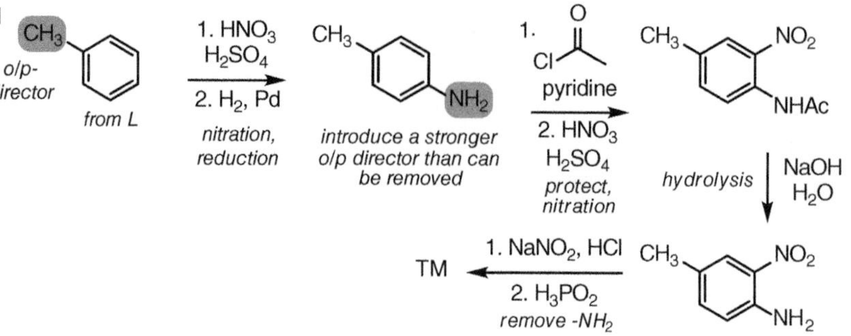

CH₃ (benzene with methyl, o/p-director, from L)

1. HNO₃
H₂SO₄

2. H₂, Pd

nitration, reduction

CH₃ — NH₂ (para)

introduce a stronger o/p director than can be removed

1. acetyl chloride / pyridine

2. HNO₃
H₂SO₄

protect, nitration

CH₃ — NO₂ / NHAc

hydrolysis | NaOH, H₂O

CH₃ — NO₂ / NH₂

TM ← 1. NaNO₂, HCl
2. H₃PO₂

remove -NH₂

or

CH₃ (benzene, o/p-director)

Na₂Cr₂O₇
H₂SO₄

oxidize

HOOC-benzene (meta-director)

HNO₃
H₂SO₄

nitration

HOOC — NO₂ (meta)

1. SOCl₂
2. DIBAL-H
3. NH₂NH₂, KOH

reduction (of acid chloride to aldehyde, then to alkane)

TM

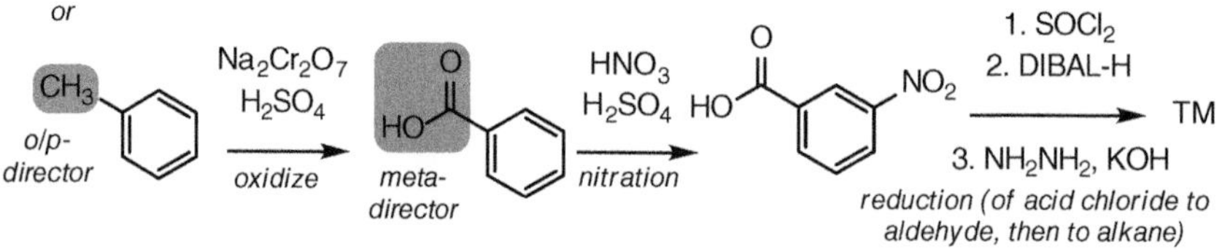

5-2. C-14 Synthesis Part A.

A

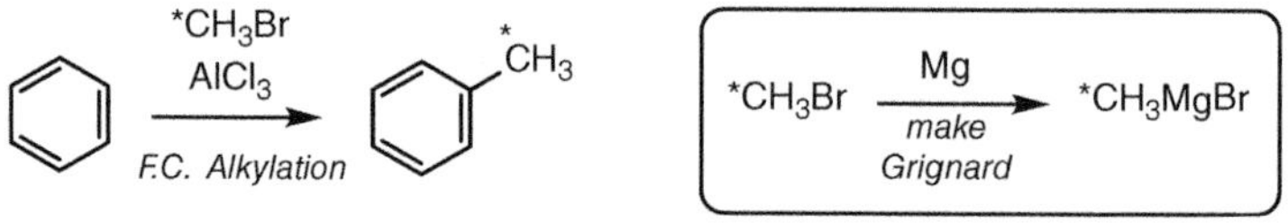

benzene

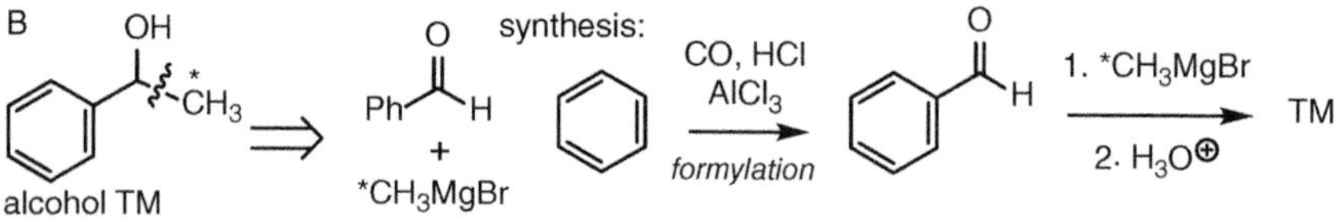

*CH₃Br
AlCl₃

F.C. Alkylation

toluene with *CH₃

*CH₃Br — Mg / *make Grignard* → *CH₃MgBr

B

alcohol TM (phenyl-CH(OH)-*CH₃)

⟹

Ph-CHO
+
*CH₃MgBr

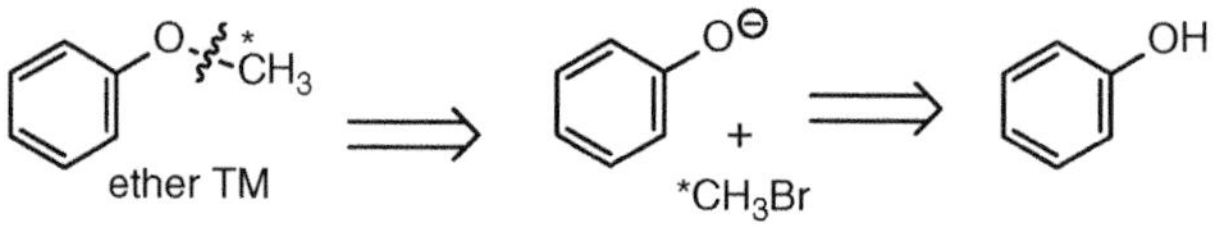

synthesis:

benzene

CO, HCl
AlCl₃

formylation

benzaldehyde (Ph-CHO)

1. *CH₃MgBr
2. H₃O⊕

TM

C

ether TM (phenyl-O-*CH₃)

⟹

phenoxide (O⊖)
+
*CH₃Br

⟹

phenol (OH)

synthesis:

benzene

1. HNO₃
H₂SO₄
2. H₂, Pd

aniline (NH₂)

1. NaNO₂
HCl
2. H₂O

phenol (OH)

1. NaOH
2. *CH₃Br

deprotonate, S_N2 (Williamson ether)

TM

install -OH via diazonium salt

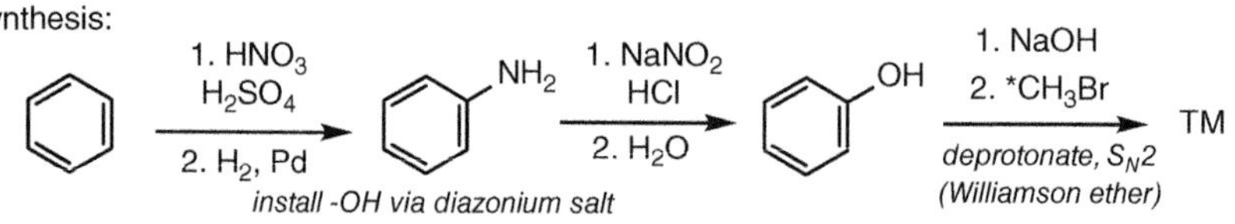

D

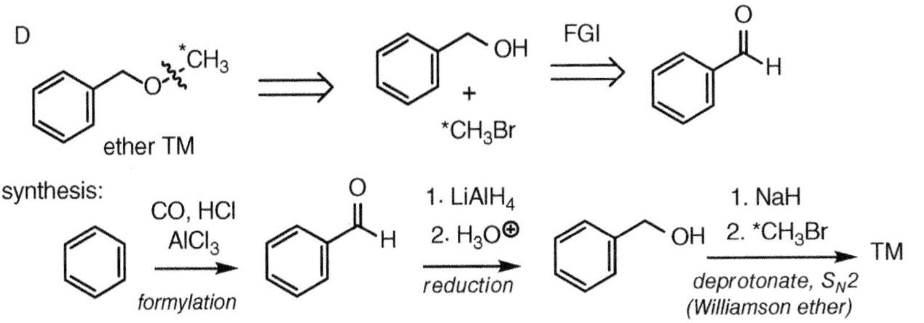

ether TM

+

*CH$_3$Br

FGI

synthesis:

CO, HCl
AlCl$_3$

formylation

1. LiAlH$_4$
2. H$_3$O$^\oplus$

reduction

1. NaH
2. *CH$_3$Br   TM

deprotonate, S$_N$2
(Williamson ether)

E

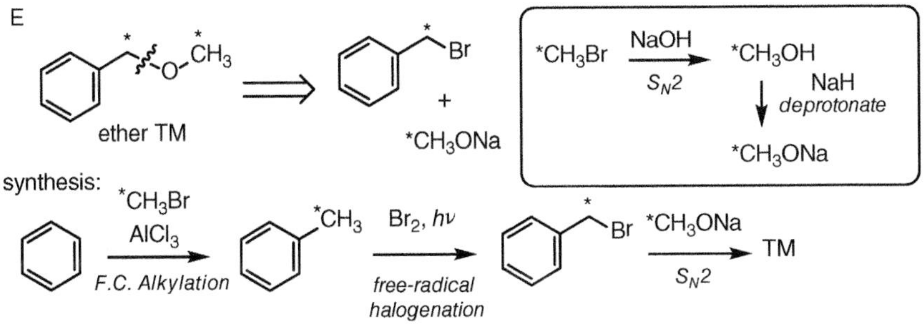

ether TM

+

*CH$_3$ONa

*CH$_3$Br   NaOH   *CH$_3$OH
           S$_N$2
                        NaOH
                        deprotonate

*CH$_3$ONa

synthesis: *CH$_3$Br

AlCl$_3$

F.C. Alkylation

Br$_2$, hν

free-radical
halogenation

*CH$_3$ONa

S$_N$2   TM

F

+ H$_2$N⤳*CH$_3$

*CH$_3$Br
+
"H$_2$N:"$^{\ominus}$

synth. equiv.
needed

synthesis:

1. KOH
2. *CH$_3$Br

deprotonate,
S$_N$2 (Gabriel)

NaOH
H$_2$O

hydrolysis

H$_2$N–CH$_3$

1. Br$_2$, FeBr$_3$
2. Mg

1. CO$_2$
2. H$_3$O$^\oplus$

1. SOCl$_2$
2. *CH$_3$NH$_2$ TM

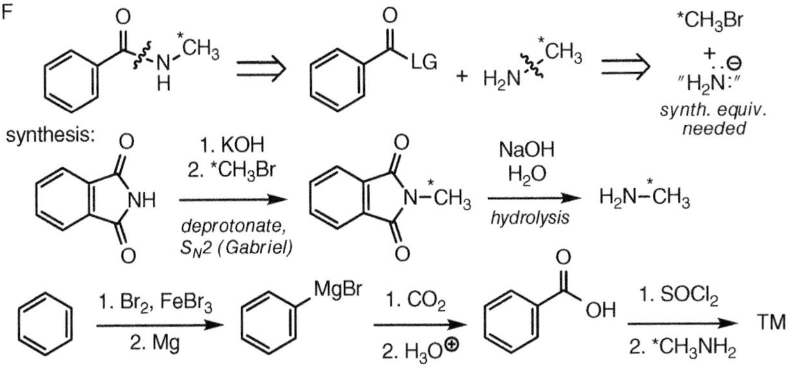

5-3. C-14 Synthesis Part B.

A

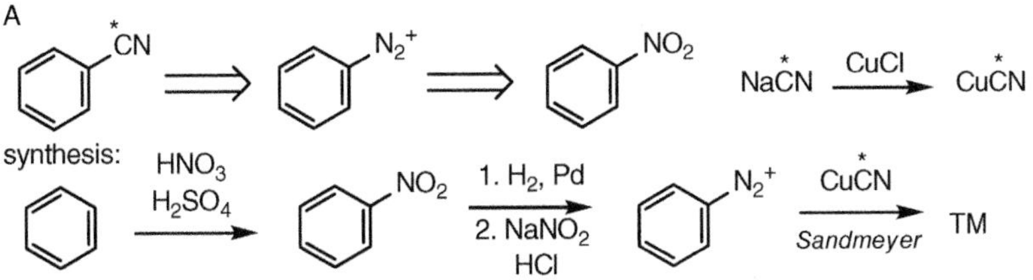

NaCN   CuCl   CuCN

synthesis:

HNO$_3$
H$_2$SO$_4$

1. H$_2$, Pd
2. NaNO$_2$
HCl

CuCN

Sandmeyer   TM

B

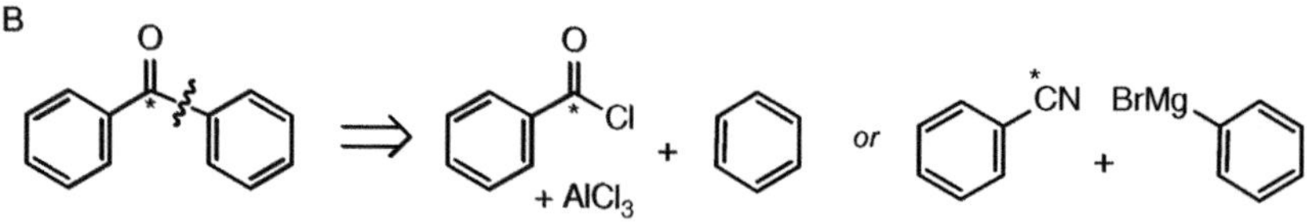

synthesis:

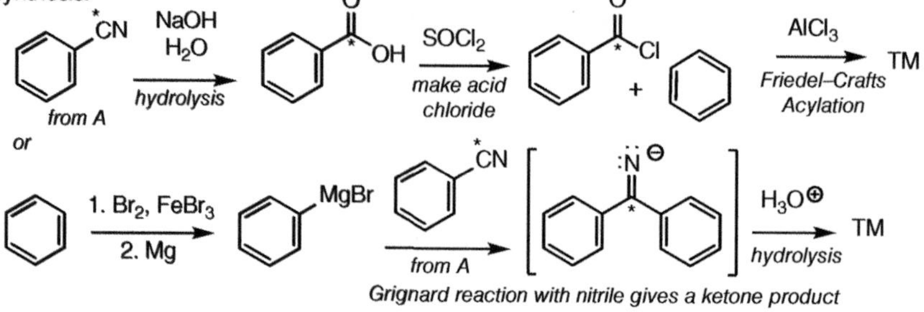

from A

or

Grignard reaction with nitrile gives a ketone product

C

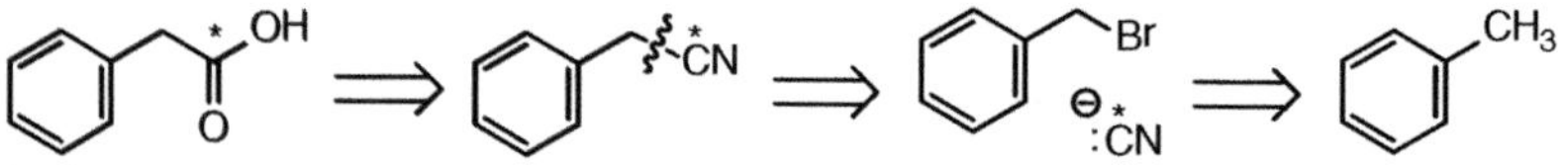

synthesis:

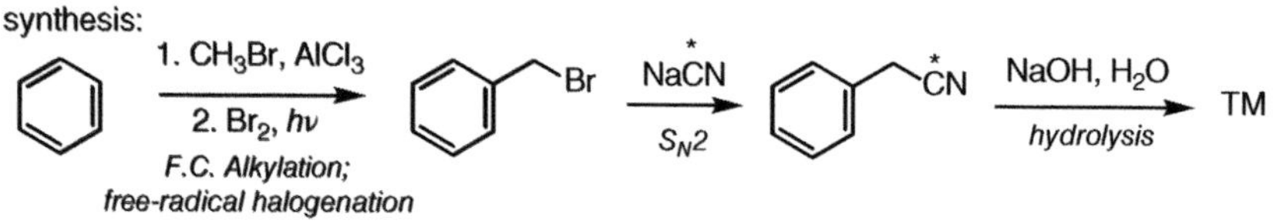

F.C. Alkylation;
free-radical halogenation

hydrolysis

D synthesis:

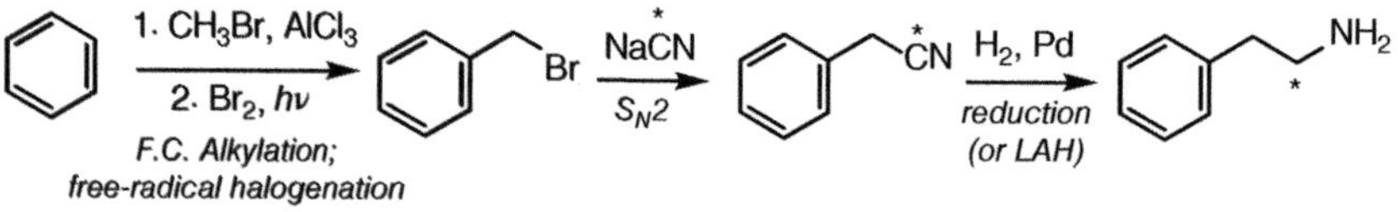

F.C. Alkylation;
free-radical halogenation

E

*this ketone disconnection
can come from nitrile*

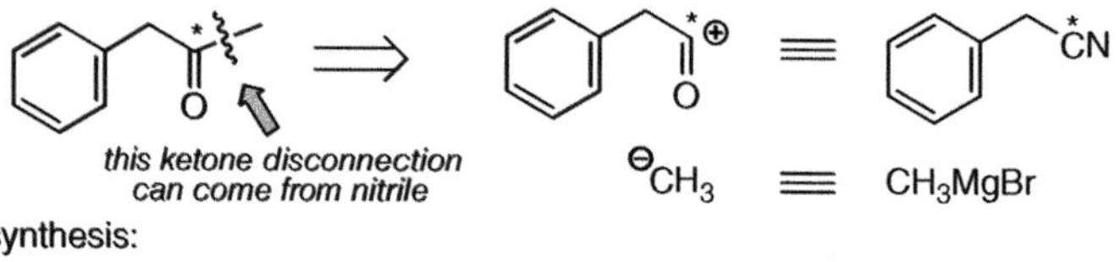

synthesis:

from C

Grignard reaction with nitrile gives a ketone product

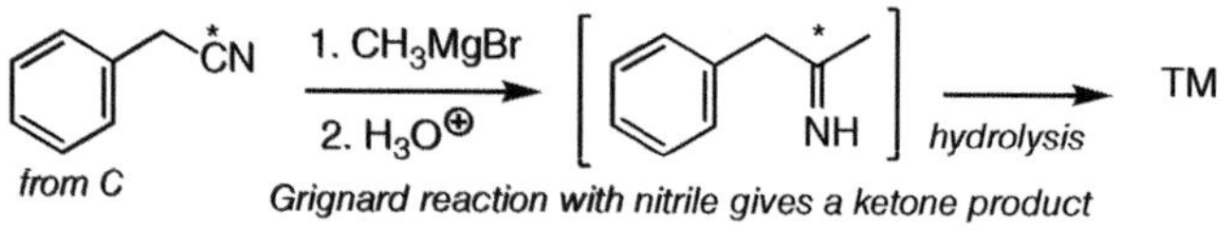

F

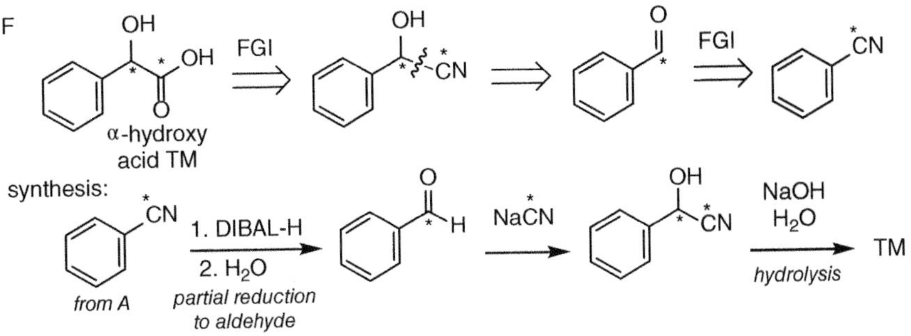

α-hydroxy
acid TM

synthesis:

1. DIBAL-H
2. H₂O

from A   *partial reduction
to aldehyde*

NaCN

NaOH
H₂O

hydrolysis

TM

SOLUTIONS TO CHAPTER 6: CYCLIC TMs

6.4 Predict the major product(s) or provide the missing reactants.

A

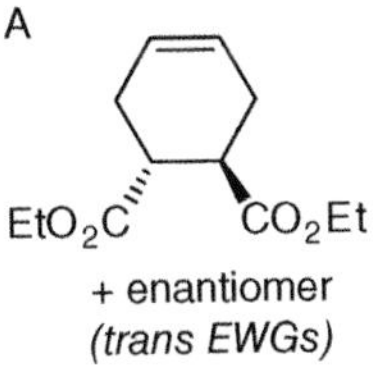

EtO₂C   CO₂Et

+ enantiomer
(trans EWGs)

B

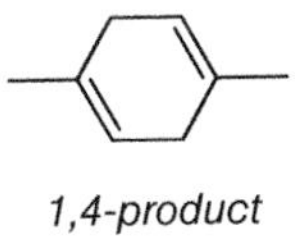

1,4-product

C

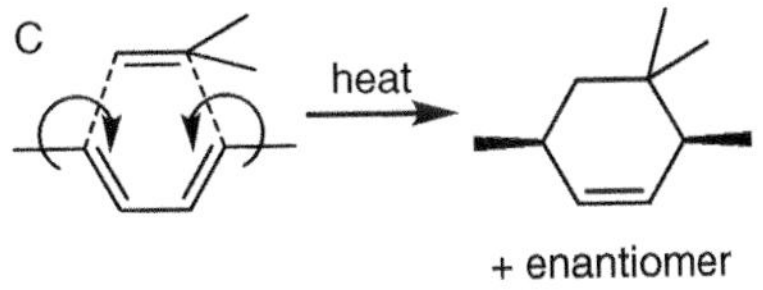

heat

+ enantiomer

D

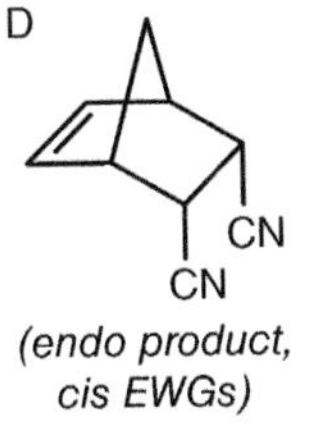

CN
CN

*(endo product,
cis EWGs)*

E

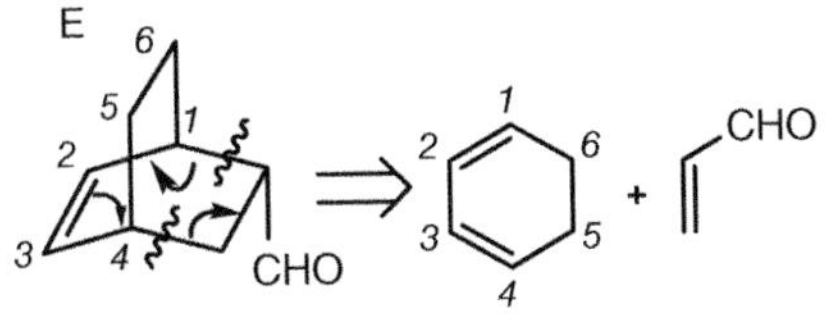

CHO

F

HC≡CH +

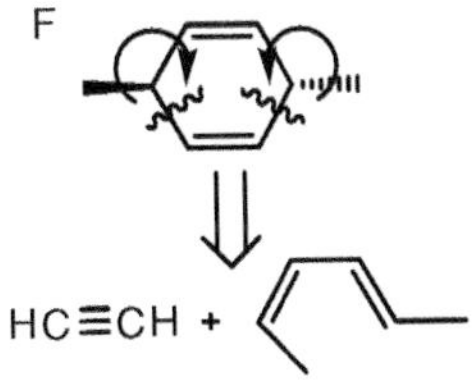

End-of-Chapter Problems

6-1. Provide a retrosynthesis.

A

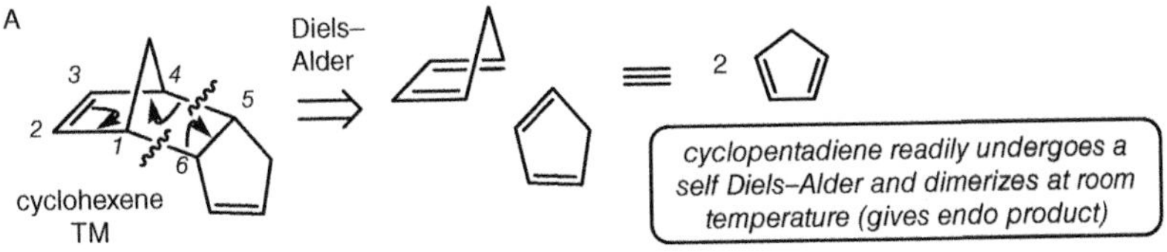

Diels–
Alder

≡ 2

cyclohexene
TM

*cyclopentadiene readily undergoes a
self Diels–Alder and dimerizes at room
temperature (gives endo product)*

B *the cis ethyl groups at C-1 and C-4 indicate both were on "outside" of diene**

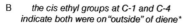

select the alkene with the EWGs to be the dienophile

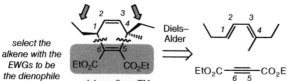

cyclohexadiene TM
(EWGs are on pi bond so use Diels–Alder)

EtO$_2$C≡CO$_2$Et (numbered 6 5)

Diels–Alder ⟹

EtO$_2$C (1) ... (2 3) ... (4) with methyl branch

**the ethyl groups can't be on the inside because the s-cis conformation of this diene is too crowded (so no Diels–Alder)*

C

cyclopentane TM

FGI ⟹ ⟹ ≡

+ Bu$_3$SnH, AIBN

D

cyclohexadiene TM
(EWG not on pi bond so use Birch reduction)

CO$_2$CH$_3$ Birch ⟹ CO$_2$CH$_3$ FGI ⟹ CO$_2$H

E

trans stereochemistry is retained (start with trans alkenes)

HO$_2$C ... CO$_2$H, Ph ... Ph
cyclobutane TM

[2+2] ⟹

HO$_2$C ... Ph + Ph ... CO$_2$H

photodimerization of trans-cinnamic acid gives TM

F

O ... cyclohexene TM (positions 6 1 2 5 4 3), O

Diels–Alder ⟹

O ... O (positions 6 5) + (positions 1 2 3 4)

G

CO$_2$Et ... CO$_2$Et (positions 1 6 2 3 5 4)
cyclohexene TM

cis stereochemistry is retained from cis dienophile

Diels–Alder ⟹

(positions 1 2 3 4) + CO$_2$Et ... CO$_2$Et (positions 6 5)

H

H O ... O, O (positions 1 2 6 3 4 5) ≡ (positions 2 1 6 3 4 5)
cyclohexene TM

Diels–Alder ⟹

O (furan) + O ... O (maleic anhydride)

I

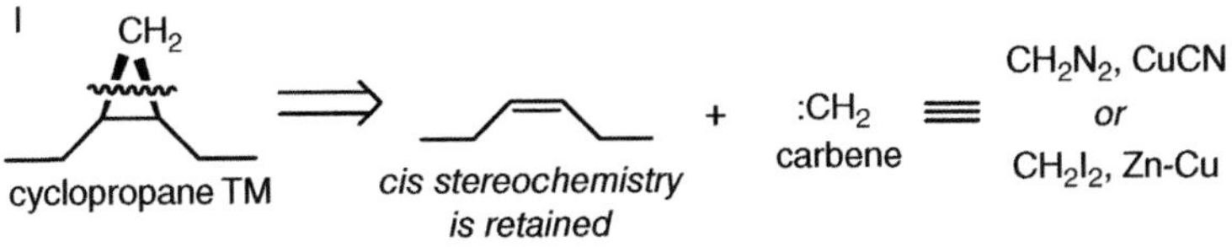

cyclopropane TM   cis stereochemistry is retained

+   :CH$_2$ carbene   ≡

CH$_2$N$_2$, CuCN
or
CH$_2$I$_2$, Zn-Cu

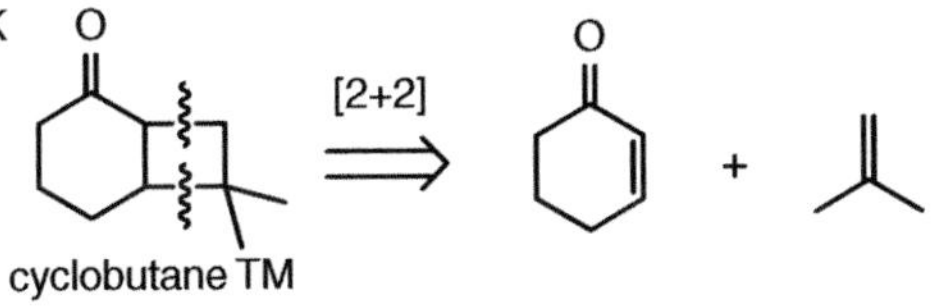

J

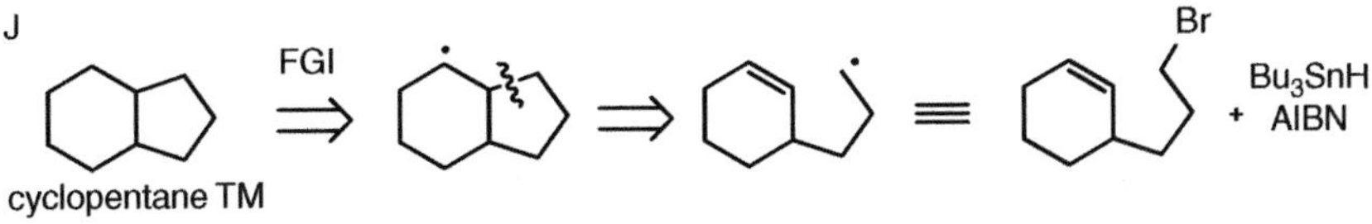

cyclopentane TM

FGI ⟹ ⟹ ≡

Bu$_3$SnH
+ AIBN

Br

K   O

cyclobutane TM

[2+2] ⟹

O

+

L   trans stereochemistry is retained from trans dienophile

cyclohexene TM

⟹

≡

NC   CN
5
6

when the central ring of anthracene acts as a diene, the outside rings retain their aromaticity

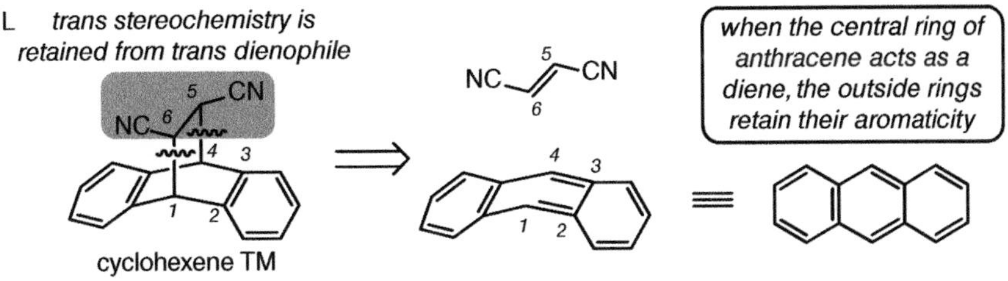

6-2. Provide a synthesis.

A   *trans stereochemistry is retained from trans dienophile*

was Nu:
(Wittig)

Diels–Alder ⟹

was E+
(C=O)

Wittig ⟹

Ph$_3$P   CO$_2$Et

O

OEt

Nu:

E+

synthesis:

O

Ph$_3$P   OEt

Wittig (or HWE)

CO$_2$Et

heat

Diels–Alder

TM

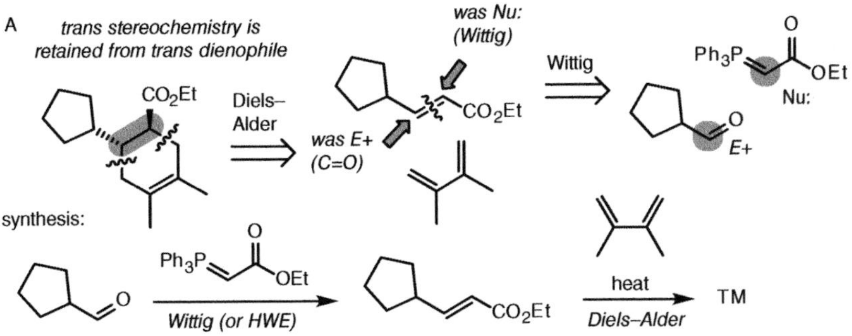

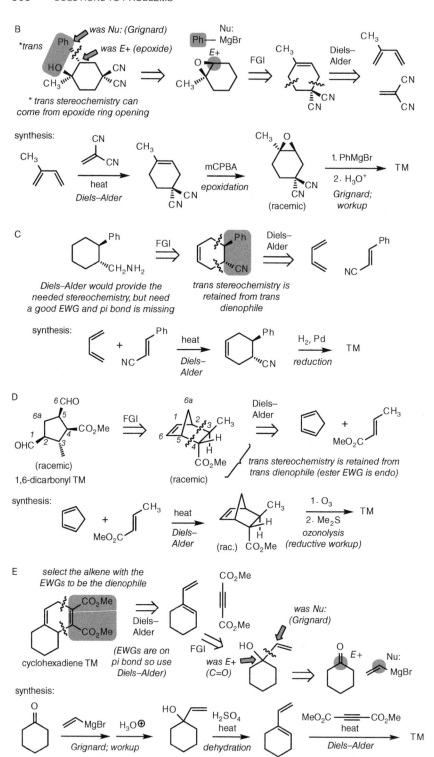

B

*trans

was Nu: (Grignard)

was E+ (epoxide)

* trans stereochemistry can come from epoxide ring opening

Nu:
Ph— MgBr
E+

FGI

Diels–Alder

synthesis:

heat
Diels–Alder

mCPBA
epoxidation

(racemic)

1. PhMgBr
2. H₃O⁺
Grignard; workup

TM

C

FGI

Diels–Alder

Diels–Alder would provide the needed stereochemistry, but need a good EWG and pi bond is missing

trans stereochemistry is retained from trans dienophile

synthesis:

heat
Diels–Alder

H₂, Pd
reduction

TM

D

6 CHO

(racemic)
1,6-dicarbonyl TM

FGI

Diels–Alder

(racemic)

trans stereochemistry is retained from trans dienophile (ester EWG is endo)

synthesis:

heat
Diels–Alder

(rac.)

1. O₃
2. Me₂S
ozonolysis (reductive workup)

TM

E   select the alkene with the EWGs to be the dienophile

Diels–Alder

(EWGs are on pi bond so use Diels–Alder)

FGI

was E+ (C=O)

was Nu: (Grignard)

E+

Nu:
MgBr

cyclohexadiene TM

synthesis:

MgBr   H₃O⊕
Grignard; workup

H₂SO₄
heat
dehydration

MeO₂C——CO₂Me
heat
Diels–Alder

TM

F

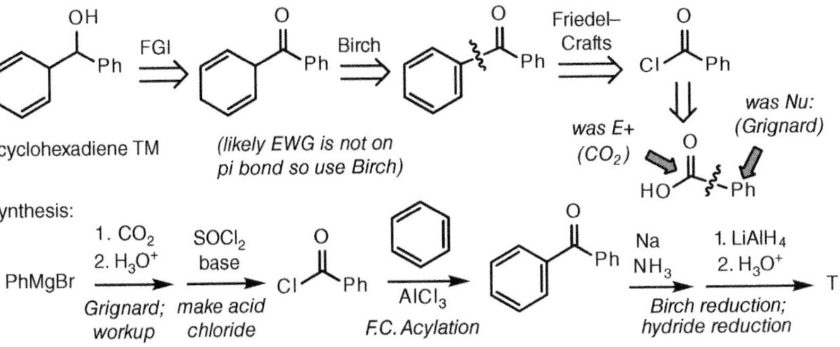

OH
Ph
FGI
O
Ph
Birch
O
Ph
Friedel–Crafts
O
Cl
Ph
cyclohexadiene TM
(likely EWG is not on pi bond so use Birch)
was E+ (CO_2)
was Nu: (Grignard)
O
HO
Ph
synthesis:
PhMgBr
1. CO_2
2. H_3O^+
Grignard; workup
$SOCl_2$
base
make acid chloride
O
Cl
Ph
$AlCl_3$
F.C. Acylation
O
Ph
Na
NH_3
1. $LiAlH_4$
2. H_3O^+
Birch reduction; hydride reduction
TM

G

CH_2OH
CH_2OH
FGI
CO_2Et
CO_2Et
cis stereochemistry is retained from cis dienophile
Diels–Alder
CO_2Et
CO_2Et
Diels-Alder would provide the needed stereochemistry, but need a good EWG
synthesis:
+
CO_2Et
CO_2Et
heat
Diels–Alder
CO_2Et
CO_2Et
1. $LiAlH_4$
2. H_3O^+
reduction
TM

H

O
CO_2Et
EtO_2C
O
1,6-dicarbonylTM
FGI
CO_2Et
CO_2Et
Diels–Alder
CO_2Et
EtO_2C
trans stereochemistry is retained from trans dienophile
synthesis:
+
CO_2Et
EtO_2C
heat
Diels–Alder
CO_2Et
CO_2Et
(racemic)
1. O_3
2. Me_2S
ozonolysis (reductive workup)
TM

I

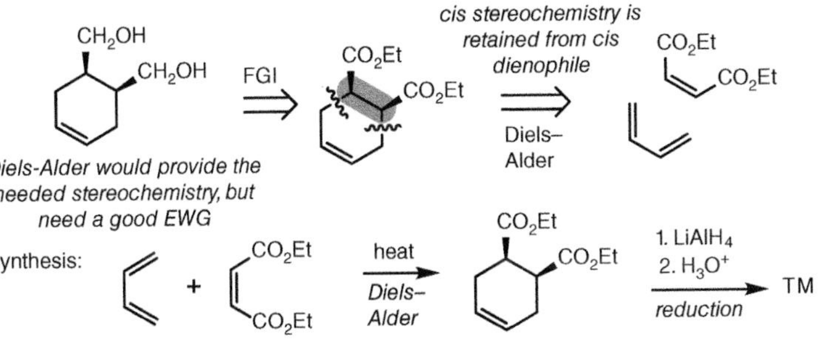

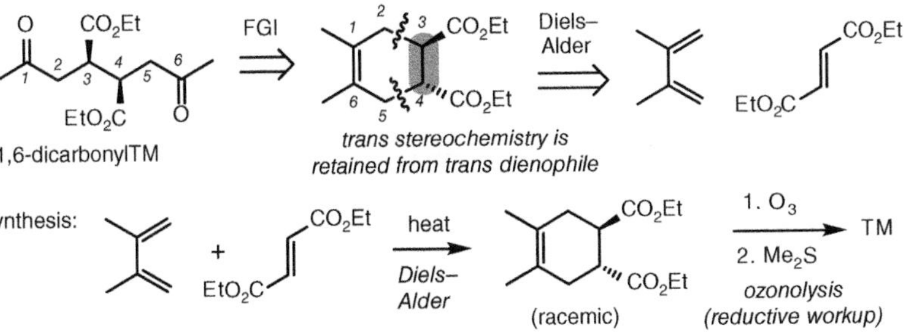

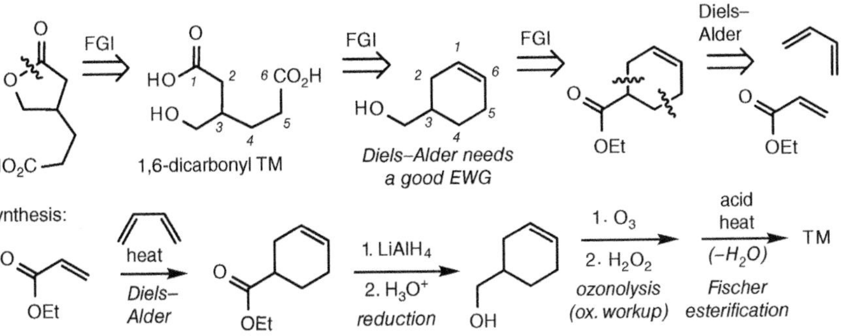

O
O
HO_2C
FGI
O
HO
CO_2H
HO
1,6-dicarbonyl TM
FGI
HO
Diels–Alder needs a good EWG
FGI
O
OEt
Diels–Alder
O
OEt
synthesis:
O
OEt
heat
Diels–Alder
O
OEt
1. $LiAlH_4$
2. H_3O^+
reduction
OH
1. O_3
2. H_2O_2
ozonolysis (ox. workup)
acid
heat
($-H_2O$)
Fischer esterification
TM

6-3. Predict the product and explain the regiochemistry and stereochemistry.

regiochemistry:
(consider resonance to identify dienophile, and best E+/best Nu: positions)

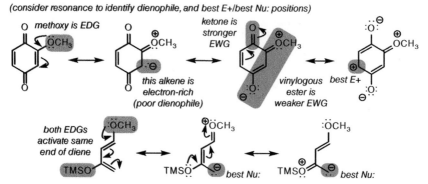

regiochemistry:
(properly align diene and dienophile)

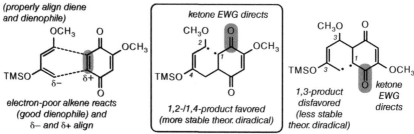

stereochemistry:
(endo EWGs)

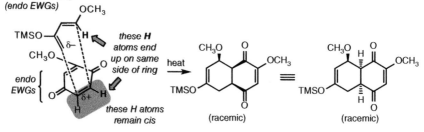

6-4. Provide products and mechanism.

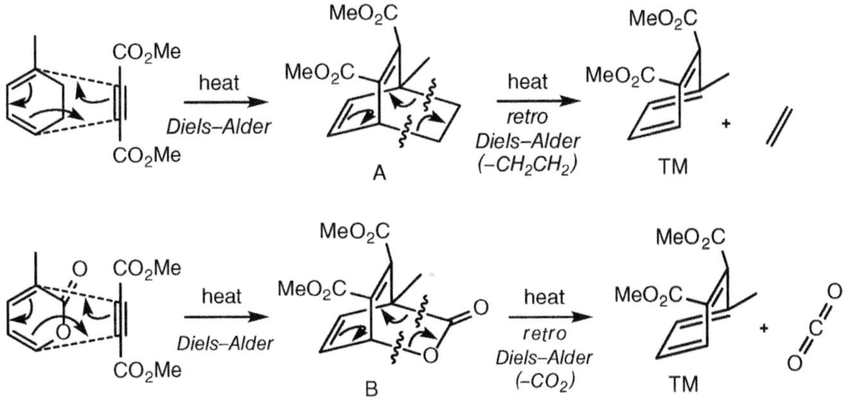

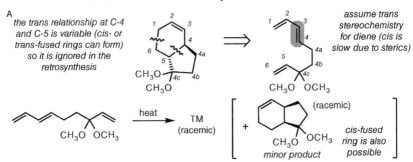

6-5. Intramolecular Diels–Alder retrosynthesis.

A

the trans relationship at C-4 and C-5 is variable (cis- or trans-fused rings can form) so it is ignored in the retrosynthesis

assume trans stereochemistry for diene (cis is slow due to sterics)

heat → TM (racemic)

(racemic)

cis-fused ring is also possible

minor product

M. E. Jung and K. M. Halweg, *Tetrahedron Lett.*, **1981**, *22*, 3929.

B

the cis relationship at C-1 and C-4 indicates trans-trans diene (both groups are on "outside" of diene)

the cis relationship at C-4 and C-5 is variable (cis- or trans-fused rings can form) so it is ignored in the retrosynthesis

the cis relationship at C-5 and C-6 was retained from a cis dienophile

heat → TM (racemic)

W. R. Roush and H. R. Gillis, *J. Org. Chem.*, **1982**, *47*, 4825.

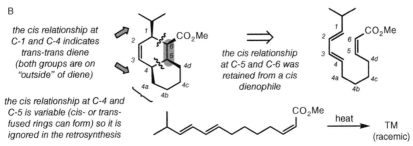

C

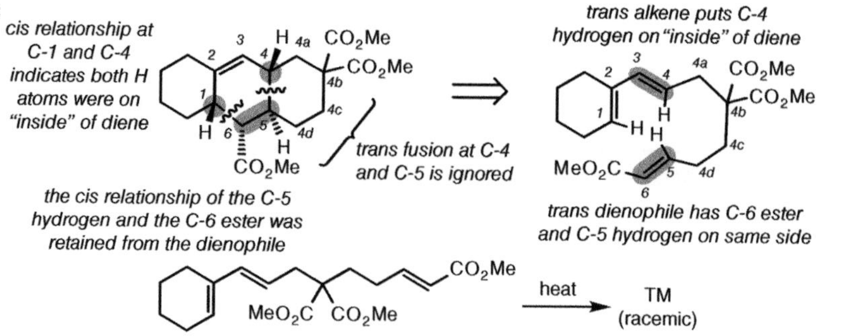
cis relationship at
C-1 and C-4
indicates both H
atoms were on
"inside" of diene
the cis relationship of the C-5
hydrogen and the C-6 ester was
retained from the dienophile
trans fusion at C-4
and C-5 is ignored
trans alkene puts C-4
hydrogen on "inside" of diene
trans dienophile has C-6 ester
and C-5 hydrogen on same side
CO2Me
MeO2C
heat
TM
(racemic)

B. M. Trost, M. Lautens, M. H. Hung, and C. S. Carmichael, *J. Am. Chem. Soc.*, **1984**, *106*, 7641.

6-6. Mechanism.

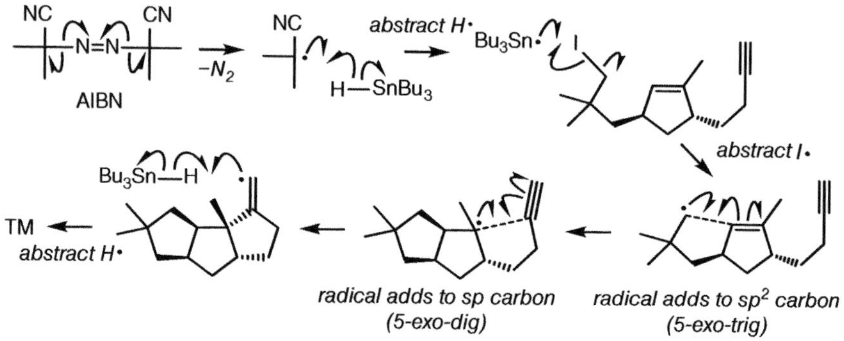
NC
CN
AIBN
−N2
abstract H•
H—SnBu3
Bu3Sn•
abstract I•
radical adds to sp carbon
(5-exo-dig)
radical adds to sp2 carbon
(5-exo-trig)
Bu3Sn—H
TM
abstract H•

D. P. Curran and D. M. Rakiewicz, *J. Am. Chem. Soc.*, **1985**, *107*, 1448.

6-7. Mechanism.

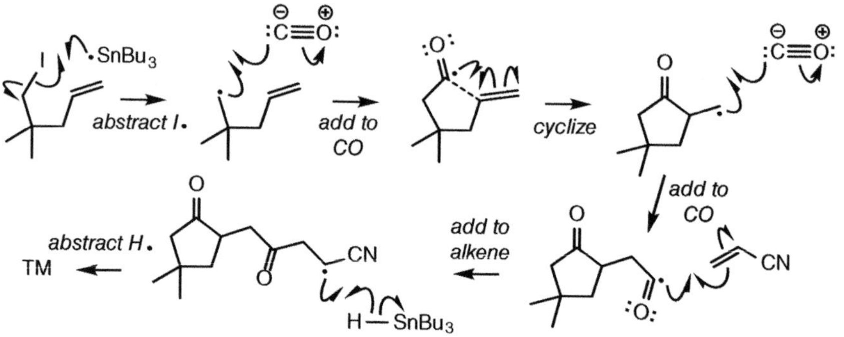
•SnBu3
abstract I•
:C≡O:
add to
CO
cyclize
add to
CO
CN
add to
alkene
CN
H—SnBu3
abstract H•
TM

L. Set, D. R. Cheshire, and D. L. J. Clive, *J. Chem. Soc., Chem. Commun.*, **1985**, 1205.

6-8. The regiochemistry of the hetero Diels–Alder is directed by polarity, with the electron-rich end of the diene aligning with the electron-deficient end of the carbonyl dienophile. Hydrolysis of the Diels–Alder product results in an α,β-unsaturated ketone, and an intramolecular radical addition to the alkene forms a five-membered ring.

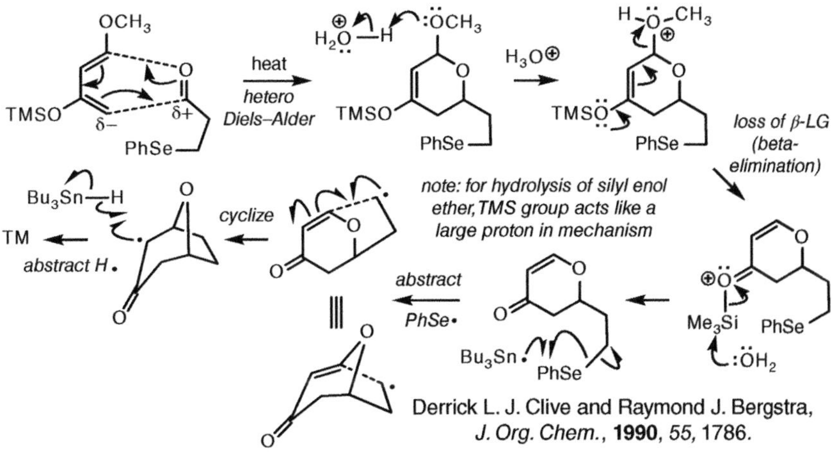
heat
hetero
Diels–Alder
loss of β-LG
(beta-
elimination)
note: for hydrolysis of silyl enol
ether, TMS group acts like a
large proton in mechanism
abstract
PhSe•
cyclize
abstract H•
TM

Derrick L. J. Clive and Raymond J. Bergstra, *J. Org. Chem.*, **1990**, *55*, 1786.

6-9. Mechanism.

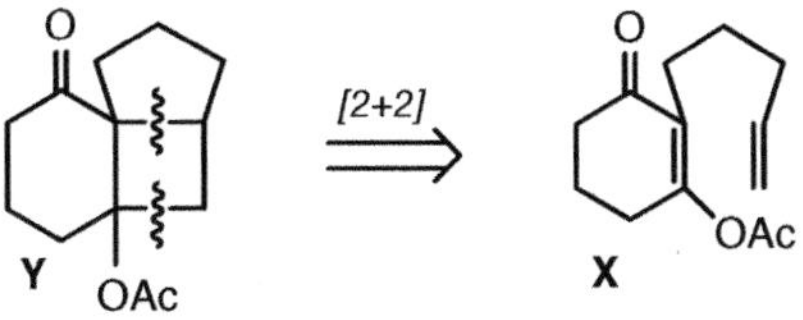
[2+2]
Y
X
OAc

Michael T. Crimmins, *Chem. Rev.*, **1988**, *88*, 1453.

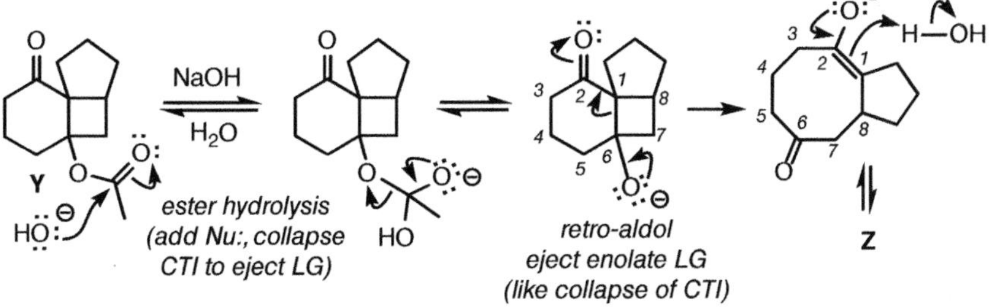
NaOH
H2O
Y
ester hydrolysis
(add Nu:, collapse
CTI to eject LG)
retro-aldol
eject enolate LG
(like collapse of CTI)
Z

6-10. Mechanism.

6-11. Provide missing structures.

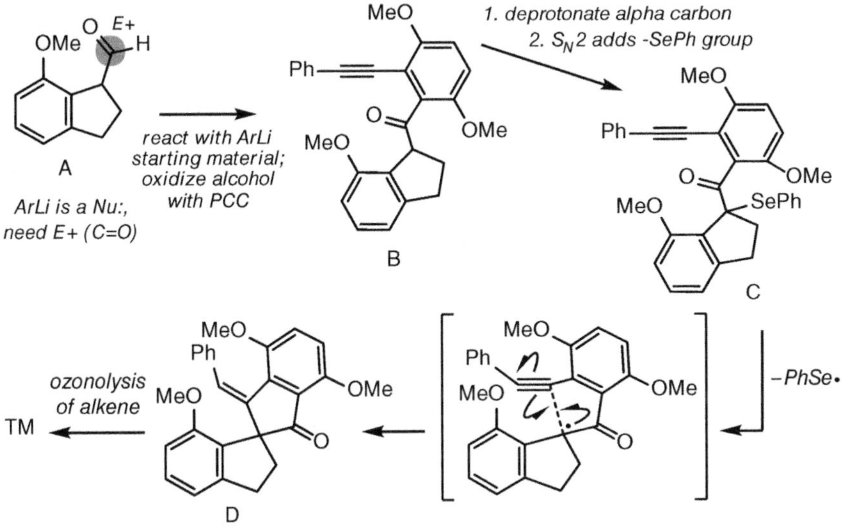

MeO

1. deprotonate alpha carbon
2. S_N2 adds -SePh group

OMe
E+
H

react with ArLi
starting material;
oxidize alcohol
with PCC

A

*ArLi is a Nu:,
need E+ (C=O)*

Ph—≡—

MeO

OMe

B

MeO

Ph—≡—

OMe
SePh

MeO

C

ozonolysis
of alkene

MeO
Ph

OMe

MeO

TM

D

MeO
Ph

OMe

MeO

−PhSe•

D. L. J. Clive, Y. Tao, A. Khodabocus, Y.-J. Wu, A. G. Angoh, S. M. Bennett,
C. N. Boddy, L. Bordeleau, D. Kellner, G. Kleiner, D. S. Middleton, C. J. Nichols,
S. R. Richardson, and P. G. Vernon, *J. Am. Chem. Soc.*, **1994**, *116*, 11275.

6-12. Provide the reagents.

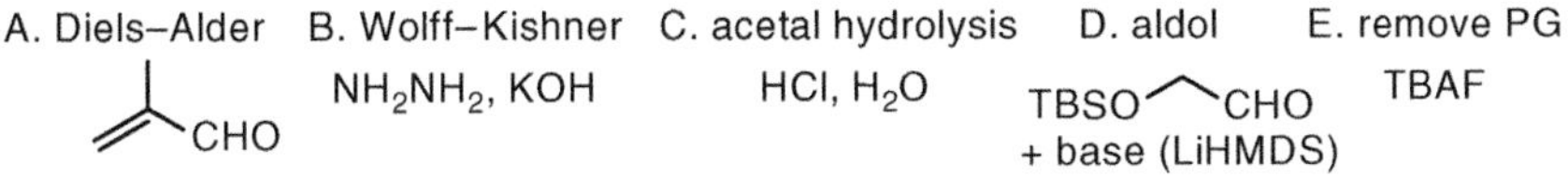

A. Diels–Alder   B. Wolff–Kirkner   C. acetal hydrolysis   D. aldol   E. remove PG

CHO

NH_2NH_2, KOH

HCl, H_2O

TBSO⌒CHO
+ base (LiHMDS)

TBAF

SOLUTIONS TO CHAPTER 7: STEREOCHEMISTRY

7.4 Predict the major product(s) or provide the missing reagent(s).

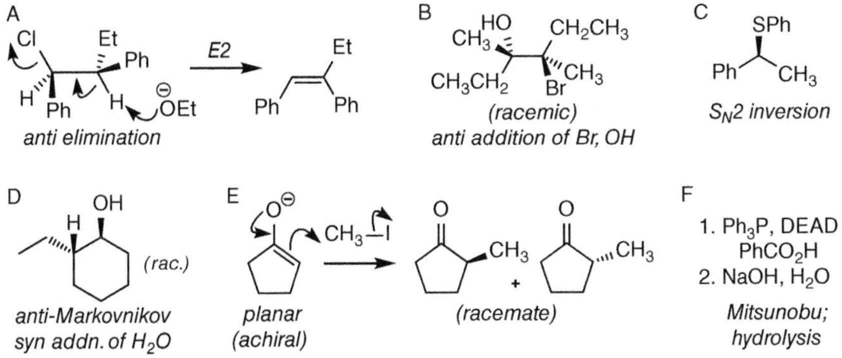

A

Cl
Et
Ph
H
Ph
H
OEt

E2

Et
Ph
Ph

anti elimination

B

CH_3
HO
CH_2CH_3
CH_3CH_2
CH_3
Br

*(racemic)
anti addition of Br, OH*

C

SPh
Ph
CH_3

S_N2 inversion

D

OH
H
Et

(rac.)

*anti-Markovnikov
syn addn. of H_2O*

E

O⁻
CH_3—I

*planar
(achiral)*

O
CH_3
+
O
CH_3

(racemate)

F

1. Ph_3P, DEAD
   $PhCO_2H$
2. NaOH, H_2O

*Mitsunobu;
hydrolysis*

7.7 In each case, imagine making a substitution for each of the groups being compared (indicated below with *). To determine *pro-R* or *pro-S*, substitute with a heavier isotope, such as replacing a hydrogen atom with a deuterium atom, and determine if the resulting structure has the *R* or *S* configuration.

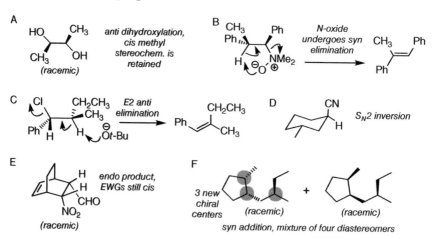

A

enantiomers
H_A & H_B are enantiotopic

same compound
Me_C & Me_D are homotopic

enantiomers
H_B & H_F are enantiotopic

B

same compound
H_A & H_B are homotopic

enantiomers
H_C & H_D are enantiotopic

#3 → #2 → #1
is clockwise (R)
H_C is pro-R

note: rotation
is reversed
because #4
is pointing
toward us

C

these structures are diastereomers (cis vs. trans)
so H_A and H_B are diastereotopic

End-of-Chapter Problems

7-1. Predict the major product.

A   HO   CH_3   anti dihydroxylation,
     CH_3   OH   cis methyl
     (racemic)   stereochem. is
                  retained

B   CH_3   Ph   N-oxide
     Ph   Ph   undergoes syn   CH_3   Ph
     H   NMe_2   elimination
         O   Ph

C   Cl   CH_2CH_3   E2 anti
            CH_3   elimination   CH_2CH_3
     Ph H   H   Ot-Bu   Ph   CH_3

D   CN
          S_N2 inversion
        H

E   H   endo product,
     H   EWGs still cis
         CHO
     NO_2
     (racemic)

F   3 new   +
     chiral
     centers   (racemic)   (racemic)
     syn addition, mixture of four diastereomers

G

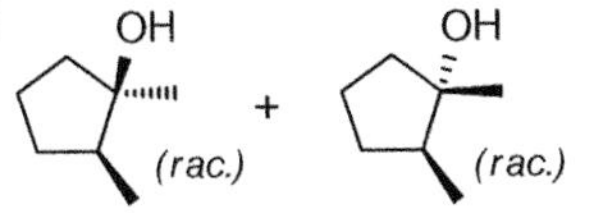

both anti and syn addition occur,
giving mixture of four diastereomers

H

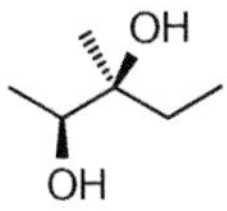

syn dihydroxylation

(racemic)

I

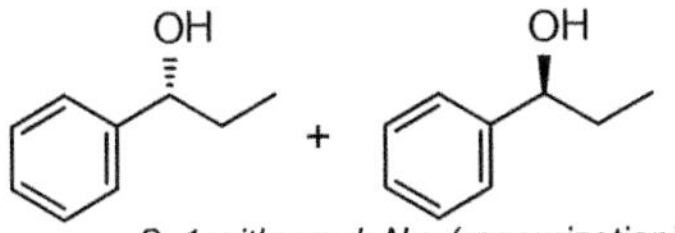

S$_N$1 with weak Nu: (racemization)

J

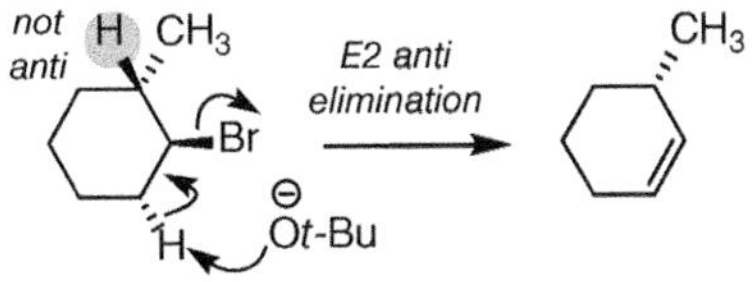

not anti

E2 anti elimination

K

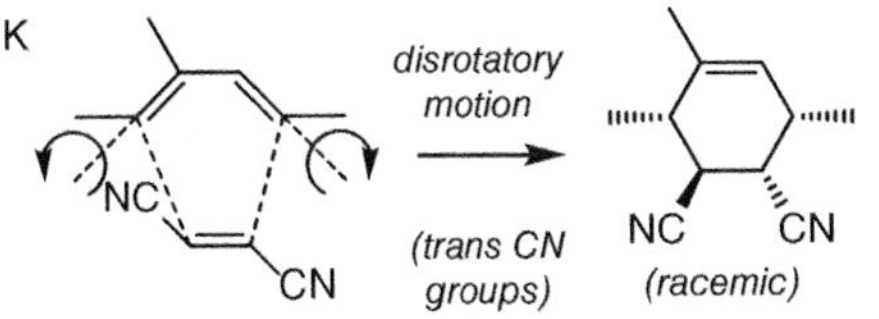

disrotatory motion

(trans CN groups)

(racemic)

L

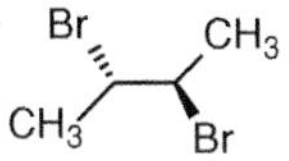

anti bromination,trans methyl stereochem. is retained

(meso)

7-2. Predict the monochlorination products. Consider all regioisomers and stereoisomers.

A (6 possible products)

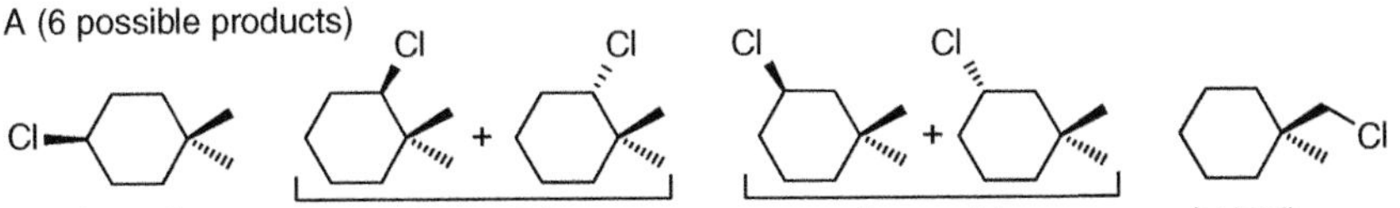

(achiral)   *(racemate)*   *(racemate)*   *(achiral)*

B (7 possible products)

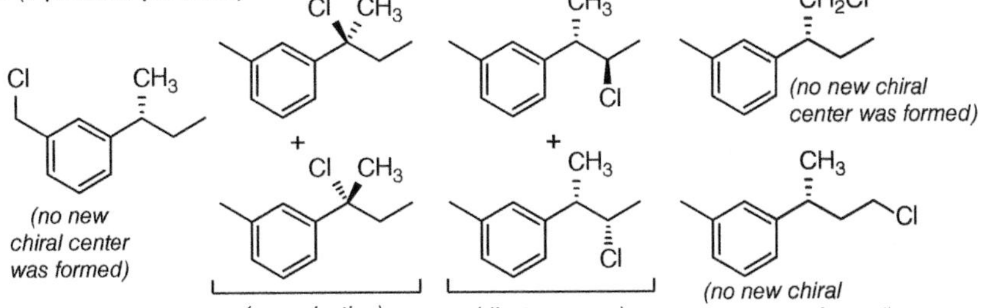

(no new chiral center was formed)

(no new chiral center was formed)

(racemization)   *(diastereomers)*   *(no new chiral center was formed)*

C (7 possible products)

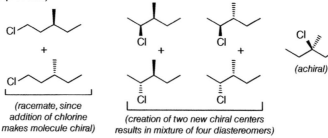

(racemate, since addition of chlorine makes molecule chiral)

(creation of two new chiral centers results in mixture of four diastereomers)

(achiral)

D (6 possible products)

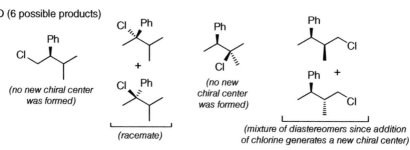

(no new chiral center was formed)

(racemate)

(no new chiral center was formed)

(mixture of diastereomers since addition of chlorine generates a new chiral center)

7-3. Predict the major product(s), paying close attention to stereochemistry.

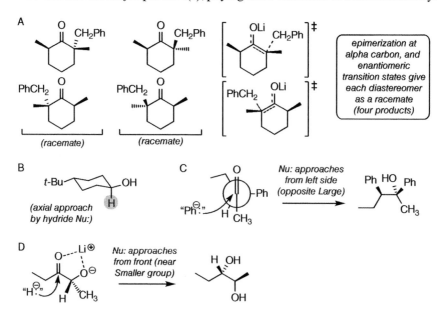

A

(racemate)

(racemate)

epimerization at alpha carbon, and enantiomeric transition states give each diastereomer as a racemate (four products)

B

t-Bu—⟨⟩—OH

(axial approach by hydride Nu:)

C

Nu: approaches from left side (opposite Large)

"Ph:⊖"

D

Nu: approaches from front (near Smaller group)

"H:⊖"

E

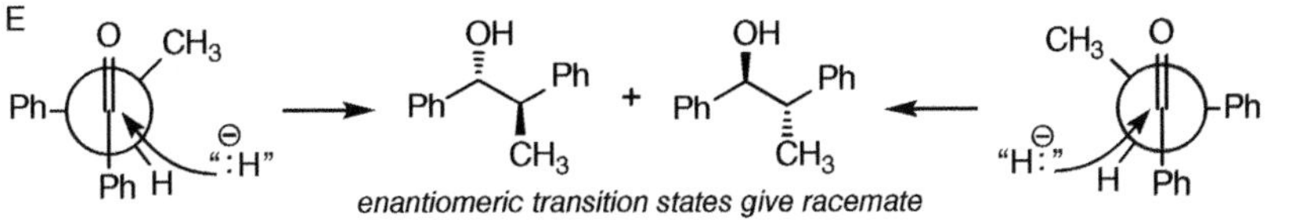

enantiomeric transition states give racemate

F

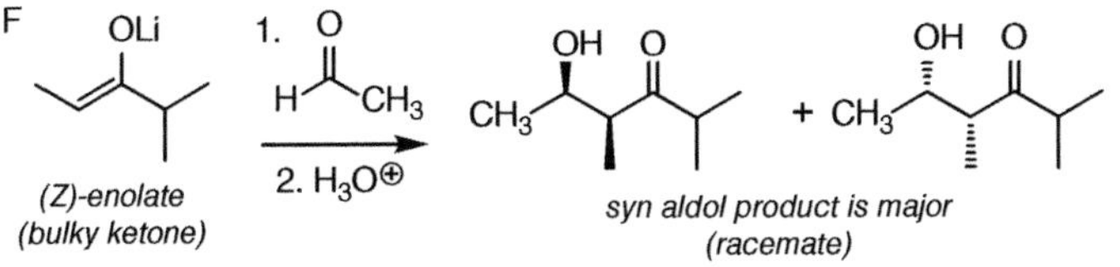

(Z)-enolate
(bulky ketone)

1. O
H⁀CH₃
2. H₃O⊕

syn aldol product is major
(racemate)

7-4. Predict (E)- or (Z)-enolate.

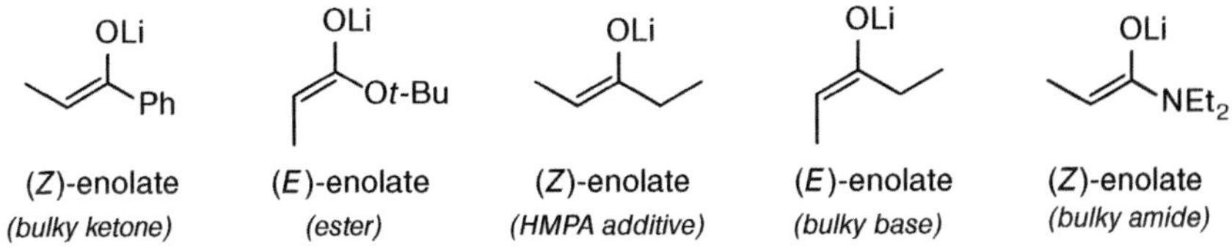

(Z)-enolate	(E)-enolate	(Z)-enolate	(E)-enolate	(Z)-enolate
(bulky ketone)	(ester)	(HMPA additive)	(bulky base)	(bulky amide)

7-5. To determine the appropriate starting material, begin by analyzing one enantiomer of the desired product and consider the two possible retrosyntheses.

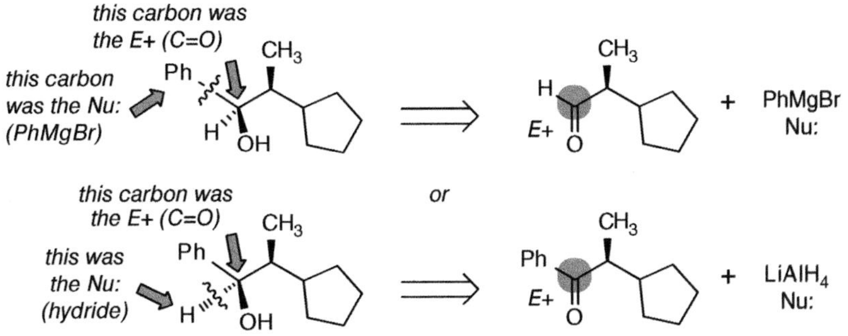

this carbon was
the E+ (C=O)

this carbon
was the Nu:
(PhMgBr)

+ PhMgBr
Nu:

or

this carbon was
the E+ (C=O)

this was
the Nu:
(hydride)

+ LiAlH₄
Nu:

For the aldehyde starting material, a PhMgBr nucleophile is needed. This reaction would give the desired stereochemistry.

desired stereochemistry

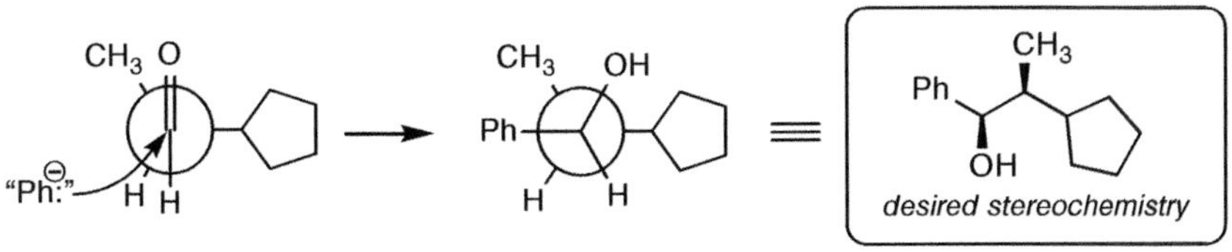

For the ketone starting material, a LiAlH$_4$ nucleophile is needed. This reaction would NOT give the correct stereochemistry.

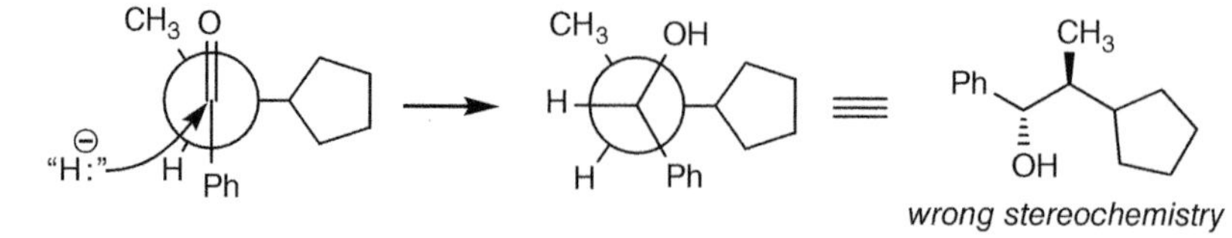

wrong stereochemistry

Therefore, the correct synthesis begins with the aldehyde.

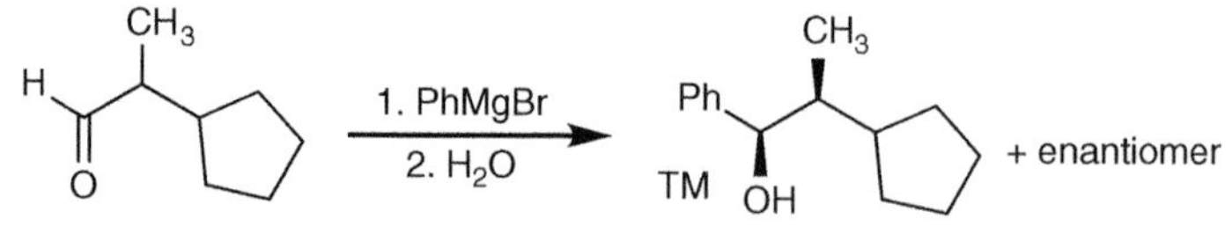

7-6. Upon treatment with acid, the chiral carboxylic acid undergoes tautomerization to give an achiral enol.

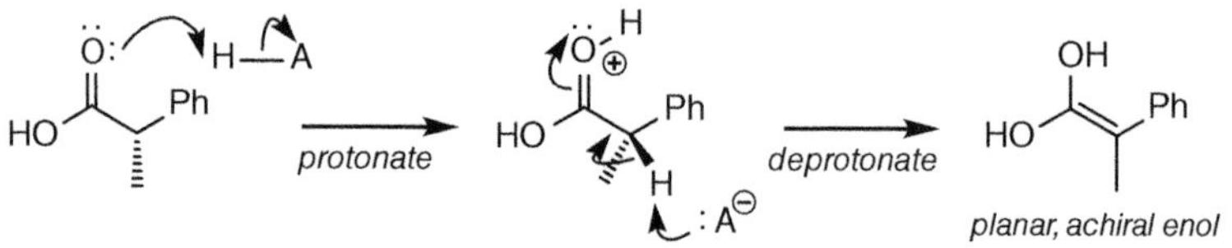

planar, achiral enol

When this enol undergoes the reverse tautomerization mechanism, protonation of the planar enol can occur on either face, resulting in the formation of both enantiomers of the carboxylic acid (i.e., racemization).

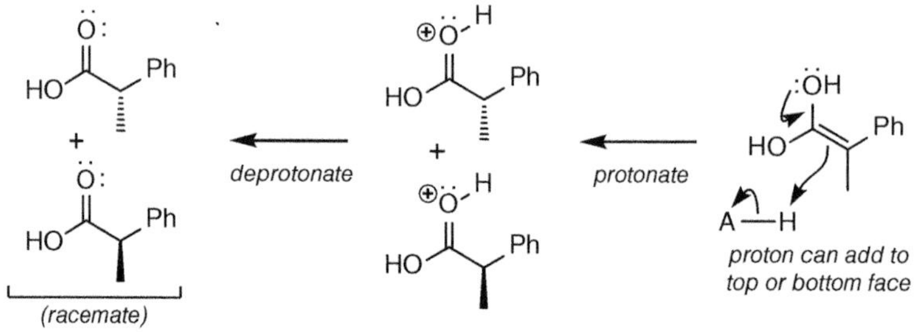

(racemate)

proton can add to top or bottom face

This racemization is expected to be slower in base because the acidic carboxylic acid functional group would be deprotonated under basic conditions and this protects the alpha carbon from deprotonation.

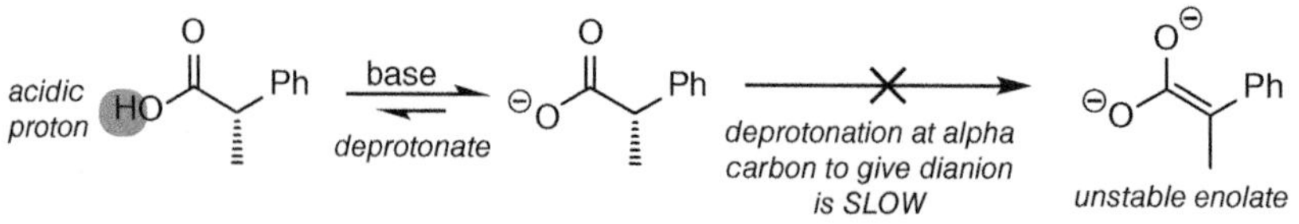

acidic proton

deprotonation at alpha carbon to give dianion is SLOW

unstable enolate

7-7. Upon treatment with acid, 3-methylcyclohexanol undergoes epimerization at the carbon bearing the hydroxyl group, since the hydroxyl will be converted to a good leaving group (H_2O). The cis diastereomer is expected to be the major product formed since that is the more stable product with both substituents in equatorial positions.

carbon bearing
LG is epimerized

more stable 1,3-cis
(diequatorial)
diastereomer is favored

The mechanism involves protonation of the hydroxyl group, loss of water to give a planar carbocation, and then addition of water to reform the alcohol.

+ enantiomer

cis is major
product

+ enantiomer

$-H_2O$

loss of LG

protonate

deprotonate

$H_2\ddot{O}:$

add Nu:
S_N1

7-8. Bromine can attack either face of the alkene to give the bromonium ion intermediate. Attack from the bottom face gives intermediate A (**Int-A**) and attack from the top gives intermediate B (**Int-B**). Because of the existing chiral center, the transition states for this step of the reaction are diastereomeric and not equal energy. As a result, the intermediates are formed at different rates and will not be formed in equal amounts.

Energy

Progress of Reaction

alkene

TS-B

TS-A   Int-B

Int-A

Br_2 attacks from top

TS-B (higher E)

Int-B
minor intermediate

Br_2 attacks from bottom

TS-A (lower E)

Int-A
major intermediate

diastereomeric transition states
are not equal in energy

Int-A leads to product **A**, when the bromonium ion is attacked by bromide at the more substituted, and therefore more electrophilic carbon with an S_N2 mechanism (backside attack results in trans bromine atoms). The same mechanism gives rise to product **B** from **Int-B**. Since the intermediates are formed in unequal amounts, the diastereomeric products are also expected to be formed in unequal amounts.

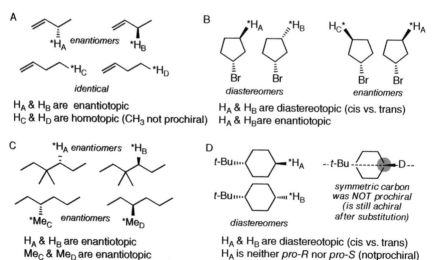

7-9. In each case, imagine making a substitution for each of the groups being compared (indicated below with *). To determine *pro-R* or *pro-S*, substitute with a heavier isotope, such as replacing a hydrogen atom with a deuterium atom, and determine if the resulting structure has the *R* or *S* configuration.

A

enantiomers

identical

H_A & H_B are enantiotopic
H_C & H_D are homotopic (CH_3 not prochiral)

B

diastereomers   *enantiomers*

H_A & H_B are diastereotopic (cis vs. trans)
H_A & H_B are enantiotopic

C

*H_A *enantiomers* *H_B

*Me_C *enantiomers* *Me_D

H_A & H_B are enantiotopic
Me_C & Me_D are enantiotopic

D

t-Bu····   ····*H_A

t-Bu····   ····*H_B

diastereomers

--*t*-Bu------D--

symmetric carbon was NOT prochiral (is still achiral after substitution)

H_A & H_B are diastereotopic (cis vs. trans)
H_A is neither *pro-R* nor *pro-S* (notprochiral)

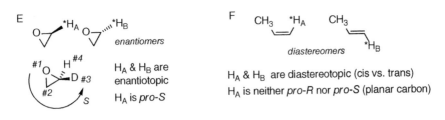

E enantiomers

H_A & H_B are enantiotopic

H_A is pro-S

F diastereomers

H_A & H_B are diastereotopic (cis vs. trans)

H_A is neither pro-R nor pro-S (planar carbon)

7-10. Label each ester group as either *pro-R* or *pro-S* and draw a structure of the product.

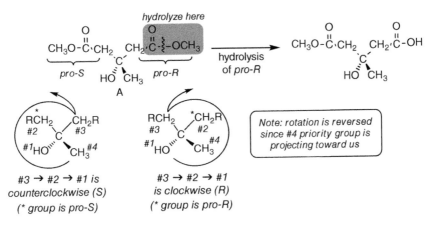

hydrolyze here

hydrolysis of pro-R

Note: rotation is reversed since #4 priority group is projecting toward us

#3 → #2 → #1 is counterclockwise (S)
(* group is pro-S)

#3 → #2 → #1 is clockwise (R)
(* group is pro-R)

7-11. This dihydroxylation of naphthalene has occurred at the *si* faces of both C-1 and C-2.

C-1 *si*
C-2 *si*

enzymatic oxidation

counterclockwise so the front/top of C-1 is the si face

counterclockwise so the front/top of C-2 is also the si face

7-12. The target molecule must first be rotated to fit the Sharpless asymmetric epoxidation model (need to have allylic alcohol group in bottom right position). The retrosynthesis of the epoxide leads to an alkene with the stereochemistry shown. Since the oxidation has occurred at the top face, the (+) enantiomer of diethyl tartrate (DET) is required. Note that only the alkene with the allylic hydroxyl group is oxidized in the Sharpless epoxidation.

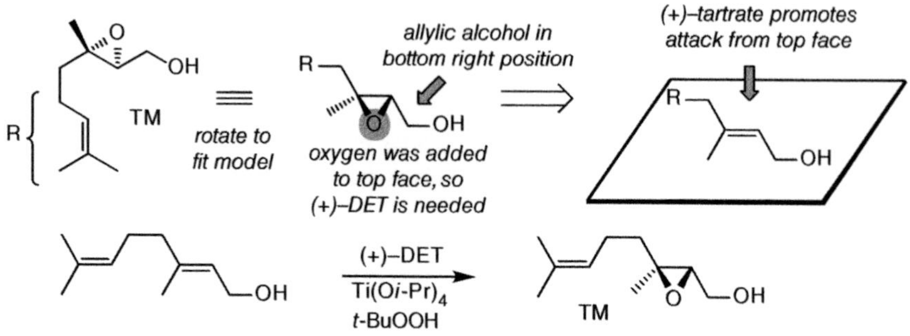

7-13. Since 1-amino-2-cyclohexene is basic, a chiral acid is needed. When (+)-mandelic acid is the resolving agent used, a mixture of diastereomeric salts is formed as shown. These diastereomers have different physical properties and can therefore be separated (often by recrystallization). Treatment with base neutralizes the amine and allows it to be separated from the charged mandelate salt (typically by extraction), to give the neutral amines as pure enantiomers.

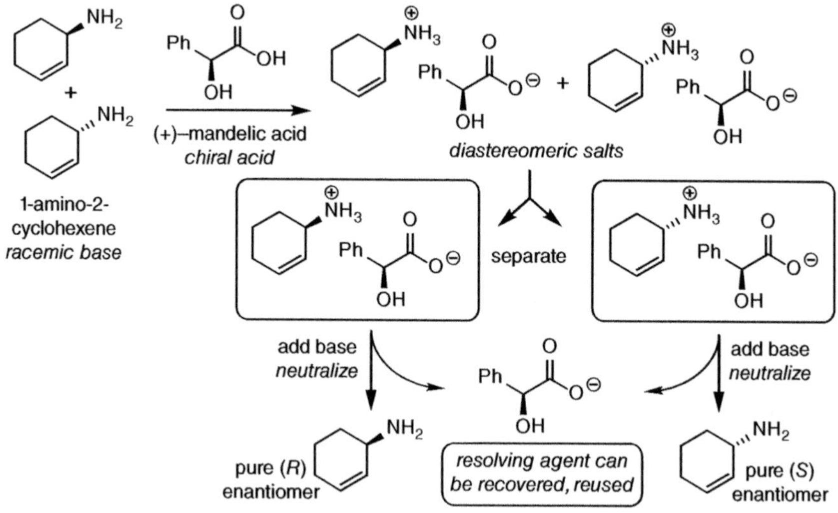

SOLUTIONS TO CHAPTER 8: TRANSITION METAL CHEMISTRY

8.1   Give the electron count, oxidation state, and d^n configuration.

A

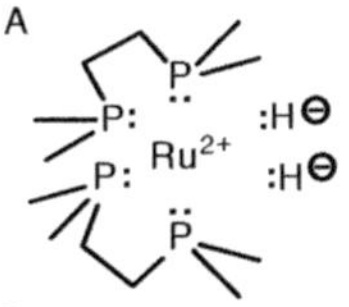

Net charge:
$+2(Ru^{2+}) + (-2)(H^- \times 2) = 0$
Electron count:
$Ru(II)d^6 + 4e^-(H \times 2) +$
$8e^-(dmpe \times 2) = 18e^-$

D

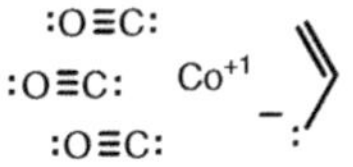

Net charge:
$+1(Co^{1+}) + (-1)(allyl^-) = 0$
e^- count:
$Co(I)d^8 + 6e^-(CO \times 3)$
$+ 4e^-(allyl) = 18e^-$

B

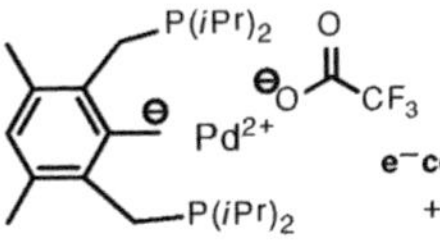

Net charge: $+2(Pd^{2+}) +$
$(-1)(RO^-) + (-1)(R^-) = 0$
e^- count: $Pd(II)d^8 + 4e^-(R_3P \times 2)$
$+ 2e^-(RO) + 2e^-(R) = 16e^-$

E $[Rh(NH_3)_5Br]^{2+}$ $Br^- Br^-$

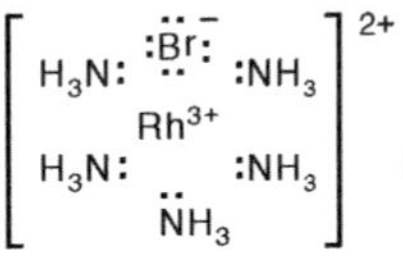

Net charge:
$+3(Rh^{3+}) + (-1)(Br^-) = +2$
Electron count:
$Rh(III)d^6 + 10e^-(NH_3 \times 5)$
$+ 2e^-(Br) = 18e^-$

C

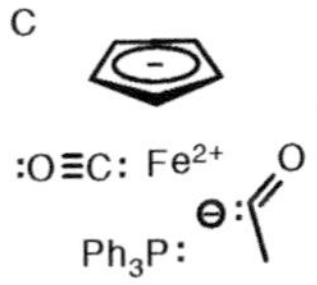

Net charge:
$+2(Fe^{2+}) + (-1)(Ac^-) + (-1)(Cp^-) = 0$
Electron count:
$Fe(II)d^6 + 2e^-(Ac) + 6e^-(Cp) +$
$2e^-(CO) + 2e^-(PPh_3) = 18e^-$

8.4   Predict the product or provide the missing reagent in the following Heck reactions.

A

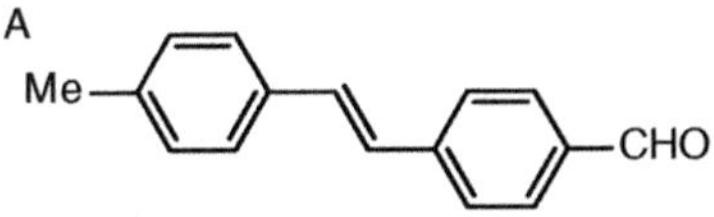

Me—⟨ ⟩—CH=CH—⟨ ⟩—CHO

B   CO₂Me

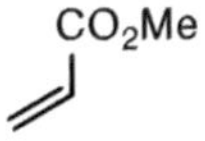

8.5   Predict the product(s) and provide the name of the coupling reaction.

A   OH
⟨ ⟩—⟨ ⟩—CO₂H

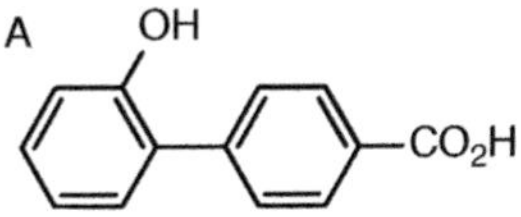

Suzuki–Miyaura

B
⟨ ⟩—⟨ ⟩

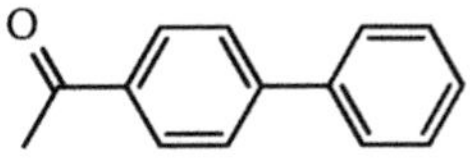

Stille

C

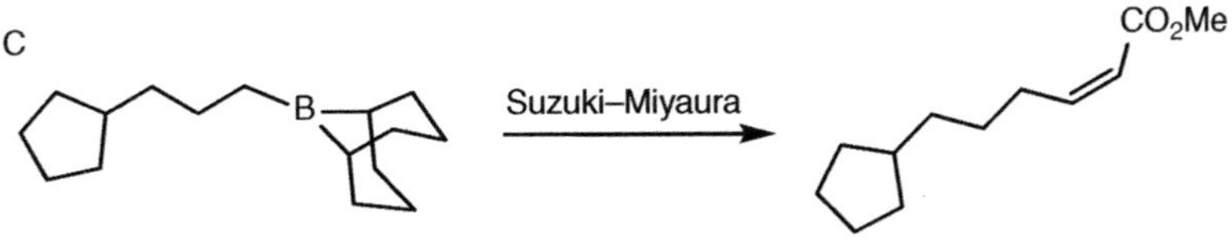

Suzuki–Miyaura →   CO₂Me

D

Suzuki–Miyaura →   Ph

End-of-Chapter Problems

8-1. Give the electron count, oxidation state, and d^n configuration.

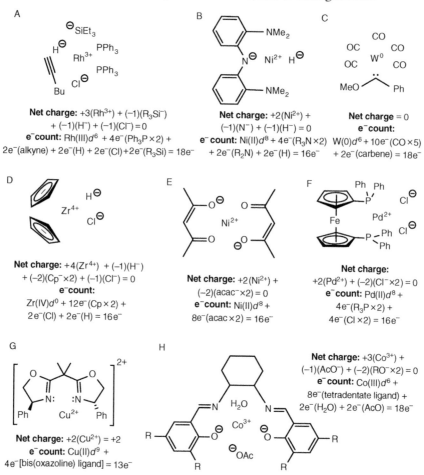

A

Net charge: $+3(Rh^{3+}) + (-1)(R_3Si^-)$
$+ (-1)(H^-) + (-1)(Cl^-) = 0$
e^-**count:** $Rh(III)d^6 + 4e^-(Ph_3P \times 2) +$
$2e^-(alkyne) + 2e^-(H) + 2e^-(Cl) + 2e^-(R_3Si) = 18e^-$

B

Net charge: $+2(Ni^{2+}) +$
$(-1)(N^-) + (-1)(H^-) = 0$
e^- **count:** $Ni(II)d^8 + 4e^-(R_3N \times 2)$
$+ 2e^-(R_2N) + 2e^-(H) = 16e^-$

C

Net charge $= 0$
e^-**count:**
$W(0)d^6 + 10e^-(CO \times 5)$
$+ 2e^-(carbene) = 18e^-$

D

Net charge: $+4(Zr^{4+}) + (-1)(H^-)$
$+ (-2)(Cp^- \times 2) + (-1)(Cl^-) = 0$
e^-**count:**
$Zr(IV)d^0 + 12e^-(Cp \times 2) +$
$2e^-(Cl) + 2e^-(H) = 16e^-$

E

Net charge: $+2(Ni^{2+}) +$
$(-2)(acac^- \times 2) = 0$
e^-**count:** $Ni(II)d^8 +$
$8e^-(acac \times 2) = 16e^-$

F

Net charge:
$+2(Pd^{2+}) + (-2)(Cl^- \times 2) = 0$
e^-**count:** $Pd(II)d^8 +$
$4e^-(R_3P \times 2) +$
$4e^-(Cl \times 2) = 16e^-$

G

Net charge: $+2(Cu^{2+}) = +2$
e^-**count:** $Cu(II)d^9 +$
$4e^-[bis(oxazoline) ligand] = 13e^-$

H

Net charge: $+3(Co^{3+}) +$
$(-1)(AcO^-) + (-2)(RO^- \times 2) = 0$
e^- **count:** $Co(III)d^6 +$
$8e^-(tetradentate ligand) +$
$2e^-(H_2O) + 2e^-(AcO) = 18e^-$

8-2. Give the electron count, oxidation state, and d^n configuration.

Net charge:
$+2(Ru^{2+}) + (-2)(Br^- \times 2) = 0$
Electron count:
$Ru(II)d^6 + 4e^-(Br \times 2) +$
$4e^-(BINAP) = 14e^-$

8-3. Give the electron count, oxidation state, and d^n configuration.

a. The oxidation state of each Rh atom is +1.
b. Each Rh atom has a total of 16 e⁻ so it is
   coordinatively unsaturated.
c. "cod" stands for "**C**yclo**O**cta**D**iene"
   This complex is described as the
   1,5-cyclooctadienerhodium(I) chloride dimer.

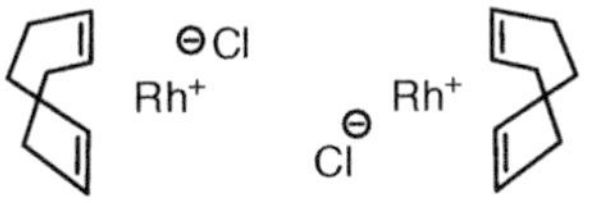

Net charge: $+1(Rh^+) + (-1)(Cl^-) = 0$
e⁻count: $Rh(I)d^8 + 4e^-(cod) + 4e^-(Cl \times 2) = 16e^-$

8-4. Describe each step of the Negishi coupling mechanism.

A. oxidative addition *[Pd(0) → Pd(II)]*
B. transmetalation *[R group is transferred from Zn to Pd]*
C. reductive elimination *[Pd(II) → Pd(0)]*

8-5. After formation of a pi-olefin complex with allyl alcohol, the catalytic
cycle involves an insertion followed by a beta-elimination. The isomer-
ized vinyl alcohol then undergoes enol-keto tautomerization to generate the carbonyl group in propionaldehyde.

alkene isomerization: **tautomerization:**

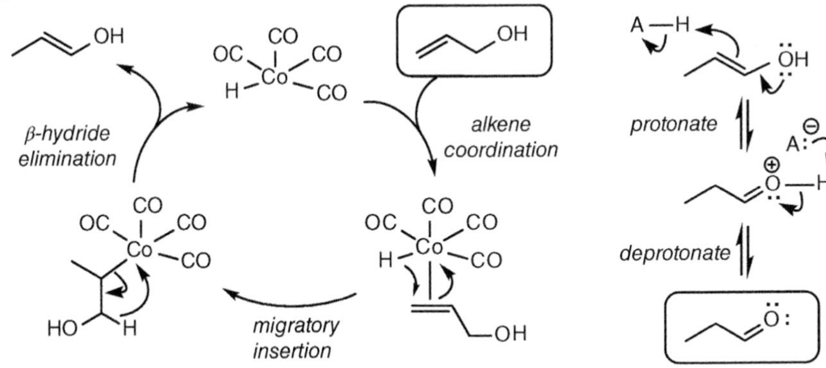

8-6. Predict the product(s) or provide the reactant and give the name of
each reaction.

A   Suzuki-Miyaura

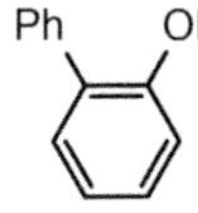

B   Heck

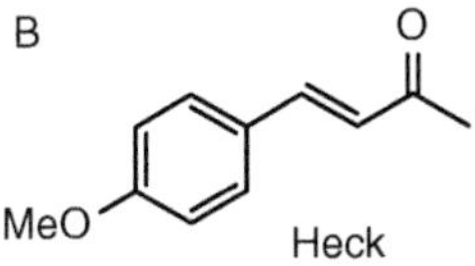

C   Stille

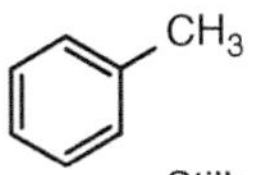

D   Heck

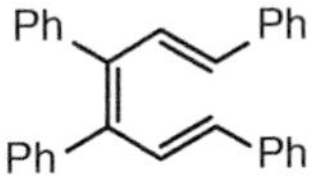

E   Pd(dppf)Cl₂   NaOH   Suzuki-Miyaura

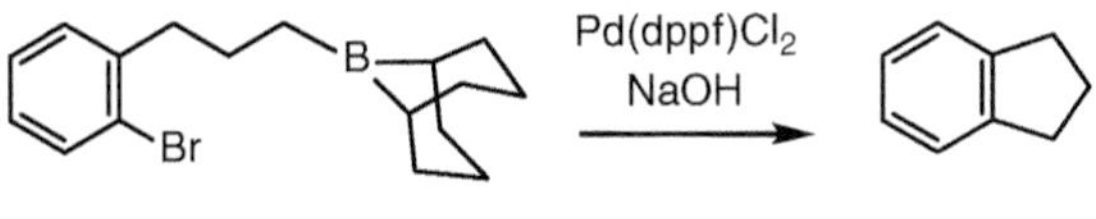

F   Stille

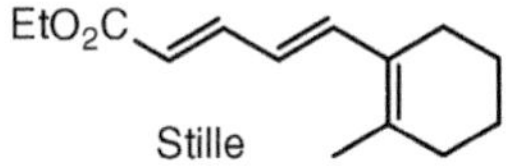

G   Suzuki-Miyaura

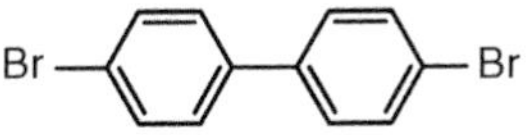

8-7. For each reaction, name the catalyst, describe the reaction, and predict the product.

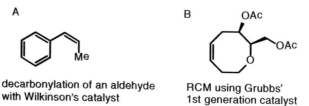

A

B   OAc

decarbonylation of an aldehyde
with Wilkinson's catalyst

RCM using Grubbs'
1st generation catalyst

8-8. Provide the missing reactant and reagent.

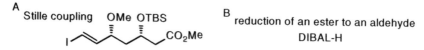

A
Stille coupling   OMe OTBS

B
reduction of an ester to an aldehyde

DIBAL-H

8-9. Provide the missing reagents.

A   NBS, *hν*
   (free-radical halogenation)

B   (deprotonate, alpha   EtONa + CO₂Et
   alkylation via S$_N$2)

C   Grubb's catalyst (ring-closing methathesis, RCM)
(A higher yield was obtained with Mo-based catalyst *J. Org. Chem.*, **1997**, *62*, 7310.)

8-10. Provide a synthesis. A trans alkene can be prepared with a Heck reaction.

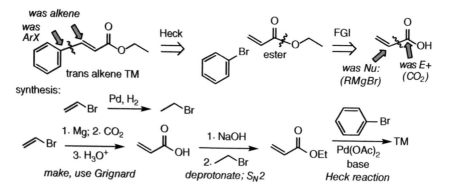

Introduction to Strategies for Organic Synthesis, Second Edition. Laurie S. Starkey.
© 2018 John Wiley & Sons, Inc. Published 2018 by John Wiley & Sons, Inc.

Printed and bound by CPI Group (UK) Ltd, Croydon, CR0 4YY

27/10/2024

14580474-0001